Psychology of Technology

V.K. Kool • Rita Agrawal

Psychology of Technology

Springer

V.K. Kool
State University of New York
Polytechnic Institute
Utica, NY, USA

Rita Agrawal
FMT, HCPG College
Varanasi, India

ISBN 978-3-319-45332-3 ISBN 978-3-319-45333-0 (eBook)
DOI 10.1007/978-3-319-45333-0

Library of Congress Control Number: 2016956563

Printed on acid-free paper

This Springer imprint is published by Springer Nature
The registered company is Springer International Publishing AG
The registered company address is: Gewerbestrasse 11, 6330 Cham, Switzerland

Preface

It would be extremely difficult to think of any segment of human behavior that is not influenced by technology, be it classroom learning, the gospel that is presented over the microphone in the background of soothing lights in the church, or while packing for a vacation and making sure that the iPhone and/or computer has been packed and is fully charged. And yet, as we peruse through currently available textbooks of psychology, it is apparent that the effects of technology on human behavior have received scant attention. We were disappointed to notice that even books on social psychology, and more importantly, those on applied social psychology, have failed to pay due attention to the multifarious ways in which human behavior has been affected by a wide variety of technological interventions in every field of life and work.

Historically speaking, we can see that the study of behavior with reference to technology has found some place in certain subfields of psychology. Having its roots in applied experimental psychology, the phenomenon of behavior in the context of technology has been addressed by several subfields which, in turn, led not only to the expansion of the scope of psychology but also to the growth of psychology as a science. What immediately comes to mind are the areas of industrial psychology, organizational psychology, human performance, human factors engineering, and ergonomics. Because of the exponential growth in technology, the last 50 years has seen human behavior expanding to levels hitherto unthought-of and reaching vistas which are not only new but also much beyond the canvas of normal evolutionary changes in behavior. Despite such overarching effects of technology, has psychology as a science attempted to analyze its impact? We have reached the juncture where it has become imperative to develop a relatively nascent subfield of psychology, namely psychology of technology, and to delineate its scope and content as a comprehensive and independent subfield of psychology, with biology on one side and technology on the other. We anticipate that psychology of technology, perched, as it would be, between the symbiosis of biology and technology would then enrich our understanding of behavior in general and human cognition and emotion in particular. Psychology, as a science, will have a very special role to play, albeit it may sometimes lead to situations wherein such roles of psychology could be undermined by other disciplines. As psychologists, it would then become even more crucial to make attempts to disentangle behavior and its biological substrates from the throes of technology.

Predicting what technology would do to us or what would be its future has never been easy. From Galileo and Edison to Martin Cooper, and, from Hubble to Hawking and to the many other famous contributors to technology, they have all been great innovators and thinkers. However, have they been able to predict the effects of technology in advance? For example, Edison doubted the use of the AC current and Martin Cooper concluded that cellular phones would never replace wire systems of telephony. As we look back to the times when the first generation of computers was introduced, were we able to comprehend what it and the World Wide Web would do to us? Did many professors who started teaching in the 1960s have any inclination of how technological developments would change their teaching methods through the use of Internet and PowerPoint presentations? Even the National Science Foundation has been skeptical, being able to foresee the effects of technology on humanity for a time slot of not more than 20 years (Roco & Bainbridge, 2002). To what extent will we remain human? Will we become a cyborg? What about man as a social animal? What about the limits to our vision and audition and even our ability to think, to solve problems—in fact, to cognize? These and many such questions continue to plague us, making them even fiercer, in the face of technology, which is becoming not only omnipresent but also almost omnipotent. And, in our daily lives the effect of technology has become so integral that we tend to forget information because it is readily available on the Internet or loaded in our iPhone (popularly known as the Google Effect or digital amnesia).

Simply put, it is often not easy to grasp the gap between the creation of a technological product and its use in our daily life. If you have tried to assemble "Do It Yourself (DIY)" kits after reading the instructions given by the engineers, you would be well aware of the frustrations. In other words, there is a big gap between the creation of a thing and its use, between the features of a gadget and its applications.

The above uncertainty about the effects of technology also calls for an understanding of its impact in the building of relationships, a core topic of study among social scientists. Many scholars have raised doubts about how technology will help in connecting people to each other. Robin Dunbar, a prominent evolutionary psychologist and author of *How many friends does one need* (Harvard University Press, 2010), is skeptical about the peaceful interactions of humans in the context of meeting new people during the course of evolution. As per Robin Dunbar's calculation, our neocortex sets a limit for personalized relationships with not more than 150 people. Can technology change this course? Can we afford to play with the course of evolution? With technology moving at a pace faster than our capacity to adapt to it, we have already begun to have fears regarding our own creations, say, robots. Will man remain master of his creations or will he become a slave to them? Human beings have always been fascinated by the special abilities of other living beings, for example, the capacity of birds to fly, and this has led to the use of human intellect to acquire such skills through artifacts. With the complexity of such creations growing considerably, would such artifacts begin to control us? A movie, based on Pierre Boulle's novel, *In the planet of apes* (translated from the French, *La planete des singes*, published by "Le cercle du nuveau livre," 1963) focuses on what would have happened if the other creatures on earth had intelligence similar to that of human beings? Given our understanding of the survival of the human being through the forces of evolution, the answer to the above, in all likelihood, would be pessimistic, namely that "chaos" will reign for the simple reason that the peaceful coexistence of human beings with other living beings has been questionable throughout the history of mankind. That raises another question: is intelligence (or too much of it) a good thing? We might argue that the above novel and movie were mere fiction based on the imagination of the writer, Pierre Boulle, but with robots around, smarter than us and outsizing us in our biological abilities to process information, are we heading for "chaos" in the real sense? We can only guess that if a situation akin to that of Boulle's fiction does occur, the answers would be provided through co-evolution as a result of and through technology, although the issues of peaceful coexistence might well lead to connotations, different from what we have been used to in the past.

At the outset, we want to let the readers know that this book has been written with its own inherent limitations. First, take a look at the human brain which is the source of all creativity including that leading to the growth of technology. Technology, which originated outside our bodies, is gradually, but surely, becoming a part of ourselves and our bodies. A technologically modified and augmented cognition with its ever increasing capacity to handle and store information, erupting emotions in relations to artifacts such as "my robot is my friend," and the functioning of our two limbs—one of which may be natural and the other artificial—truly give many of us a sense of being an extension much beyond our biological inheritance. Second, borrowing and integrating knowledge from such diverse fields as technology and biology has not been easy and this book has attempted, in a straightforward and simplified form, to illustrate how the two disciplines converge together, to enrich the subfield of psychology of technology. And, as we cruise along through the growth of this field, we foresee its future course, with its interdisciplinary nature being its major strength, and contributions from sister social science disciplines, such as sociology, anthropology, economics, and many more, leading to its robust development.

While surveying the courses offered in psychology programs by universities around the globe, it became clear that there are not many which offer a course on this burgeoning field of psychology of technology. There may be several reasons for this. A primary need while designing and offering a course, as we have learnt from instructors across a wide range of disciples, is the availability of a suitable textbook on the subject. Diverse as it is, a course on psychology of technology requires integration of material from multiple disciplines. As a result, specialists from different areas of study have occasionally converged to publish a few edited books on this subject, many of which are proceedings of conferences culminating in the form of edited volumes. The problem with such publications is that they lead to a potpourri of material, difficult to glue together and leaving many topics untouched or unconnected to each other. While a few scholars have made solo efforts, their volumes cover only some specific and limited area/s of this subfield. A comprehensive textbook is missing.

Our approach for the writing of this book has emanated from a focus traditionally used in describing core psychological processes, that is, sensory, perceptual, cognitive, and affective processes. The introductory chapter begins with raising the question about the growth of psychology of technology in the context of philosophy of science and the emerging field of philosophy of technology. Obviously, a subfield needs to identify clear goals which will help to distinguish it from its sister disciplines—a focus that has been raised in the opening chapter of the book. Thereon, in the next chapter, we move to explore the ways in which technology emerged in the history of mankind, aiding the human species in its search for mastery over the ravages of an unfriendly environment, and, how, in contrast today, our bid for greater advancements in technology rarely enhances survival, being used more to gratify our wants than our needs. We also focus on how, as technology starts being used for the masses, we learn that one size does not fit all, but nevertheless we seek to do so for commercial and other reasons. It generates issues concerning anthropometrics and the operation of the human body in a restricted environment—man-made or otherwise, with long-term sobering or harmful effects and the chances of its acceptance or rejection at the societal or cultural levels.

In the next two chapters of the book, we move on to the sensory, perceptual, cognitive, and motor processes, focusing on the limits of human behavior and how technology has helped us to enhance our performance and also, how a greater understanding of the human apparatus has led to growth in usable technology. In the chapter on how technology bears on our affective and emotional processes, we began with a description of neuro-aesthetics and continue to show how the design of things is so very important in taking decisions (rational or otherwise) for using them. The next chapter on virtual reality (VR) builds on the complex ways in which VR has changed our very persona and the breadth of our experiences. We have probed into how technology contributes to the formation of our "second self" as aptly argued by Sherry Turkle (2011), along with the need to understand how robotics is bringing about changes in various aspects of life, including education, therapy, and industry.

The last chapter of the book focuses on our concerns with technology, including fears emerging from the future impact of technology. While the course of human evolution is natural, the chart of technological growth would depend on what we think about our own existence in the context of who we were, who we are, and what we want to be—our understanding at the phenomenological level, our ability to seek connections with others, and overall, how we develop a sense of "cognitive collectivity" for embracing the depth and range of available technology and laying down the course of its future growth.

In writing this book, we have made several compromises. The reader might notice that while some topics are included, others have been given scant or no attention. We did not plan to prepare a compendium of psychology of technology. Rather, we focused on the need for a simple book, making the integration of material across disciplines unembellished and easy to understand. While we acknowledge that this book could have been written in many different ways, our focus has been to align the material in terms of our understanding of the core psychological processes such as perception, cognition, and emotion in the context of technology.

We would like to thank our friends, colleagues, and students, and last but not the least, our families, who supported us throughout this venture and also provided feedback at various stages of the preparation of this book. We appreciate the patience and continued support from our publisher and, particularly, from Sylvana Ruggirella of Springer (USA), in steering this project smoothly and making it available as soon as possible.

Utica, NY, USA — V.K. Kool
Varanasi, India — Rita Agrawal

Contents

List of Figures

List of Tables

List of Boxes

Chapter 1
The Emerging Nature of Psychology of Technology

1.1 An Introduction to Psychology of Technology

"Life was much easier when apple and blackberry were just fruits!"

Do we really think so? Imagine life without the Apple's iPhone, or the Blackberry that has been the hallmark of most people around the world in the last decade; life without a mobile phone, a computer, the internet, and the many other gizmos surrounding us? It certainly seems difficult, to say the least.

We still remember the evening when we sat down and began to wonder about the times before we started writing letters predominantly through the email rather than through the traditional postal mail (or snail mail as the i-Generation would call it). It brought back memories of painstakingly writing the letter (choosing the paper was often equally important as the contents), going to the post office, buying and affixing stamps, and making sure that the receiver's address was correct. Then came the email: simple and fast. However, in the beginning, the thought uppermost in our minds was whether it was appropriate to communicate via such a novel technological method—especially, if it was a culturally significant event, such as a wedding invitation or the birth of a child. It soon dawned upon us that the use of technology in sending letters had slowly changed our very style of communication. It seems as if technology had walked into our lives in a very subtle, albeit calm, way. Yes, many of the effects of technology are simply happening, without much ado. However, when each of us started enlisting the sorts of behavior that have been affected by technology and the ways in which it had conspicuously changed our lifestyle, we differed in our preferences. But, we did agree on one thing: we had just got hooked onto technology with little or no fuss. Additionally, we also noticed that this predominant influence of technology had its own sweet rewards that had been, hitherto, unknown to us. For example, as we began to browse our own emails that had been written over the years in multiple contexts, ranging from personal to professional correspondence, little did we realize that we had created a virtual Avatar for our self by using various portals—starting from Microsoft Windows to Twitter and

V.K. Kool, R. Agrawal, *Psychology of Technology*,
DOI 10.1007/978-3-319-45333-0_1

Facebook, of late. It was more than a collection of memories. It was about how we had changed with the times not just as people but also as learners of technology as our mainstay of communication, as consumers of technology for intervention in our well-being and health care needs, and as enablers of technology to understand our self beyond the given biological limits. Had we foreseen some 25 years back, the ways through which we are teaching today or even writing this book? Whatever we do, where ever we go, technology seems to be omnipresent, whether we like it or not—ranging from the virtual customer service agent to the insecurity of mobile banking and virtual banks.

The pace at which technology is growing is more than obvious. No longer do we need to look at the stars and the sky to read signs about impending weather. Our smart phone has it all ready for us at any time of the day. Nay, even the night or the week before, one can get weather forecasts and plan life accordingly. Going to work? Having a fear of being caught in traffic snarls? Your GPS guides you through the quickest route, avoiding all the traffic, blocked roads, etc. Once at work, gone are those days when one had to rely on a personal secretary to bring in voluminous files on which decisions had to be taken. You may not even have a personal secretary. The work place has changed and work rules have been reengineered. From CAD-CAM and robotics leading the way in manufacturing; graphic designing in advertising; e-commerce, e-banking, and CRM in marketing; ERP, SAP, and the use of big numbers and analytics in finance and enterprise management, to name just a very few, work has taken on a new shape.

And, it is not just at work. Our very lifestyles have changed. Schools boast of smart classes, homes are highly automated, from the lowly microwave oven to the *I-robot* that vacuums your home without your help and even when you are not there and vehicles change settings according to the customized key of the driver. Equipped with an Apple watch and a Fit-bit, riding a Harley Davidson bike and earphones tucked into the ears, the youth of the twenty-first century is ready to take on the world.

However, our capacity to judge the impact of technology has not grown proportionately, for our response to the technological change is much slower. It takes time to accommodate to such changes at the cognitive level and we often realize its significance only when the technology is lost or withdrawn. Imagine the day you lose your mobile phone!

An incident such as that of losing one's phone and the ensuing emotional reaction shows very clearly that technology is not just about creating tools: it is about understanding human nature and how it reacts to technology. This is where psychology of technology can come to the rescue.

But first things first: let us get to the crucial question: what is technology? Also, has technology always had such a pervasive impact on our lives? Were our ancestors also using some kind of technology to understand and predict the ravages of nature? Were they using technology to predict when it would rain, or when there would be a thunderstorm? Was technology being used to move from one place to another, or to communicate with people far away?

Coming to the first question raised above, our initial reaction is that technology constitutes the world of material possessions. It is about how we make things and use those things. On second thought, and looking at it more seriously, it starts becoming clear that some kind of organization of knowledge must have taken place to initiate the development of such things. When a chimp feels hungry and craves for protein in his diet, he uses a small twig to burrow into the soft soil to dig out insects to satiate his need. The size and diameter of the twig has to be such that it is strong enough to penetrate the soil, or else, the chimp will be unsuccessful in his task. So the chimp organizes his knowledge regarding the soil, the twigs around him, and that digging would lead to the finding of food.

As human beings dealt with the forces of nature, including predictable or unpredictable conditions such as weather, they gathered knowledge which enabled them to fabricate material needed to promote their survival. From a practical point of view, technology refers to a form of human activity geared to the making of things. Such activity involves using components of nature, typically called raw materials, to develop a product. Therefore, technology is rooted in the practice of how we engage in manipulating our environment so as to gain mastery over it. Additionally, as we realize that the forces of nature keep changing, our adjustment to the use of things also varies and our activities become technological in nature. For example, as human beings learned how to increase food production through improved gadgets for cultivation, they also began to devise things to preserve their cultivated products. In this sense, technology is not merely something which is a product or an object. Rather, its use may result in several forms of other activities, totally dynamic in nature. Similarly, we cannot think of technology merely as a possession or a collection of tools; it is also a source of finding something new and may result in various forms of transformations. A bicycle has been used for mobility for years, but now, in its static form, it is also used as an important tool for enhancing health and fitness.

Before we go into the complexities of psychology of technology, we will introduce some basic aspects about technology and our understanding of the role of technology in our lives. An attempt will be made to answer many of the questions raised above, as also, to raise other questions which will be elaborated upon in the chapters to follow.

1.2 Defining Technology

Let us start by trying to answer a very basic question, what do we mean by the term, technology? Amazing though it may seem (because we generally take technology to be synonymous with the last couple of centuries) the first discussions on **technology** can be found in the works of Aristotle and Plato and hence the word has a Greek origin, coming from

> **The word technology** ***originates from the Greek word "techne" meaning art, skill and "logia" referring to the underlying laws***

the Greek word "*techne*," meaning "art, skill, cunning of hand" and "*logia*," of course, referring to the underlying laws or principles (Liddell & Scott, 1980, *The Greek-English Lexicon*). With the coming of the twentieth century, we witness more formal definitions of the term. Probably, the definition most social scientists accept and adhere to is the one given by the American sociologist, Read Bain, in the *American Sociological Review*. According to Bain, "technology includes all tools, utensils, weapons, instruments, housing, clothing, communication and transporting devices and the skills by which we produce and use them" (Bain, 1937, p. 860). Another popular definition is the one advanced by Bernard Stiegler. He defines it in two ways: firstly, as "the pursuit of life by means other than life" and secondly, as "organized inorganic matter" (Stiegler, 1998, p. 17). By the twenty-first century, the range of technologies has increased in manifold ways, requiring an even more encompassing definition. The **Merriam-Webster Dictionary** gives the following meaning for the term technology: "the practical application of knowledge especially in a particular area" (2007) and W. Brian Arthur defines technology "as a means to fulfill a human purpose" (Arthur, 2009, p. 28).

> ***Definition of technology****: all tools, utensils, weapons, instruments, housing, clothing, communication and transporting devices, and the skills by which we produce and use them*

> ***Merriam-Webster Dictionary definition of technology****: the practical application of knowledge especially in a particular area*

Another way of looking at technology is in terms of its structure and functions. From the point of structure, it stands before us in the form of a product or a machine. But for most people, its system of operation and the intricate nature of its formation remain a mystery. It can be likened to the black box of an aircraft about which we normally do not know anything except its operation. So, for the vast majority of us, the primary interest in technology is only in the functions or the uses of that technology. However, when we deal with technology, we need to focus on both the issues: the creation of a product and its operation. Both aspects are related to each other, but ironically, as technology becomes highly complex, the ordinary human being starts feeling helpless and restricts himself to its functional or utilitarian aspect. Just think of the very many of us who have been driving cars with a great deal of perfection and for years together, without ever knowing about the complex ways in which its engine operates.

Thus, there seems to be no singular way by which we can define technology because of the very nature of the concept itself. As human beings engaged in various forms of activities to enhance their adaptation to the environment, not all forms of behavior, albeit goal oriented, resulted in the development of a product. Similarly, many technological achievements have taken place and are still taking place not with any particular purpose, rather, they simply happened. However, let us look at the various ways in which technology can be defined. As discussed above, its meaning has been shrouded by various connotations, some of which have been summarized below:

- Technology is goal directed; it is purposive in nature.
- Technology involves manipulation of environment to promote our adjustment.
- Technology involves making things and therefore it is a form of human activity.

- Technology is resource based.
- Technology involves integration of raw material in some form to fabricate a product
- Technology is restricted by the availability of knowledge and resources.

Putting it all together, one can say that a **comprehensive definition of technology** would include the use of matter that has been organized in some way so as to help us to encounter our lives and solve its problems effectively. The range of items that come under the rubric of technology is vast, as aptly described by Bain and, therefore, a simple hammer is technology and so is the complex space station.

Comprehensive definition of technology*: use of matter that has been organized in some way so as to encounter our lives and solve its problems*

1.2.1 Science and Technology

Another aspect that often confuses the common man is the differences between technology and some related, albeit, independent terms. When we talk of medical technology, transportation technology, food technology, communication technology, textile technology, or cooking technology, or even space and satellite technology, we are adding prefixes to the word technology to specify the use of technology in that particular field. But, how about the generic term known as technology? Is it synonymous with the term "science" or "engineering" for that matter? What does this technology accomplish for us and even more so, what does technology do to us? Have we become different from our ancestors of yore, who lived in the jungles and caves? Is technology useful for us? Is it good for us? Should we allow it to proliferate, ad infinitum? Why is it that we like certain technologies and are ready to adopt them, while there are others that we tend to abhor from the very start? These and many more such questions continue to perplex us once we start thinking about the whole issue of the role of technology in our lives.

As a child, you have probably often been asked to solve the riddle regarding the hen and the egg: which came first, the hen or the egg? Trying to differentiate between science and technology is very much akin to this riddle. Scientists and engineers are convinced that technology is nothing but applied science, but did we not have technology even when there was no science or scientific principles per se? In the next chapter, we will describe how even subhuman species such as dolphin, elephants, and chimpanzees have been seen to devise tools to aid them in the digging of food. We will also describe some of the earliest known tools used by man, dating to some 3.5 billion years ago. Did animals use principles of science to construct their tools? Did science exist 3.5 billion years ago? Though scientists will insist that it is science that paves the way for technology, there are enough examples to clarify that technology can exist even without formal science. The chimpanzee uses a twig to dig up worms while the elephant uses a branch of a tree to drive away mosquitoes from its back. Similarly, we had tools used by early man to crack nuts, or the pointed tool used to pull away flesh from the bones of a cadaver, The twig, the branch, the stone hammer, the pointed tool—they are all tools and, thus, they can all be classed as technology that we have constructed to solve a problem—they have all made life easier for us, but did science, as a set of principles, exist at that time? And then, what would be your conclusion regarding science and technology when you think of the fact that science has advanced often because of technology? Galileo would not have been able to advance his theory without the invention of the telescope and modern science is dependent upon a whole host of technologies in order to gain insight into the phenomenal world. At the same time, there are other examples to show that technology is a consequence of science. The telephone could be invented only after it was discovered that sound travels from one point to another given a medium through which to travel. The airplane is a consequence of the scientific principles of aerodynamics being put to use by engineers. As can be seen, there is no one answer to whether science preceded technology, did technology exist before science came into being, or whether they have advanced, each in their own way, independent of one another. But, one can try to analyze the differences between the two.

Difference between science and technology: *science deals with reality that is given while technology creates reality according to a design*

By and large, science "**investigates the reality that is given**," whereas technology "**creates reality according to a design**" (Skolimowski, 1966, p. 44). Later, Herbert Simon (1969) contended that whereas science is concerned with how things are, technology is concerned with how things ought to be. Let us examine this difference with an example. Science refers to the development of systematic knowledge that gives us insight in order to establish a cause–effect relationship. A strong science offers firmly tested theories that have high predictive value in different conditions. Take objects of different weights, toss them up and you will find that all of them fall down. The common underlying cause is gravity. Using this knowledge, we learn to give support to objects we use or else they would be rolling down on the surface. However, with no gravity in outer space, the same gadgets would still float irrespective of the support provided to them. Presently, science is unable to deal with this problem of weightlessness and lack of gravitational forces. However, will we wait until such time as when science is able to offer information on restraining objects from flying in a weightless environment? The answer is certainly "no." Instead, we have developed technology to restrain flying objects and astronauts have been managing the weightlessness with the use of appropriate technology developed for that purpose.

In fact, technology has been in existence long before human beings understood science. Before we created the formal nature of science, commonly known as experimental science, technology, in the form of various human crafts, was already a part of human endeavor. Explaining behavior in terms of adaption to the environment, as long back as in the early part of the last century, Dewey (1935) argued that there is ample anthropological evidence to show that human activity can be viewed as being synonymous with technology. For Dewey, there is no need to stipulate that science should be a precursor of technology, for, technology was there when humans were continuously making efforts to develop products to augment their adjustment to their changing environment (Hickman, 1990). When an ethnologist found a bird creating the tallest nest of about 6 ft to attract a female in an island in Java, it was a technological marvel based on the capacities of this bird and comparable to the technological marvel of those craftsmen who built the Taj Mahal without having received any training in the field of architecture. In the entire history of mankind, a lot of technology evolved as a result of human activity, and it occurred more by accident than because of any purposeful design.

__Science__ comes from Latin word "scientia" meaning knowledge

Another way to understand differences between the two terms, science and technology, is through their respective etymology. As described above, technology originates from the Greek word "techne." Science on the other hand emanates from the Latin word, "**scientia**" meaning knowledge. As the *ScienceDictionary.com* (2007) states

> "Science is the reasoned investigation or study of phenomena, aimed at discovering enduring principles among elements of the phenomenal world by employing formal methods such as the scientific method."

Thus, while science attempts to unravel the mysteries of the world, technology helps us solve some of the day-to-day problems. Science is epistemological while technology is pragmatic. It would probably not be wrong to say that while science studies the world as it is, technology aims at changing that world to suit human adjustment. In order to be useful, technology must satisfy our requirement in several ways: it must not only be useful, it must also be usable and at the same time, safe. Any knife would be useful, but it becomes safe only when its blade is protected by a handle and usable only when the handle fits into the palm of the user. In order to satisfy all three requirements, technology is often not exclusively a product of science. It is based on many fields of knowledge, science being just one of them. Technology has to draw heavily from disciplines such as ergonomics, mathematical sciences, linguistics, and even historical knowledge and culture, so as to construct something that would be of use to us.

1.3 Philosophy of Technology

While the origins of philosophy of science can be said to date back to Aristotle's *Organon* and many of the writings of Plato, philosophy of science emerged as a distinct discipline only in the mid-twentieth century with the works and writings of Thomas Kuhn (1962, *The Structure of Scientific Revolution*). Thereafter, of course, there was a proliferation of philosophies, one for every discipline. So, one finds a philosophy of physics, that of chemistry, biology, economics, social sciences, and even a philosophy of psychology. What we are specifically interested in is technology and whether there exists anything akin to the philosophy of science as far as technology is concerned. Specifically, we would like to know whether there is a philosophy of technology, and if there is, what are its aims and objectives.

There is a basic difference between the philosophy of science and that of technology. As far as the former is concerned, it has laid emphasis upon what constitutes science, what should be the subject matter of science, its methods and implications. As far as the impact of science upon society is concerned, such questions have been left to social scientists and historians. However, when we start reflecting upon the writings of scholars of **philosophy of technology**, the main question has been the impact of technology upon the human race.

Philosophy of technology: *tries to delineate the impact of technology upon the human race*

Philosophers have reflected upon technology from the times of the Greeks and the Romans. The writings of Plato and Aristotle clarify the ways in which technology was conceptualized at that time. Plato considered technology to be basically an imitation of nature (so man watched spiders weave their webs and this gave him the idea that the weaving of cloth is possible and a technology emerged for that purpose). Aristotle, however, was of the view that while man did imitate nature, it also added to what he saw in nature. "Generally art in some cases completes what nature cannot bring to a finish, and in others imitates nature" (*Physics* II.8, 199a15; see also *Physics* II.2, Schummer, 2001, cf. *Stanford Dictionary of Philosophy*).

However, it was not until the middle of the twelfth century and the writings of Francis Bacon that due emphasis was laid upon developing a philosophy of technology, and it was still later, in the nineteenth century that the first text on the philosophy of technology appeared on the scene. This was *Eine Philosophie der Technik*, written by a philologist and historian Ernst Kapp in 1877. As technology became more pervasive in the twentieth century, we find a large number of social scientists and scholars from other disciplines becoming concerned with the role of technology in the sociocultural fabric of the world. The more prominent among these are Martin Heidegger, Arnold Gehlen, Lewis Mumford, Jacques Ellul and Albert Borgmann, Don Ihde, Bernard Stiegler, and Bruno Latour. While a number of important individual works were published in the second half of the twentieth century, two books published at the turn of the century marked the development of the philosophy of technology as an academic subdiscipline. These were, *Technology and the Good Life* (2000), edited by Eric Higgs, Andrew Light and David Strong, and *American Philosophy of Technology* (2001) by Hans Achterhuis.

Let us now focus on some of the important aspects of the philosophy of technology. First and foremost, we find that there are two broad groups of thinkers who have tried to work out a philosophy of technology. The first group constitutes those who have been called the **humanities group** because they are generally from the humanities disciplines and more importantly because they have been concerned with the role of technology in our lives. The second group, known as the **analytical**

Humanities group: *from humanities group and concerned with role of technology in our lives*

Analytical group: *concerned with the design and engineering aspect of technology*

group is concerned with a totally different set of questions, namely, with the design and the engineering aspect of technology, as also with methodological issues. These include Henry Skolimowski, Herbert Simon, Mario Bunge, Ian Jarvie, and Michael Polanyi. While some defend the similarities between science and technology, others argue for their relative independence.

The turning point in the growth of philosophy of technology was the rejection of the Cartesian model having its roots in the works of the philosopher Descartes who believed, "I do not recognize any difference between the machines made by craftsmen and various bodies nature alone composes" (Cottingham, Stoothoof, & Murdoch, 1984, p. 99). This approach made an enormous impact on physical and biological sciences and most scientists began to investigate body functions as if they were functions of a machine. As far as the domain of psychology is concerned, such an approach implied that we tend to entertain ideas to be true (or false) and to stay so by default until we make an attempt to change them (Gilbert, 1991).

Unlike the commonly known philosophy of biology or the philosophy of physics, the specific subfield of philosophy of technology remained relatively unknown until German philosopher Martin Heidegger (1962) argued that technology is about more than merely developing instruments. He concentrated on two aspects of technology:

- "Technik," referring to older forms of technology concentrating on making things as we see in the conventional areas of mechanical and electrical engineering
- "Technologie," which is instrumental in nature and aims at uncovering and discovering, for example, what we now know as biotechnology, nanotechnology, etc.

> ***Hervorbringen****: to make things*

> ***Herausforden****: to change nature*

Sensing the changing role of technology and its impact on human lives, Heidegger argued that in our pursuit of a deeper understanding of technology we need to move beyond ***hervorbringen*** (to make things) to ***herausfordern*** (to change nature). While a clear example of the former is our ability to develop the windmill which is dependent on the flow of air to obtain maximum effects, the latter is embodied in technology that challenges nature and modifies it to obtain results, for example, creating dams and managing the flow of water in them.

While philosophy of technology is a relatively new subject, it has considerable significance for psychology. For example, while most of us believe that technology makes us modern, French scholar Bruno Latour (1993) raised the question "What does it mean to be modern" in his famous book, *We have never been modern*, and addressed issues regarding our mental landscape. In developing his Actor-Network Theory, popularly known as ANT, he questioned our understanding of our own society and how it influences our relationship with technology. If we view society as a bundle of ties, it is confined to its material domain and remains a product for us manifested in our normative behavior or in the established and stable state of affairs. On the other hand, when we view society in the context of its original meaning, that

Table 1.1 Our understanding of technology and society

	Classification of environment	
Human conception	Evolutionary (society)	Unnatural (technology)
View as a product	A bundle of ties/product	Objects/artifacts
View as a process	Seeking connections/ties	Seeking possibilities

is, how it was assembled or what was assembled, we engage in, what Latour called, "seeking trace connections." Essentially, he was referring to the fact that a better understanding of relationships can be had only in the context of anthropology.

We believe that the impact of the views of both Heidegger and Latour are important for the understanding of technology and human behavior in more than one way. First, both invite us to think beyond society or technology in static forms or products, and that they would be better understood in terms of processes rather than as fixed materials. Second, in the context of our conception of nature, both technology and society have been playing many significant, and often difficult to measure, roles. The role of philosophy of technology or for any science, for that matter, is to help us develop an understanding of how and why the boundaries between science, technology, humanities, and social sciences remain unconnected or become blurred. Any scientist, including the psychologist, will then be forced to face the questions of ethics that are so relevant for our own existence as humans. Table 1.1 illustrates the interrelationship between technology and society in the context of such a conceptualization.

With the amazing technological developments of this century giving us the feel of living in a natural environment despite the fact that the environment has been artificially produced (as in virtual reality or VR) or for that matter, parts of our body may be artificial (e.g., the pace maker in the heart or the artificial limb), the boundaries between technology, society, and human beings are fading and becoming blurred at both the individual and the social levels, which is, in itself, overwhelming for modern social sciences. The ways in which technology, society, and humans will connect is best summarized in a recent issue of the *Time* magazine (August 17, 2015). Quoting Jaron Lanier, VR entrepreneur, in the article, *Inside the box*, *Time* reported: "Virtual reality," he says, "is a means of spontaneous, improvisational visual expression, the same way that talking is a means of aural communication; it's the next logical step from written language to printing press to photographs to audio recording to film. It can blur the distinction between you and the rest of the world," (p. 49).

In his article, *Toward a philosophy of technology*, Hans Jonas (1979) had reflected on several aspects of technology. **First**, according to him, the traditional view of technology as a mere state of human activity or possession is no longer valid. Such a static view of technology goes to undermine the comprehensive nature of its impact. Technology has now become a part of human enterprise forcing human aspiration to raise the bar and to strive for success as never before. It has helped create new laboratories that have expanded the role of science and paved the way for indefinite progress. Technology is no more considered simply a means to an end; it is now an art to complement our knowledge that is seemingly fathomless and unending. It even seeks to replace our cognition with technology in the form of robots, enabling it to function beyond present human limits.

Secondly, technology may be viewed in terms of its material form. In this case, it is forced to depend on science for its growth. A discovery in science is of no value if it cannot be applied. The discovery of electricity fails to have much significance unless and until it can be used to light homes and operate numerous gadgets. It is this material form of technology which also represents its ugly nature. When we generate electricity with the help of atomic energy, we create a risk for our environment. The malfunctioning of nuclear reactors and dumping of its waste has led to catastrophic effects. Even with our current knowledge, what we now know about such adverse effects may be regarded as merely the tip of the iceberg.

This material connotation of technology has undergone rapid changes during the past century. While scientific knowledge of mechanical and chemical components was the basis of technological developments in the past few decades, it is developments in electronics, today, which have led to technology becoming an integral part of our life. Take away the various gizmos from college going youth, and they would begin to feel as if they have regressed to some primitive form of life. Similarly, a power failure or the loss of an internet connection for even a very short period of time has a debilitating effect on us.

The **third** impact of technology, according to Jonas, is related to moral issues. Technology does not distinguish between the good and the evil or between what is fair and unfair. The use of surveillance gadgets could be viewed by some as good for the security of community, but for others, it amounts to invasion of their privacy. One of the greatest dangers that technology produces is the way in which it empowers the user. The cold war between the USA and the Soviet Union may be considered to be a good example of this. Both had nuclear power and continued to accelerate their technical arsenal to dominate the world, each one believing that it would be powerful enough to prevail upon the other. We all know that if the use of technology is not monitored carefully, it may cause catastrophic effects (see Box 1.1). It is no wonder that we are living on the edge of moral failure in our use of dangerous technology.

Box 1.1: Disaster on the Three Mile Island (TMI)

The Three Mile Island accident was a nuclear meltdown that occurred on March 28, 1979, in reactor number 2 of the Three Mile Island Nuclear Generating Station in Pennsylvania, USA. It is said to be the worst accident in US commercial nuclear power plant history.

The accident began with failures in the nonnuclear secondary system, followed by a stuck-open pilot-operated relief valve in the primary system, which allowed large amounts of nuclear reactor coolant to escape.

(continued)

Box 1.1: (continued)

The mechanical failures were compounded by the initial failure of plant operators to recognize the situation due to inadequate training and other human factors. Despite the valve being stuck open, a light on the control panel supposedly indicated that the valve was closed. As a result, the operators did not correctly diagnose the problem for several hours. Even more so the operators had not been trained to understand the ambiguous nature of the pilot-operated relief valve indicator and to look for alternative confirmation that the main relief valve was closed.

What really happened? Blockages are common with the resin filters in the nuclear generators and are usually fixed easily, but in this case the usual method of forcing the stuck resin out with compressed air did not succeed. At 6:56 a.m., a plant supervisor declared a site area emergency, and less than 30 min later station manager Gary Miller announced a general emergency, defined as having the "potential for serious radiological consequences" to the general public. The worst part was the fragmentary, ambiguous, or contradictory statements made to government agencies and to the press, particularly about the possibility and severity of off-site radioactivity releases. At a press conference held thereafter, officers were reassuring, yet confusing, about this possibility, stating that though there had been a "small release of radiation… no increase in normal radiation levels" had been detected. These were contradicted by another official, who claimed that no radioactivity had been released. Later that day, the official changed his statement, saying that the situation was "more complex than the company first led us to believe," and schools were closed, residents were urged to stay indoors and farmers were told to keep their animals under cover and on stored feed. The Governor, on the advice of NRC chairman, advised the evacuation "of pregnant women and pre-school age children…within a five-mile radius of the Three Mile Island facility." The evacuation zone was extended to a 20-mile radius on Friday, March 30. Within days, 140,000 people had left the area. More than half of the 663,500 population within the 20-mile radius remained in that area. According to a survey conducted in April 1979, 98 % of the evacuees had returned to their homes within 3 weeks.

The accident caused intense concerns for safety among activists and the general public, resulted in new regulations for the nuclear industry. Cleanup started in August 1979, and officially ended only in December 1993, with a total cleanup cost of about $1 billion.

Post-TMI surveys have shown that less than 50 % of the American public were satisfied with the way the accident was handled by Pennsylvania State officials and the NRC, and people surveyed were even less pleased with the utility (General Public Utilities) and the plant designer.

(continued)

Box 1.1: (continued)

The Three Mile Island accident inspired the now well-known Normal Accident Theory by Charles Perrow (1984), in which an accident occurs, resulting from an unanticipated interaction of multiple failures in a complex system. TMI was an example of this type of accident because it was "unexpected, incomprehensible, uncontrollable, and unavoidable." Perrow concluded that the failure at Three Mile Island was a consequence of the system's immense complexity. Such modern high-risk systems, he realized, were prone to failures however well they were managed. It was inevitable that they would eventually suffer what he termed a "normal accident." Therefore, he suggested, we might do better to contemplate a radical redesign, or if that was not possible, to abandon such technology entirely. "*Normal Accidents* contributed key concepts to a set of intellectual developments in the 1980s that revolutionized the conception of safety and risk. It made the case for examining technological failures as the product of highly interacting systems, and highlighted organizational and management factors as the main causes of failures. Technological disasters could no longer be ascribed to isolated equipment malfunction, operator error or acts of God" (Pidgeon, 2011, p. 404).

Source: Nuclear Regulatory Commission: Backgrounder on the Three Mile Island Accident (2014)

Morally responsible design: *designing technology keeping in mind that it should not harm the user*

More recently, the thinkers from the analytical school have been showing concern regarding the issue of **morally responsible design** and value-sensitive design (Roeser, 2012; van den Hoven & Weckert, 2008; Zwart, Poel, Mil, & Brumsen, 2006; Friedman, 2004). A very telling example is provided by Burg and Gorp regarding the design of a safer trailer. By empathizing with potential victims of a suboptimally designed trailer, an engineer can use his imagination and knowledge of engineering to avoid such mishaps even if the customer had not asked for such features. After all, it is the designer of the technology who knows about or who can think of the possible consequences of the technology, either negative or positive. In other words, technology should not be value neutral for even the designer. Engineers should not be like unemotional calculators; rather they should be morally responsible engineers (Burg & Gorp, 2005). It is not just a mere tool, its design will determine the ways in which it will be used and whether it will be not just usable but also safe to use.

A very different conceptualization has been posited by Ihde and Selinger (2003), who argue that we would be able to understand technology by using the notion that some form of human action precedes human cognition. As we interact with technology, it becomes a part of our self-concept, that is, we begin to perceive and evaluate our own existence in relationship with that technology (Box 1.2). As this self-conception grows stronger, we start feeling empowered to achieve endlessly.

Box 1.2: Don Ihde: Non-neutrality of Technology-Mediated Experiences

> "For every enhancement of some feature, perhaps never before seen, there is also a reduction of other features. To magnify some observed object, optically, is to bring it forth from a background into a foreground and make it present to the observer, but it is also to reduce the former field in which it fit, and—due to foreshortening—to reduce visual depth and background" (Ihde, 1979, p. 111).

Ihde uses several examples from optical technology to explain the above. Take the wearing of simple eyeglasses. While the glasses enable you to see things more clearly, it also reduces your vision for objects that lie at the periphery of your vision. And, according to Ihde, no matter what the technology, such magnifications of or affordances for certain objects minimize our perception of other aspects. No form of technology is a simple neutral tool which does nothing more than helps us to perform certain activities; technology alters our perception and experiences of the world; technology-mediated experiences are different from those which are not technology mediated. We pay a price for every technology that we use; if nothing, one must care for it, one must carry out maintenance functions; and, we definitely pay a monetary price for it. Technology is therefore non-neutral.

Ihde also distinguishes between three types of relations that people could have with technology, namely, embodiment, hermeneutical, and alterity relations.

Embodiments: for certain technologies, our relationship becomes an embodiment, meaning that we see the world through the technology. Again, whether it is optical technology, hearing aids, a cane, the type people with visual impairments use, our perception is restricted by the technology. But the beauty of such sensory enhancing technology is that after initial adjustments, the sensory aid recedes into the background, it becomes transparent and allows the person to sense the environment through it; it is as if the technology has become a part of the person, it has become an embodiment.

Hermeneutical relations: however, all technologies do not become embodiments of the user. Rather, a relationship opposite of the above can form. Take a glucometer, such that is used to measure blood glucose levels by diabetics. Or, think of a thermometer used to measure temperature, whether it is that of the human body or that of a certain city. In both these cases, the technological aid does not become transparent. The diabetic does not see his blood through the glucometer, and, neither do we feel the temperature of our body through the thermometer. Rather than receding into the background, the tool becomes the foreground and the world recedes into the background. Moreover, one must acquire skills necessary to use the tool. One not only becomes dependent on it as in the case of embodied relationships, but also, one cannot experience these aspects without the tool. Without the glucometer, the diabetic would not

(continued)

Box 1.2: (continued)
know what her blood sugar level is like, and without the thermometer, you would not be able to know what the temperature is like. One becomes dependent on the tool even though one may not know whether it is working properly or not. The instrument may be giving you wrong results but unless you compare the results from some other tool, you will not know whether you are getting true or false results. Your sugar levels may have risen but you trust your glucometer which is saying that the level is well within the normal range. Such a relationship with technology can be called a hermeneutical relationship.

Alterity relationships: whether it is your eyeglasses or whether it is a thermometer, you normally do not become emotionally attached to it. You drop your spectacles and they break, you simply go to the optician and get another pair readied for use. But there is some technology to which you get emotionally attached. There are other technologies for which there is a sense of interaction with it much as your interactions with other people or animals. But the difference is that these interactions fall somewhere in between in our continuum of experiences. They are not as strong as that with other people and animals and nor are they as weak as are relationship with objects. They fall in between. A good example is that of computer games or video games. While playing them, you get the feeling that you are playing against an opponent, when in reality there is none. The feelings on winning or losing are akin to those you undergo when playing with a real opponent, you must beat it. Your relationship with a mere technology has been altered, you start anthropomorphizing it, giving it human like attributes. Another characteristic of alterity relationships is that your emotions regarding the technology may get altered. As long as the computer or your cell phone is working, it is fine, and you love your cell phone, but just imagine that yourself losing it. It is as if *you* are lost without it.

"The quasi-love relationship" is lost, the tool itself becomes conspicuous, and the relationship transforms into frustration and "quasi-hate," a kind of alterity human–technology relation (Ihde, 1990, p. 106).

"To summarize, technologies mediate our experience of the world. They appear in between humans and the world and change our experiences, amplifying some aspects while reducing others. While a technology may enable one to act on the world, it simultaneously limits other ways of acting, enabling new experiences while closing down the potential for others. It is however important to realize that "no technology is 'one thing,' nor is it incapable of belonging to multiple contexts" (Ihde, 1999, p. 47). Technologies are thus multi-stable, as they can be embodied in various ways for various purposes."

Source: Fallman (2011)

As we can see from the above, the relationship between human beings and technology has been changing in the course of human history. As mentioned earlier, the role of technology in the past was confined to its subservient nature, but now this has changed such that it transforms our relationship with nature even to the extent of creating a paradox. Technology is not only being used to enhance our adaptation to nature, but with the developments in genetic engineering, our own body has become an object for technology. In other words, we have become so dependent on technology that it is helping us to explore the very nature of our existence.

While no one doubts that technology has deeply and irreversibly affected the social character and social fabric, there is considerable controversy between scholars as to whether technology drives society (technological determinism) or whether it is vice versa (social determinism). These have also been known as "push" and "pull" theorists, respectively. Linda Green in her book, *Technoculture*, (2001) attempts to draw distinctions between these two sets of thinkers. According to her, those who believe in technological determinism, firmly believe that "it was features of technology that determined its use and the role of a progressive society was to adapt to (and benefit from) technological change," (p. 2) while those who talk about social determinism, "looks upon society being at fault for the development and deployment of technology" (p. 3).

We are all familiar with the works and thoughts of Karl Marx who elaborated on this very idea when he wrote that changes in modes of production are the primary reason for change in social structure and cultural practices. So, with the advent of the industrial revolution, mass production became possible, bringing in its wake many a social change, including the migration of people from rural to urban manufacturing hubs, changes in the family structure and the intermingling of cultural practices from a variety of regions. This led, in turn, to the development of a new culture, the culture and traditions of the migrant worker, who is physically present in the factory but his heart still burns for his family and home in the faraway village.

A more recent proponent is Thomas Friedman, who, though a journalist, changed the mindset of many through his highly acclaimed book *The World is Flat* (2005). In other words, we organize ourselves and our society in a manner that is dictated by technology. For example, with advancements in communication technology has not our mode of communication, and especially that of the youth, undergone a sea change? Did we SMS earlier? Did we use Whatsapp or Facebook to share photos not only with people across continents but also with the person who is sitting very close to you, say in office, but whom you fear to disturb directly? As luggage with wheels is becoming more and more common and technologically more advanced, that genre of people, known as porters, is fast disappearing. The same is the case with the stenographer in office. With more and more people typing their own correspondence on their computers, one does not need to dictate letters, memos, notes, etc., and as a result one hardly sees stenographers in offices, leave alone seeing advertisements for courses on typing and stenography. Communication technology has also enabled concepts such as flexi-work times and video conferencing to become a reality. Technology seems to lead to a process of natural selection much akin to the Darwinist natural selection in the evolution of species, causing some societal processes to become extinct and at the same time, retaining those that are conducive to the advancement of that technology.

But this is only half the truth. Yes, technology is important and does affect our lives in very important ways but there are other factors that are also important, sometimes to a lesser degree but at other times, to an equal degree. This realization is what brought people to talk of a "soft" technological determinism, as contrasted to the "hard" deterministic role of technology discussed above. A soft deterministic view point believes that though technology is important, it interacts with the socio-political situation. While technology continues to be the guiding force, we have a chance to take decisions regarding the adoption of the technology. A prime example of this is what has been termed as a cultural lag with regard to technology adoption.

While the above viewpoints were true during the early phases of technology development, for example, when inventions were taking place at great speed during the industrial revolution, the intertwining of technology with other social, cultural, economic, and political factors is apparently more complex.

> ***Social shaping of technology school:*** *bidirectional relationship between technology and society with each shaping the other*

There is a third school of thought, namely, the **social shaping of technology school**, which believes that rather than there being a unidirectional relationship between technology and society, there is a bidirectional interaction between the two, with each shaping the other. As stated by Murphie and Potts (2003), "technology does not determine but operates, and is operated upon, in a complex social field." In complete contrast, the social constructionist believes that the path of innovation is shaped by society and various societal forces, allowing only those aspects of technology to become reality as fit in with the social setup. Why is the technological divide so clear in some countries but not in others? And even this technology divide is not equal for all types of technology. In India, for example, penetration of internet is still limited to urban and semi-urban areas and also follows a pattern dictated by socioeconomic class, but mobile telephony does not follow this pattern. One finds people from all socioeconomic classes and from even remote rural areas using a cell phone very smartly. One reason is that cell phones are cheaper than computers. The same can be said for labor saving devices used in the kitchen. With low incomes, these devices are more of a luxury than a necessity as far as India and other developing countries are concerned but they become a necessity where household help is far more expensive than the price of gadgets, as in the case of the developed Western countries.

Summarizing from the above, it is clear that the role of technology in human life and endeavor has been undergoing widespread change. Some of these have been delineated below:

1. Initially, technology operated at a very basic level. For example, look at the tools of primitive human beings found in Ethiopia some 3.5 million years ago.
2. As human beings entered the era of land cultivation and agriculture, a more complex set of tools was needed. Tools became a medium through which man could

unravel the mysteries of nature. During the era which saw the growth of the agricultural society, nature was adored and any technology that brought a change in our relationship with nature was abhorred and perceived as being unsafe.

3. With the onset of the machine era, the role of technology transformed enormously. It took over several functions of the gifts of nature, for example, an automobile replacing the animal driven buggies. Human skills gradually surrendered to that of mechanical gadgets. No longer was the pride in human handwriting considered practical in large-scale transactions and for formal and large-scale communication, typewriters were obviously a better option. By and large, several chores performed with the help of the human body were taken over by human-made devices, that is, technology.
4. With the invention of computers—the greatest technological achievement of the twentieth century—each and every part of our life has changed. We no longer look at a book for an answer but open a computer and search. We do not remember telephone numbers of friends and acquaintances but look at the list of contacts in our mobile phone. Technology has offered us artifacts that have the ability to supplement the functioning of our brain. Its role has expanded from controlling functions of the body to the control of mental operations.

Consider the long white cane of a blind man. It is not simply a tool. Over a period of time, it has become a part of human operation much the same way as he used his hands for navigation. This is the physical part. With advancing technology, human beings have found an alternative to their cognition. No longer do we need to think about and solve complex numerical operations; we simply feed them into a computer and almost immediately we receive an output. Technology now provides a substitute to mental operations and more. It has already begun to inform us about the functioning of our brain through a product that has been invented by our own brain. As such the relationship between human beings and technology has become interactive; the more we invent, the more we learn about our own brain.

The pervasive influence of technology has led us to perceive the resemblance between our body and a machine. Our nerves are projected as an electrical system and our arms as levers, and so on. With the advancement of genetic programming, human beings have become an object of technology. Can we imagine where technology will lead us once the script of our genetic code has been identified? There would be an unending door of applications for which we might not even be prepared. In essence, the role of technology will shift from being a means to becoming an end. Rather than watching the movie *Shreck*, the viewer would want to be the *Shreck*. In short, the contemporary view of technology seeks answers for what technology is versus what it is not; and even more, what it can and should do (Latour, 2013, 1987; McGinn, 2010; Martin & Schinzinger, 2005; Mitcham, 1994).

It is clear that as technology became more pervasive our conception of the relationship between technology and society has also changed. This has been presented in the flow chart below and detailed further in Box 1.3.

Technological determinism → hard and soft determinism → social construction of technology → technology as a system

Box 1.3: Albert Borgmann and the Device Paradigm: Our Subservience to Technology

"In this rising tide of technological devices, disposability supersedes commanding presence, discontinuity wins over continuity, and glamorous thrills trump centering experiences," (Strong & Higgs, 2000, p. 24).

According to Albert Borgmann, as far as technology is concerned, a thing that is useful may not always be good. So a refrigerator is useful but it also exudes gases that lead to global warming.

Borgmann's device paradigm: Borgmann differentiates between focal things and technological artifacts that he calls devices. Giving the example of the traditional wood fire place, he explains how this demanded not only patience but also the presence of people who are both present and also continuously engaged in keeping the fire going. It thus engages not only the mind but also the body. There is a continuous connection between the means and the end, possible only if means (i.e., the wood), is there, making up the fire, lighting it and tending it and the end of procuring heat and warmth from it. This wood fire place afforded a place where people would gather around, gossiping till the last embers remained, collectively tending the fire and, thereby, connecting people and creating a social bonding between the people who are present. Such things are focal things that engage not only the mind but also the body. In the words of Borgmann, "a focal thing is not an isolated entity; it exists as a material center in a complicated network of human relationships and relationships to its natural and cultural setting" (Strong & Higgs, 2000, p. 23).

Borgmann then compares the traditional fire with the modern heating system. While it is very convenient, all you have to do is switch it on and turn it to the temperature you want and sit back to relax with a book or with your laptop. Does it require your constant presence? Does it require collective effort to keep it going? Does it require physical effort or any special skill? No, none of these are required and these are the supposed advantages of the heating technology. But by so doing, the technology makes us passive users. At the same time, it is useful only so far that it provides us warmth; that camaraderie that characterized the wood fireplace is also missing, making us passive recipients of the technology, but removing us further from the world around us. As Fallman (2011) puts it, "At the heart of Borgmann's philosophy of technology is thus the notion that modern technology tends to operate to deconstruct things and reconstitute them into devices, and that this transformation is accelerating with recent advances in information technology." The irony of technology is that its devices take us further and further away from people, which require less and less of human input, disengaging us from each other and in so doing frustrating the higher and deeper aspirations of the person. He is of the view that technology is gradually creating us into passive creatures and fears that rather than being the master of that technology, we may become subservient to it.

Source: Strong and Higgs (2000)

1.4 Psychology as a Science

Compared to the many other sciences that have been taught for centuries, the teaching of psychology as an independent science of behavior and mental processes does not have a long history. Whereas, on the one hand, philosophy and religion accounted for the conceptual growth of psychology for a long time, physical sciences, on the other hand, expanded its growth through explorations in brain functioning, mostly through improvized tools of research. The German scholar Helmholtz, known for his pioneering work in vision and audition, believed that psychology would never gain the status of a science owing to its vulnerability in not being able to employ objective tools for the study of mental processes.

As psychology began employing tools that were scientific in nature, more universities and colleges started offering undergraduate and graduate programs in their curriculum. The first laboratory of psychology was established in Leipzig, Germany in 1879 by Wilhelm Wundt who was both a philosopher and a physiologist. Around the same time momentum gained and led to the starting of the teaching of psychology at Harvard and other schools in the universities of the USA. Subsequently, psychology became a popular subject of study in North America and around the globe. For example, in 1985, there were only five universities teaching psychology in China. By the end of the century, this number rose to 40.

The growth of psychology is generally described on the basis of what are known as the four main schools of psychology: psychoanalytic, behaviorist, gestalt, and humanistic schools. Led by Freud, the **psychoanalytic school** was primarily concerned with the study of different aspects of mental life ranging from the unconscious to the conscious mind and the impact of early childhood experiences. **Behaviorism**, on the contrary, focused more on the observable form of behavior and using various scientific techniques, demonstrated how our behavior is linked to the environment. **Gestalt psychology** primarily offered principles regarding the ways in which we organize information to create a unified whole known as a gestalt and laid the foundation of what we now know as the "cognitive revolution" in psychology. Cognitive psychology studies how we perceive and think and tend to adapt to our environment. Finally, led by Rogers and Maslow, **humanistic psychology** offered insights away from negative or neutral forms of human nature focusing on our ability to seek growth and healthy development. While each school has contributed to the development and exploration of several aspects of behavior and mental processes, they seem to have disappeared from contemporary psychology. However, the impact of these schools has led to the

> ***Psychoanalytic school:*** *study of mental processes such as the unconscious and conscious mind and role of early childhood experiences*

> ***Behaviorism:*** *focus on observable behavior and scientific methods to study behavior*

> ***Gestalt psychology:*** *offered principles for the ways in which we organize information and create a unified whole*

> ***Humanistic school:*** *focus on or ability to seek growth and healthy development*

growth of several subfields of psychology, for example, abnormal and clinical psychology—profoundly influenced by psychoanalysis; cognitive psychology—initially rooted in gestalt school; and **positive psychology**—emerging from the basic tenets of humanistic psychology.

> ***Positive psychology****:* *focus on the positive side of human nature, rather than on the negative*

As our behavior varies in different settings of our life ranging from the religious to the industrial setting, or say from the social to the educational setting, the need for the study of behavior in each setting led to the growth of many subfields of psychology: social psychology, educational psychology, industrial psychology, health psychology, etc. For our purpose here, we are presenting an argument for a specific, albeit interdisciplinary, subfield of the psychology of technology, believing that, as discussed in the opening paragraphs of this chapter, technology has influenced our behavior even in those subfields that have long been studied in the domain of psychology. Has not our teaching and learning within the classroom or outside it been influenced by technology? With our plan to travel in space, will not technology bring about major changes in our lifestyle? In the context of virtual environments, how are we going to develop a renewed sense of social life? In the company of robots around us, what type of relationships would we develop with them? Why was a robot given a funeral in Japan, much in the same way that a human is given a funeral? Will we feel and value the artificial left leg in a manner similar to how we feel and value the real right leg? The impact of technology has opened up a plethora of questions regarding our behavior in personal, social, and human-made conditions. In fact, it has called for a reevaluation of our existence as human beings.

Keeping the above scenario in mind, let us explore the subfields of psychology that have so far addressed these and many similar forms of behavior that have been influenced by technology (summarized in Box 1.4). In a conventional sense, subfields such as industrial psychology, applied experimental psychology, engineering psychology, human factors engineering or ergonomics have been addressing the impact of technology on our behavior (Box 1.4). By and large, **industrial psychology** attempts to seek "the application of the methods, facts, and principles of psychology to people at work" (Schultz & Schultz, 2009). Its goal is to study the behavior of people in the work setting. In most conditions of work, people are involved with the use of machines, which should be, ideally speaking, compatible to them or, at a minimum, understandably so. As machines became more complex and sophisticated, the issues of compatibility, safety, and optimum performance became salient in analyzing and understanding several key psychological processes such as cognition, perception, motivation, and emotions. With a diverse group of scholars from engineering, biological sciences, and social sciences including sociologists, this subfield of psychology became a specialized area in itself. Within the leading organization of psychologists in the world, the *American Psychological Association*, it emerged as a separate entity known as *Division 21: Applied Experimental and Engineering Psychology* and it coexists with *Division 14: Industrial Psychology*.

> ***Industrial psychology****:* *application of the methods, facts, and principles of psychology to people at work.*

Box 1.4: Precursors of Psychology of Technology

Applied experimental psychology: concerned primarily with the applications of empirically drawn psychology in the context of perceptual, cognitive, and motivational processes

Industrial psychology: study of behavior in the work setting

Human factors engineering: focus on man–machine systems and related issues

Ergonomics: ergon = work; nomos = natural laws

At the same time, there exist other subfields with similar or equivalent titles, for example, engineering psychology, psychology at work.

Box 1.5: Eindhoven University of Technology, UG Program, Psychology, and Technology

In Psychology and Technology, you learn how technology influences people and how you can use knowledge of psychology to let people use technology as effectively as possible. But you also learn how you can use psychology to design completely new technology applications such as motivational rehabilitation technology in healthcare, sociable robots, or brain–computer interfaces.

People and technology

Teaching staff and students of Psychology and Technology program deal with a wide range of different subjects, all of which focus on the relationship between people and technology. For example:

- Robots that help in the household
- Intimate technology for showing affection at a distance
- Online auctions you can trust
- Lighting that makes you healthier and smarter
- Avatars that help you save energy
- Intelligent street lighting
- Persuasive technology or how your mobile phone helps you to lose weight. And how can your shower make you use less water.

(Reproduced from the Eindhoven University web site) https://www.tue.nl/en/university/departments/industrial-engineering-innovation-sciences/education/undergraduate-programs/psychology-technology/

During the last decade, many psychology programs have become increasingly specialized in their curriculum and include technology as their primary focus even at the undergraduate level. They contain a large number of topics that are barely paid attention to in most colleges and universities around the globe. To make readers familiar with this scenario, Box 1.5 illustrates the psychology and technology program of a university in Europe.

1.5 On Relating Psychology with Technology

When we make an attempt to relate technology with psychology, there are at least three facets that can be considered. These have been detailed below:

1.5.1 Psychology in Technology

Psychology IN technology*: role of psychology to determine which types of technology the human apparatus can handle*

The entire gamut of technology and its development is highly complex, involving various academic disciplines ranging from physics to biology to engineering. With the increasing role of communication engineering, the range and breadth of information and communications technology far exceeds the capacities of the human being's limited sensory apparatus. Thus, when we speak of **psychology IN technology** we are limiting the role of psychology to the extent to which it can tie up with other disciplines to show which types of technology the human anatomy and physiology can handle and what it cannot. Other aspects of technology can be left for other disciplines to handle. In other words, if we take the example of cloning, the role of psychology would be limited to the understanding of the effects of cloning. Other aspects of cloning such as creating a clone could safely be left to the biological sciences. At the same time, increasing our pool of knowledge and moving towards this goal of cloning in terms of human capabilities definitely falls within the range of psychology. In this sense, our cognitive capabilities for innovation and entrepreneurship define the growth of technology. In other words, technology, at times, needs psychology, just as it needs various other disciplines, for its growth and development.

1.5.2 Psychology and Technology

Psychology AND technology*: the two disciplines remain independent entities, coming together only when they need each other.*

Second, we can consider **psychology AND technology** as independent entities coming together only when we require them simultaneously. For example, psychology uses technology when psychologists feel that they can gain from the use of technology. In a similar vein, we keep on manufacturing new products for business purposes until we discover its positive or negative impact. Some teachers still ban the use of calculators and computers in learning basic mathematics. We walk on the road instead of using a treadmill, believing that natural exercise is better for us.

While almost all sciences owe their growth to sophistication in technology, as of date, the role of technology in psychology is negligible. This argument is based on the fact that mental concepts require an interpretation of data at a level other than that required by other sciences. Whereas concrete findings in physics and chemistry revolve around a material component seeking an answer for another set of matter, much of behavior is intangible and for its interpretation, it is imperative to go beyond pure experimental data. Based on statistical validity, that is, significance of results based on the rejection of a hypothesis of zero difference (i.e., the null hypothesis), as experimental findings are, is not a sure guarantee for establishing the relationship between the observable form of behavior and its concurrence with human cognition. Sometimes, concurrence may be found between the two, at other times, there may be none.

Take the study of the role of reinforcement (basically, rewards and punishments) in influencing behavior. When Skinner began training his pigeons to track silhouettes of ships and reinforced them to learn to peck at an image, his primary motivation was to offer an application of this study for managing combat operations in WW II (www.bfskinner.org/org/project-pigeon). He believed that behavior is only what organisms do and refused to consider any mentalistic conception of behavior.

Sensing the above and being fully aware that interpretations of psychological data based on experimental evidence could be problematic, Skinner remarked:

> "My treatment of human behavior was largely an interpretation, not a report of experimental data. Interpretation was a common scientific practice, but scientific methodologists had paid little attention to it" (p. 27, *A matter of consequence*, 1983).

Most students of psychology know that Skinner's reinforcement theory is one of the most popular theories of psychology. Its applications have been equally popular. Through reinforcements, Skinner was able to change the probability of occurrence of behavior. By inventing, what he called *Teaching Machines*, Skinner offered a means of presenting reinforcement. With each success the student kept going ahead, recording his level of achievement. However, when he made a mistake, he was provided feedback so that he could take corrective action (based on the feedback). Such programmed learning opened the door for a very heavy dependence on technology for determining the outcomes of learning. The above example goes to clarify how **psychology and technology** can be related to each other as per the needs of psychology and psychologists (see Box 1.6).

Box 1.6: Skinnerian Reinforcement in the Technology-Driven World

It is generally believed that ease of use and positive experiences are two factors that will make people adopt new technology. In other words, they will act as positive reinforcers for change in behavior. But is this always so? But the reality is that technology can be a tale of two heads; one where positive reinforcement increases behavior and the other where the lack of reinforcement

(continued)

Box 1.6: (continued)

decreases desired behavior. A case in this direction is the new technology, Yammer, which was introduced by organizations as the organization's own social networking site over which people could discuss various work-related issues. Through this program, coworkers can collaborate and share ideas across an infinite number of people and locations instantaneously. Can you imagine that there would be some people who would not like this innovation? Yes, there were people who did not take to it. These were people who were slower in their thinking, who were more wont to weigh the loss and benefits of every action. The reason was that others who were faster or who were less risk aversive often took decisions even before this slow person had weighed all the alternatives and who may actually have the right solution, with the end result that such people would soon stop thinking. Thus, organizations that look to these new technologies must ensure that the reinforcement it provides to those using it generates the positive, desired behavior that they seek.

Thus, reinforcements have to be considered in congruence with the needs of the people being reinforced. Many a time, the same reinforcement may work in different ways for different people. We are drawing closer towards a psychology of technology based on the symbiosis of different schools of psychology.

Source: Daniels (2000)

1.5.3 Psychology of Technology

The goal of the present volume, however, goes far beyond the above-mentioned ways in which psychology has been related with technology. What we are interested in is **psychology OF technology** which attempts to unravel the ways in which we, as humans, engage with technology or are affected by it.

> ***Psychology OF technology****: it attempts to unravel the ways in which humans engage with or are affected by technology*

It is the scenario which develops when psychological principles are applied to the study of human behavior in the context of technology so as to provide insights regarding how we adapt under such varying conditions. This is the theme of this book through which we will explore the multifarious ways in which our cognitive, motivational, affective, and personality attributes are observed, explained and understood in relation to technology and how this knowledge helps in the prediction of human behavior in several domain-specific technology environments. It helps us to understand how we emote with technology, how attitudes towards technology are formed and changed as well as why some people adapt to technology very easily while others do not. In other words, it focuses on the deepening nexus between psychology and technology. An understanding of what psychology of technology entails would be better gained by looking at its goals.

1.6 The Goals of Psychology of Technology

By and large, this new field of psychology of technology deals with the kind of activity that is involved when technology and its content have an influence on human behavior. In a broader sense, technology becomes the benchmark for the capabilities of human beings for manipulating their environment and for exploring their own capacity of living in this technologically modified environment. Functioning in a virtual environment not only tests our physical skills, but also presents a challenge to us as far as managing our cognitive, motivational, and affective processes are concerned and evaluating our own existence in such an environment. If you have seen the movie *Avatar*, you may easily connect with the amazing range of human imagination that has been used in carving out this futuristic world. Just think about living in this way and then try to evaluate your behavior and yourself.

As with all fields in which the principles of psychology have been used to understand behavior in an applied realm (for further details, see Kool & Agrawal, 2006), there are **four goals of psychology of technology**, namely,

- The goal of theory building
- The applied goal
- The goal of application
- The goal of applicability

The primary goal of psychology of technology is the **goal of theory building**, that is, to seek, test and broaden psychological principles in the use of things and adapting to them. As mentioned above, when Skinner was developing his theory of reinforcement by testing his pigeons to pick up at the site of an image of a ship, he was conceptualizing an operation in war. Many military generals were not impressed by his theoretical arguments, but as we all know, Skinner's theory of reinforcement was later applied in diverse situations ranging from learning in the classroom to the modification of behavior in clinical settings. With greater understanding of human cognition, for example, of short-term and long-term memory systems and their neuropsychological bases, to the operation of our affective processes, including our response to choices around us and visceral responses to designs, psychological theories have been found useful in understanding human behavior in the context of technology (Norman, 2004). For details, the reader is referred to Chaps. 3–5.

> ***Goal of theory building:*** *to seek, test, and broaden psychological principles in the use of things and adapting to them*

The second goal of psychology of technology is its **applied aspect**, which focuses upon obtaining greater insight into the degree of usability of any technology and its consequences. If the steering wheel of my car has to be moved in the left direction to navigate the car to the left side, it demonstrates compatibility between the physical and mental operations used in driving my vehicle. Unfortunately, if you have a four-knob stove without a knob set right

> ***Applied goal:*** *focuses upon obtaining greater insight into the degree of usability of any technology and its consequences*

beneath it, the chances are that you might have, occasionally, used a wrong knob to start the desired burner. Have you ever thought of designing a lever for an up-down operation on a flat surface? What would be your configuration: moving in the right direction means going up, whereas a left movement is indicative of a downward movement? Wouldn't it be hard to learn and remember that? Traditionally, such issues have been studied in an area of psychology known as human factors engineering or ergonomics. In Chap. 3, we will focus on how human beings process information through their senses, organize it and then execute action as per the demands of the situation. In the absence of such knowledge, the development of any machine is likely to have a negative consequence for human performance. In Chap. 2, we have presented issues at the interface of technology, biology, and psychology by citing the applied nature of anthropometry (systematic measurement of physical human variation) and comfortability, or the degree to which a certain technology can or cannot be used by the population for whom it has been designed. While some answers have been provided by ergonomics and human factors engineering, a more comprehensive understanding can be had by looking at the ways in which we cognize technology. At the same time, psychology of technology in the applied context would also invite answers not commonly addressed in the current human engineering psychology: our relationship with robots, artificial limbs substituting a part of our body, and many other questions of humanistic origin. In other words, technology will be addressed in the context of "within" us rather than being "out" there (see Chap. 7).

The third goal of psychology of technology is its **application**. While it may be alright to develop a theory based on reinforcement and show its applied value, it is imperative, at least in the fast changing world, to demonstrate tangibly and accurately the value of the use of things in a community. Skinner's teaching machines used for programmed learning were criticized, but Skinner held on to his beliefs, and he even extended his ideas to help build an ideal society. This issue is at the heart of using technology as an intervention, that is, to demonstrate technology as a solution. While we may not have clear answers to all the issues and questions (e.g., at what age should we allow students to use calculators?), the field of psychology of technology will continue to seek information and facts to update itself and be ready to demonstrate the nature and limits of application.

> ***Goal of application:*** *demonstrate the value of use of things in the community*

Finally, the fourth goal of psychology of technology is **applicability**, that is, to look at the moral and ethical aspects of the applicability of basic research. Undoubtedly, this is a very sensitive issue. Applicability involves exploring the desirability of a program and its acceptance by the community. Skinner's effort to build a society based on the contingencies of reinforcement was not only rejected by his critics, but when such a society was developed in defiance of the criticism, it actually failed in the face of a real world. Similarly, although technology claims itself to be value-free, its use in real life has caused a wide range of conflicts. From women's right to drive cars in

> ***Goal of applicability:*** *look at the moral and ethical aspects of the use of technology*

some countries to restricting the hours of video games of children and more, the applicability of technology has been questioned often even before its inception. Take, for example, the use of surveillance cameras and the issue of freedom and invasion of privacy. Similarly, the operation of computers is linked to organic malfunctioning such as the carpel tunnel syndrome, but breaks at work have remained unchanged. After evaluating this problem, Australian stenographers have been provided some relief but workers in other countries have not been so fortunate. The modern wonders of engineering designs, the context of engineering, and the sustainability of new technologies such as those based on nanotechnology are surely widening our technological environment but in doing so are also making it more complex. As technology progresses, so will be the range and breadth of ethical questions (Winston & Edelbach, 2014; Mitcham, 1994; Latour, 1993).

We believe that the greatest concern regarding the issue of applicability of technology lies in its capacity to give the user a sense of power. Such power can be seen in the military of a country or an individual's capacity to deprive benefits to others. An example is the way computer hackers are notorious for abusing their knowledge of computers for harming others. Sensing such problems posed by the use of technology, Einstein had suggested a long time back that there should be accountability in the progress of technology in order to avoid potentially harmful consequences. This issue is very well illustrated in a recent book, by French thinker Bruno Latour (2013). When an industrialist raised the question, "But why should we believe you any more than the others?" in response to a scholarly lecture by a scientist on global warming, Latour wrote, "I'm astonished. Why does he put them on the same footing, as if it were a simple difference of opinion between this climate specialist and those who are called climate skeptics (with a certain abuse of the fine word "skeptics")? Could the industrialist possibly have access to a measuring instrument superior to that of the specialist?" (p. 2, quoted from, *An inquiry into modes of existence*, translated by Catherine Porter, 2013).

Many a tool may be usable but yet it has not been used. Tools much like entities in nature undergo a process of natural selection, akin to the Darwinian selection of species. Evolution of technology, too, takes place over the years and each successive generation of that technology is better than its predecessor. Over the years, the mobile phone has evolved into a genre very different from the first cell phone, in size, in shape, and also in features. The twenty-first century car is very different from those in the early twentieth century. So psychology of technology will enable us to understand what happens to the technology as we engage with it, along with what happens to us. Both change, but what are the changes? Thus, current research on virtual reality shows how the very self-concept may undergo change. It will help delineate how augmentation of our cognitive systems through neural implant technology enables us to perform mental activities otherwise not possible. While the size, shape, and weight of the neural implant have evolved, has it led to a parallel evolution of cognitive abilities? Technology has enabled us to multitask but has multitasking increased the capacity of our working memory system?

In other words, a psychology of technology will help understand the deep symbiotic relationship between human capacities and technology, which is the prime reason for the ways in which technology is becoming an irrevocable part of our psyche, our personality, and our mentality.

1.7 Conceptual Framework for Psychology of Technology in the Context of the Interaction Between Technology, Biology, and Behavior

As is evident from the above discussion, a deeper understanding of psychology of technology would not be possible without taking into account the ever changing influence of technology in our lives. At the level of our bodily functioning, the limits of our skeletal to biomechanical systems have been researched on, and we are continuing to know more about them. What we do know is that there are limits to our use of things, but at the same time technology offers opportunities for extension or alternates to our adaptation to the environment. From the invention of wheels to artificial limbs, the goal of technology has been to enhance mobility in particular and overall performance in general.

It is also apparent that there is an evolutionary basis of technology with animals using tools for a variety of purposes including, digging insects for food, swatting flies, cracking nuts, and many more. Higher up the evolutionary ladder, we come to early man, with the first discovered tools being some 3.5 million years old, namely, the Oldowan pebbles found in Ethiopia. With recent advances in imaging technology, it is now becoming increasingly clear that man's brain has been changing. The complex yet interesting ways in which the human brain has developed has been analyzed by anthropologist, John Hawks. His analysis reveals that the size of the brain of early man was very similar to that of ape today. However, it soon began to expand and during the last one-third of our history has expanded to a size nearly double of that of apes. Hawks (2013) writes,

> "As our cultural and linguistic complexity, dietary needs and technological prowess took a significant leap forward at this stage, our brains grew to accommodate the changes. The shape changes we see accentuate the regions related to depth of planning, communication, problem solving and other more advanced cognitive functions."

Thus, man's brain evolved, slowly but surely, to accommodate regions related to a host of mental functions such as planning, problem solving, and communication among others, propelling our technological prowess. The question is, would we have been able to reach the level of technological growth being seen today, without this level of growth in our mental faculties? The answer is "No." This aspect, which goes far beyond our innate sensory capacities, namely, the psychological aspects of cognition, language, communication, and decision-making in the face of complex choices can be said to be the second root of technological growth. Together, each fuelling the other, biology and psychology have led us to develop technology to never before seen heights. It is, however, important to remember that the links between biology, psychology, and technology cannot be seen as being unidirectional: not only have each grown independently, but they have also coevolved, with technology spurring developments in biology and psychology as much as the latter two have helped technological advancement.

We are apparently at a crossroad: to allow the three to coevolve, unbridled, or to rein in their growth? In May 2014, one of the world's sharpest minds, Stephen Hawking, warned us of the threats of Artificial Intelligence, while not belittling its benefits:

> "The potential benefits are huge; everything that civilization has to offer is a product of human intelligence; we cannot predict what we might achieve when this intelligence is magnified by the tools that AI may provide, but the eradication of war, disease, and poverty would be high on anyone's list. Success in creating AI would be the biggest event in human history" (Hawking et al., 2014).

Two years later, John Battele, founder, EIC and CEO, NewCo, asks,

> "Who determines what is "good"? We are just now grappling with the very real possibility that we might create a force more powerful than ourselves. Now is the time to ask ourselves—how do we get ready?" (Battele, 2016).

This is where psychology of technology has a very big role to play, helping us to get ready. While thinkers such as Ihde, Heidegger, Latour, and others have been forcing us to wake up and try to understand the role that technology is playing and could play in our lives, its good and its bad and the havoc it could cause, the role of psychology, which is, the science of behavior, becomes all the more crucial. Psychology of technology has become mandatory in the light of the fact that it is this mere 1 % difference between man and apes that has enabled the former to create tools and technology far surpassing those created by the latter. But is it not intriguing that a mere one percent can lead to such a stupendous difference in their capabilities? The net result of this one percent is the human psyche, with its unique perceptual, cognitive, and emotional aspects, which go to differentiate man from all other animals and which can be called the psychology of man. And, it is this human psychology from which emanates technology. Biology alone, without the inputs of human psychology would never have been able to bring in the kinds of inventions and discoveries we are witnessing, and even more so, the rate at which technology is advancing. Just think of Moore's law: technology doubling every 2 years, nay every 18 months.

In short, understanding psychology of technology and building it further would require continuous dialogue with growth in biological sciences on the one hand and technology per se on the other. As stated earlier, any basic research, when viewed in the context of its utility, would involve answers to the issues of applications and applicability, that is, whether technology, as an intervention, has the potential for its usefulness and is considered desirable under the prevailing social and cultural mind-set of its users and the milieu of their community. These and many other issues may form common gray areas of not only psychology of technology but also of bioethics, sociology of technology, and more. In the rest of this book, we will focus on the scientific nature of psychology of technology and its applications in the development and use of technology.

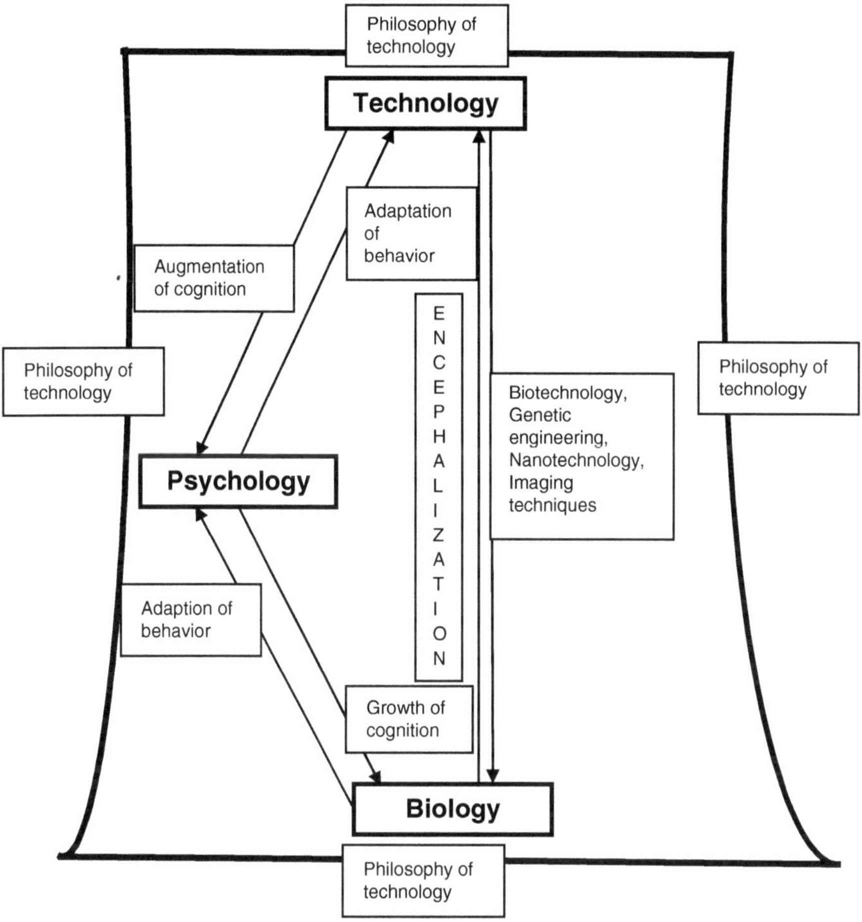

Fig. 1.1 A conceptual framework for psychology of technology

The figure above (Fig. 1.1) attempts to capture the conceptual framework for psychology of technology in terms of the relationship between biology, psychology, and philosophy of technology so as to represent their combined role in the growth of sustainable technology.

1.8 Methods of Research in Psychology of Technology

Since the study of psychology of technology involves a multidisciplinary approach, its techniques of research vary from conventional techniques of research in psychology (such as correlational analysis and the experimental approach) to simulation techniques and state-of-the-art neurological techniques such as fMRI and other scanning procedures. While such a divergence, may at times, place the study of psychology of technology in an advantageous position it may also result in some awkward growth. Consider a study of the range of comfortability or usability of a gadget. A study of this phenomenon might appear very different depending on whether it is a physiological study or one in psychology of technology. While we can classify an organic growth in fixed objective groups such as its presence or absence in the physiological study, it would be near impossible to determine the same range of threshold for psychological experiences, say of pain and comfortability, which vary considerably owing to individual differences. Therefore, what is observable at the material and organic level might not be corroborated through a related and relevant psychological report. Kool and Agrawal (2015) have pointed out how divergence in research findings could have an impact on the interpretation of results in terms of the validity of results sought in statistical terms as opposed to that sought in the conceptualization of the experimental study (Box 1.7).

Box 1.7: On Using the Experimental Method in Psychology of Technology

"By definition, the experimental method is that method in which one or more variables is/are manipulated by the experimenter in order to see its/their effect on another variable. The variable which is manipulated is known as the independent variable and since changes in the second variable are dependent on the variation in the independent variable, this latter is known as the dependent variable. Thus, to take an example from the natural sciences, one might want to study the effect of altitude above sea level on the boiling point of a substance. In order to do so, one creates conditions in which the boiling point of the substance is recorded at different altitudes above sea level. The latter, that is, altitude above sea level on which the boiling point is said to vary would be then called the independent variable while the boiling point itself becomes the dependent variable. To take another example, this time from the behavioral sciences, one might be interested in studying the effect of different training programs on the performance of employees in an automobile manufacturing unit. What would be the independent variable? The independent variable would be the one that is being manipulated by or is in the control of the experimenter, namely, the different training programs. Since performance would be a function of these different training programs, that becomes the dependent variable.

To this point, we can draw parallels between experiments in the natural sciences and those in the behavioral sciences. But a variety of differences can be pointed out, and it is these that in fact create the essence and the complexity of experimentation in the behavioral sciences. First and foremost, the subject of research, namely, the substance whose boiling point is to be measured, is lifeless. As such, no matter who conducts the research, it does not matter, and nor does the lifeless subject have moods, attitudes, personality, etc. Also, since it is without life, we can divide our total amount of solution into 10, or

(continued)

Box 1.7: (continued)

maybe, 20 or even a 100 portions and each portion would yield the same result, all other factors being kept constant. In contrast, the subject of research in the behavioral sciences is a living organism and even more so is a complex function of emotions, motivations, personality, etc. Can we divide our human subject into ten, or even two parts? These and many other similar factors, inherent in any subject that has life, call for the utmost caution while conducting research and especially experiments in disciplines such as psychology, sociology, and anthropology. It is much more difficult to obtain reliable, valid data unless due controls are exercised by the experimenter. It is primarily for this factor of control that stringent experimental designs have been created so that we may be able to establish cause–effect relationships, much akin to those obtained in the natural sciences. An experimental design can be likened to the architect's blueprint which is not only prepared after considerable thought but is also dependent upon the needs of the user of the building. So also, a variety of experimental designs are available from which the researcher can pick and choose depending upon his needs and the needs of the research problem."

Source: Kool and Agrawal (2015) p. 169–170

1.8.1 Experiments in Psychology of Technology

Experimental method: *used to study cause–effect relationships in situations where it is possible for the experimenter to vary the independent variable so as to observe differences on the dependent variable*

Experiments in psychology are conducted mainly with two considerations in mind:

1. What happens when we, as experimenter, manipulate (create changes in) a variable of interest, take for example, the effect of continuous work without rest, and
2. What is the net effect of the variable stated above, keeping all other factors in control.

Independent variable: *the stimulus, input, or the cause*

Dependent variable: *the response, output, or effect*

It is mandatory that several other considerations follow in order to establish the nature of the cause and effect relationship, usually described as the relationship between the **independent variable** (known as stimulus or input) and the **dependent variable** (broadly known as output or response). An integral part of experimentation is the random selection and allotment of subjects to each of the groups delineated in a study (e.g., the

Box 1.8: Illustration of Basic Experimental Research Method in Psychology of Technology

With advances in Information and Communication Technology (ICT), hybrid and online academic courses are on the rise. However, it remains to be established as to whether factors such as cooperative learning, known to enhance face-to-face learning, would impact computer-mediated learning in a similar fashion.

To test for the above, Roseth, Salterelli, and Glass of Michigan State University conducted an experiment on 101 undergraduates. The experimental design used was a 2 (Control: face-to-face) X 3 (medium: video, audio and text) X 2 (synchronicity: synchronous vs. asynchronous) setup.

The experimental data made it clear that cooperative perceptions declined and individualistic perceptions increased under the asynchronous condition, which in turn led to lower levels of motivation and academic achievement. The findings, thus, suggest that synchronicity of communication plays a much more important role than medium of communication in computer-mediated learning. The findings have important implications for social psychology theory too. As the authors write: "For theory, findings also suggest that social psychological theories based on face-to-face assumptions may need to be modified to indicate that predicted outcomes depend on synchronous social interaction" (Roseth, Saltarelli, & Glass, 2011, p. 804).

Source: Roseth, Saltarelli, and Glass (2011)

experimental group and the control group). Box 1.8 illustrates a typical methodology for such an experiment in psychology of technology (see Chap. 4, Sect. 4.5 for more examples).

1.8.2 Use of Descriptive Methods in the Study of Psychology of Technology: Observation and Surveys

There are many scenarios in which we are unable to conduct experiments to find the effects of technology on behavior. Or, it may be too risky or unethical to conduct an experiment in many situations. While attempting to study the effect of varying levels of radiation (independent variable), would it be ethical to study its effect on health and performance (dependent variable)? However, we can observe the current levels of radiation in the environment and then seek a relationship between the exposure levels of radiation and its consequence on health and performance of people exposed to that environment. Such correlational studies are simple, common, and efficient, but they do not ensure a cause and effect relationship, so imperative for research leading to verification and future development of theories and models in the study of psychology of technology.

Observational studies involve observing, classifying, recording, and analyzing behavior in natural or partly controlled settings. A classic study by Bandura and coworkers (1961) showed how children who watched an aggressive cartoon character later used the same tools used by the character in the cartoon to demonstrate their aggression in a free play situation. This is a classic example of observational and imitational learning and very much like what is so commonly observed in the use of gizmos around us. Another use of the observational method is illustrated in Box 1.9. See Chaps. 4 (Sect. 4.5) and Chap. 6 (Sect. 6.8) for more examples.

> ***Observational studies**: involve observing, classifying, recording, and analyzing behavior in natural or partly controlled settings*

The use of survey methods is popular in the study of psychology of technology. The bottom line for success of any product is the response of the user of a product. Therefore, surveys are very common in consumer research before, during, and after marketing a product. While writing this chapter, we saw a pop up from Microsoft, inviting us to participate in the survey on our experience with their new system "Windows 10" that we had recently loaded. For basic research purposes, the **survey method** is mostly employed for measuring user's satisfaction, comfortability, attitudes, feelings, and hedonistic attitudes, that is, mostly for subjective evaluations of the user.

> ***Survey method**: used for large-scale studies for measuring user satisfaction, comfortability, attitudes, and feelings with reference to a technology*

Box 1.9: Illustration of Observation Methods in the Study of Psychology of Technology

With widespread use of mobile phones, the hazards of talking or even texting on the phone while performing other activities is coming to the fore. Though there is ample empirical data to support the observation that using the mobile phone while driving is extremely dangerous, its effects on pedestrians still remains to be validated. It was for this purpose that Beth Ebel and her coworkers from the University of Washington collaborated with the Seattle Children's Hospital to conduct a study on the impact of social and technological distraction on pedestrian crossing behavior (Thompson, Rivera, Ayyagari, & Ebel, 2012).

We present excerpts from the abstract of the paper to show how observational studies can be performed to yield valuable insights, especially in the case of phenomena that are not amenable to laboratory conditions and variables which cannot be varied by the researcher.

> "Pedestrians were observed at 20 high-risk intersections during 1 of 3 randomly assigned time windows in 2012. Observers recorded demographic and behavioural information, including use of a mobile device (talking on the phone, text messaging, or listening to music). We examined the association between distraction and crossing behaviors, adjusting for age and gender. ... Observers recorded crossing behaviours

(continued)

Box 1.9: (continued)

for 1102 pedestrians. Nearly one-third (29.8 %) of all pedestrians performed a distracting activity while crossing...(which) included listening to music (11.2 %), text messaging (7.3 %) and using a handheld phone (6.2 %). Text messaging, mobile phone use and talking with a companion increased crossing time... Texting pedestrians were 3.9 times more likely than undistracted pedestrians to display unsafe crossing behavior (disobeying the lights, crossing mid-intersection, or failing to look both ways). Pedestrians listening to music walked more than half a second (0.54) faster across the average intersection than undistracted pedestrians.

Conclusions Distracting activity is common among pedestrians, even while crossing intersections. Technological and social distractions increase crossing times, with text messaging associated with the highest risk. Our findings suggest the need for intervention studies to reduce risk of pedestrian injury."

Source: Thompson, Rivera, Ayyagari, and Ebel (2012) p. 232–237

1.8.3 Employing Simulation

> ***Simulations:*** *use of artificial systems that resemble and follow rules and conditions similar to the real situation*

Before an operator is asked to run a machine, for example, before flying an airplane, she is involved in some practice on a system that resembles and follows rules and conditions similar to that of operating the actual machine. Broadly speaking, practicing on a model instead of the real product is a precursor to determine that the user has reached the specified levels of safety, optimum performance, and satisfaction (see Chap. 5, Sect. 5.11 for more information). Though not commonly employed by psychologists, **simulations** based on mathematical models are very helpful and useful. For example, graphical displays of positioning of human body augment our understanding of behavior in work station designs (Box 1.10).

1.8.4 Application of Methods Used in Neurosciences

> **Neuroscientific techniques***: techniques such as EEG, PET, MRI, and fMRI through which brain functioning can be understood*

It is well known that the overuse of technology has negative consequences for our body. For example, working continuously on computers has its own cost resulting in the development of the carpal tunnel syndrome. Is watching television for more than 2 h harmful to infants less than 2 years of age? How do we prove it? There are several consequences of the use of technological products that we might not know about at all without the physical examination of the functioning of the body in general and brain in particular. The use of EMG, EKG, EEG, PET, MRI, fMRI, and many other **neuroscientific techniques**,

Box 1.10: Illustration of Use of Simulation in the Psychology of Technology: COMBIMAN

One of the problems of the use of technology is to find an adequate fit between the tools, the workplace, and the human who has to operate those tools in that particular workplace. A discipline which focuses on such man–machine fit is that of human engineering or ergonomics and will be dealt with at length in Chap. 2.

Since the worker operates in three dimensions, the workplace cannot be adequately evaluated on the basis of a two-dimensional drawing. The traditional method of evaluating workplace designs in a three-dimensional setup was by creating a mock-up. Since a mock-up is hardware, it took time, money and effort to both build the mock-up and modify it. However, with limited money, manpower, and especially time, these mock-ups became unfeasible as a result of which human engineers started resorting to computer-aided simulation techniques to design and evaluate man's interactions with his working place.

The Aerospace Medical Research Laboratory has developed a computerized biomechanical man-model called COMBIMAN. It is an online interactive computer model which can be used as a manikin for workplace design and evaluation. Since its creation in the 1970s, it has undergone many changes and has also found many applications, ranging from the evaluation of existing workplaces to the creation of new workplaces and the criteria for selection of people who will work in those workplaces.

The beauty of COMBIMAN is that it does not share the handicaps of the mock-up since it exists only in the computer memory, and it does not require much expenditure in either building a workplace or modifying it. Moreover, it enables the creation of alternative designs which can then be evaluated and permanently recorded.

mostly used in biomedical research, are regarded as pertinent for demonstrating the psychological significance of technology-oriented behavior (Box 1.11). Does it mean that by doing so we have reduced our complex interaction with technology to the organic level? Not at all because the technological issues are embedded in a social context. While the prolonged use of internet may lead to social isolation (implied as a corollary for depression), it could also lead to increased socialization of a different type (virtual socialization which takes place without physical face-to-face interaction). According to Myers (2013), when psychology seeks support from neurosciences, it is not all about the under-the-skin issue (biological); we must also understand the context of between the skins (social) issues. In other words, the effects of use of technology would need to be determined at several levels—individual and collective. In short, technology will help us understand that what is good for a group may not be good for a particular individual or vice versa. See Chap. 4 (Sect. 4.10) for more examples.

Box 1.11: Neuroscientific Techniques in the Study of Psychology of Technology

One of the downsides of technological advancement is the growing addiction to the internet, video games, and social media sites. So great is this peril that many undergo sleepless nights. But what is the mechanism that makes it possible for people to remain awake despite the urge to sleep? Recent research from the field of neurosciences using modern methods of brain imaging provides an answer.

In a *ScienceDaily* report entitled, *One sleepless night increases dopamine in the human brain* research conducted by Nora Volkow, Director, *National Institute on Drug Abuse* has been provided.

> "The researchers studied 15 healthy participants who were either kept awake all night or allowed a good night's sleep. Researchers tested the same participants in both conditions. On the morning of the study, participants rated how tired they were and did cognitive tasks testing visual attention and working memory. The researchers used the imaging technique positron emission tomography to study the changes in the dopamine system that occurs with sleep deprivation. Compared to well-rested participants, sleep-deprived participants showed reduced binding of a radiolabeled compound (raclopride) that binds to dopamine receptors in the striatum and thalamus. Because raclopride competes with dopamine for the same receptors, decreased raclopride binding indicates increased levels of dopamine, according to the study authors."

Further, the report adds, "the rise in dopamine following sleep deprivation may promote wakefulness to compensate for sleep loss. 'However, the concurrent decline in cognitive performance, which is associated with the dopamine increases, suggests that the adaptation is not sufficient to overcome the cognitive deterioration induced by sleep deprivation and may even contribute to it,' said study author Volkow. Future research will examine the long-term effects of chronic sleep disturbances on dopamine brain circuits."

Source: Volkow et al. (2008))

1.9 Ethics in Research on Psychology of Technology

Conducting research in various areas of psychology of technology may be classified in two ways: first, issues of ethics commonly applied to all areas of psychology, and second, issues that are unique to the field of psychology of technology.

Regarding the former, the requirements for research in any field of psychology are stated in several sources and researchers, without any exception, are mandated to follow the guidelines enforced by their institutional review board. The following agencies/institutions offer ethical guidelines:

- The *American Psychological Association*
- The *Code of Federal Regulations HHS*, Title 45, Part 46

- The *Protection of Human Subjects, HHS*
- The *National Institutes of Health*
- The *Canadian Psychological Association*
- The *British Psychological Society*, and many more

In any empirical research in psychology, two basic ethical procedures must be followed. It is required that each participant be given enough information about the experiment before her consent is obtained for participation. This is known as "**Informed Consent**." The second is "**debriefing**," that is, after the experiment is over, the participant must be provided feedback about the goals of research and involved deception, if any, in the use of procedures applied to reach the goal of experiment. A classic example regarding the above issues of ethics and the resulting adverse effects on subjects is Milgram's famous experiment, reported in almost every contemporary introductory textbook of psychology. If you request permission to replicate this classic experiment today, the human subjects committee is not likely to accept your proposal (Box 1.12).

Informed consent*: the process of giving enough information about the research to the participant and obtaining her consent*

Debriefing*: the participant is provided feedback about the goals of the research and any deception that had been used during the research*

Box 1.12: Milgram's Experiment

Stanley Milgram's (1965, 1974) experiment on what happens when the demands of authority clash with the demands of conscience have become social psychology's most famous and controversial experiments.

Milgram, who was a professor and wrote stories and stage plays, created the following scene. Two men come to the psychology laboratory of Yale University to participate in what has been described as a pioneering study on the effect of punishment on learning. Of the two people, one has to act as "teacher" to teach a list of word pairs to the other, who is the "learner." The "teacher," who has come in response to a newspaper advertisement is administered a mild electric shock to familiarize him with the nature of the punishment to be delivered to the "learner." The "teacher" is then seated in front of the shock generator which has switches ranging from 15 to 450 V. The "learner" goes to the other room and is taught the list of words. With every mistake he is administered a shock by the "teacher," who is told by the experimenter to go on increasing the level of the shock, even though the "learner" screams and pleads to be let off.

When Milgram described this scenario to a sizable number of psychiatrists, college students, and middle class adults, people in all three groups said that they would disobey the experimenter and refuse to deliver shocks beyond 135 V and none expected to go beyond 300 V.

(continued)

Box 1.12: (continued)

But when Milgram conducted the experiment on 40 men, 63 % went clearly upto 450 V. When he re-conducted the experiment after making the learner's anguished protests even more vehement, again, 65 % fully complied with the experimenter's demands.

The obedience of the subjects disturbed not only Milgram but also social psychologists in general. A reviewer from the *New York Times* complained that the cruelty inflicted "upon their unwitting subjects is surpassed only by the cruelty that they elicit from them" (Marcus, 1974) and critics argued that there would be lasting psychological effects of such a horrifying experience.

Before Milgram started the experiment, informed consent had been taken from all the participants: that they knew the purpose of the experiment and that they were *volunteering* to participate. After such a harrowing experience, it was equally necessary to debrief the subjects and tell them about the goals of the experiment and the reason for the deception used (the learner did not actually receive any shock). When Milgram debriefed the subjects, surprisingly, 84 % said that they were glad that they had participated; only 1 % regretted the participation.

Thankfully, no permanent damage was done to the participants, as revealed by an interview conducted by a psychiatrist a year later on 40 of the participants who had supposedly suffered the most. Yet, it remains one of the most controversial of experiments in the history of psychology.

For more information on ethical and procedural issues, the reader is referred to the *American Psychologist*, 2009.

The other issues of ethics are, by and large, unique to the specific content of technology as applied to behavior. In particular, after millions of dollars are spent by engineers to develop products for human use, what would be the role of a fellow psychologist who has been teaming with them in its development? Ethical concerns require that psychologists should not socialize with other members in the team so that they are able to offer objective evaluation. However, in practice, this does not necessarily take place. The indiscriminate use of technological devices has been raising our concerns for privacy and reminds us that "racial oppression lives in bodies of color and social institutions" (Billies, 2015, p. 173). As stated earlier in this chapter and later in Chap. 7, technology has led to a major fear, known as surveillance threat. Probably, no other product more than the *Google Glass* is a clear illustration of how psychological functioning will be influenced in ways unknown to us until applied psychologists have the freedom to assess the impact of the technology in an objective fashion.

When psychologists work in applied settings along with fellow scientists from other fields, the operations that validate the use of a tool are examined with the goal of reaching the targeted purpose. In doing so, psychologists apply their knowledge and training to identify events and conditions and develop measures or tests to study them. This is known as operational psychology, a relatively new subfield of psychology that has its roots in the 1940s when the Office of Strategic Services (OSS) recruited psychologists for combating effectiveness or lowering its risk. Since then this field has been fruitfully applied in many areas of psychology including, but not limited to, clinical, industrial, and law enforcement settings (Palarea, 2007).

In contemporary research, the impact of operational psychology has been assessed in two ways: **collaborative operational psychology** and **adversarial operational psychology**. While science is considered neutral, its applied nature is not necessarily neutral and may often be viewed as being supportive or adversarial. When psychology is put into practice, the supportive role of psychologists, known as collaborative operational psychologists, is not as problematic as when they play an adversarial role, especially under high-risk conditions of human performance (Staal & Greene, 2015). Some researchers might not agree that the dangers involved in the test conditions would meet the guidelines of APA or other agencies regarding requirements of ethical concerns under dangerous conditions. Typically, we would need to ask the following questions to satisfy the collaborative nature of ethical concerns:

> ***Collaborative operational psychology***: *when psychology plays a supportive role in research in applied settings*

> ***Adversarial operational psychology***: *when psychology fails to agree with the goals of the research in the applied setting*

- Informed consent must address issues of potential harm
- Potential plan of action must be free from possible harm
- Potential plan of action must be made freely available to all stakeholders

Should we meet the above conditions, applications of research would be broadly viewed as collaborative in nature, failing which we will have to raise the bar of our scrutiny to save it from potential adversarial impact. In the new millennium, the growth of technology and its wide range of applications for human behavior pose several new questions for which we might not, as yet, have any insight, for example, genetic modification and gender change. It is expected that a number of new ethical concerns will continue to emerge and we will be addressing them using hindsight, just showing that human behavior is relatively slow in responding to the changes in technology.

Chapter Summary

The introductory chapter has been written with four goals in mind: first, to provide a working definition of technology and differentiate it from related terms such as science and engineering and introduce the reader to philosophy of technology, including the works and thoughts of Ihde, Jonas, Heidegger, Latour, and others. The second goal is to explain the basics of psychology, and especially, psychology as a science. The third goal is to relate psychology with technology with an attempt to show that since human beings and their behavior, both at the individual and social level, has become ever so intermeshed with technology, it has become imperative to not only mandate a psychology of technology, but also, its specific goals (both theoretical and applied) and methods of research (drawing from scientific methods used in psychology and modern imaging techniques used in neuroscience). The last goal is to lay emphasis on the need for an ethics of research in this emerging field, much as there is for almost all other disciplines.

Suggestions for Further Reading

Fallman, D. (2011). The new good: Exploring the potential of philosophy of technology to contribute to Human-Computer Interaction. Proceedings of the *International Conference on Human Factors in Computing Systems*, CHI 2011, Vancouver, BC, Canada, May 7–12.

Rosen, L. D., Cheever, N., & Carrier, L. M. (Eds.).(2015).*Wiley handbook of psychology, technology and society*. New York: Wiley.

Chapter 2
Technology, Psychology, and Evolution

> "Move over, King Kong. A chimpanzee named Tushi is putting up some regal resistance at the Royal Burgers' Zoo in Arnhem, the Netherlands. That's where she knocked a filmmaker's drone clear out of the sky in April using a 1.8-meter-long stick. Researchers report online today in the journal 'Primates' that this is strong evidence of planned, deliberate tool use among chimps, adding to evidence that these primates can think ahead and be creative in their toolmaking. A Dutch TV crew had set out to make a public relations documentary of the troop—part of several groundbreaking studies since it formed in 1971—and tried to use a drone to capture close-ups and overhead shots of Tushi and the other chimps. The filmmakers first tested their drone without a camera, arousing the interest of the troop. Four chimps—including Tushi—climbed several meters up the scaffolding in their enclosure. Tushi and another female carried willow switches. When the drone returned, this time with a camera, Tushi used hers to swat it out of the sky. Still filming, the drone tumbled to the ground, where it was examined by other members of the troop. Researchers say it's not clear whether Tushi was annoyed by the drone or only curious—they say her facial expression suggests she wasn't particularly afraid. But the footage is evidence that her actions were planned and deliberate, and it shows just how resourceful chimps can be at using whatever materials are available as tools."
>
> **Source: Chimps destroy documentary drone with twig tools,** Vogel (2015)

The description above is an interesting observation on the quaint ways in which animals use tools. Similarly, in the last chapter, we had mentioned two marvels, one man made, that is, the Taj Mahal in India and the other a product of nature, namely the gigantic nest created by a newly found bird in Java who has been observed collecting material from forests to craft its nest, much like its own Taj Mahal. The commonality between the two is simple. As discussed earlier, both were created by master craftsmen who were not using scientific theory for their creations, providing evidence that some form of activity had evolved among living beings, even without the cognition of the principles of science. And, even more importantly, technology, or at least a sense of technology, (that is, to use things, manipulate them, or simply indulge in mimicry), seems to manifest itself right through the evolution of behavior, among both human and other living beings. In this chapter we will focus on the forms of structures and functions that emerged in the course of evolution and made

V.K. Kool, R. Agrawal, *Psychology of Technology*,
DOI 10.1007/978-3-319-45333-0_2

living beings capable of not only displaying technological behavior but also creating artifacts of technology.

> ***Anthropomorphizing:*** *attributing characteristics of living creatures to lifeless objects*

We must also remember that, as mentioned in the last chapter, technology, per se, is interactive. The creation of some gadget and later, the possession of that gadget is not simply about having it somewhere around us. We play with it, we care about it, and we even attach our emotions to it. What do we mean when we say, "my car is my baby, I care about it the way I would care for my baby"? We refer to it as "my Mustang" and even attribute animal features to it. According to anthropologists, we are **anthropomorphizing**. This is in line with the thoughts and writings of Ihde (1979, see Chap. 1) who believes in the non-neutrality of technology and that the world is reflected back to us through technology. The things we create generate bundles of relations and carry our interests and emotions. Consider the statement of a young child who is ever so possessive of his mobile phone: "my phone is my biggest buddy because it helps me communicate with my sister living in another city." Thus, it is not attachment to just nature and its products that have been an integral part of our evolution (remember Eve and her obsession with the apples in the forbidden garden!). This attachment became even more so when such products were not only found useful but were also seen to help us to augment our adaptation to an unsympathetic environment. Over a period of time, it was but natural that such an attachment translated to other man-made artifacts which served similar purposes. In other words, not only did human structures and functions evolve to the stage that it became possible for us to create objects to aid our adaptation and survival but that we also developed an attachment for such objects, whether they originated in nature or were made by man.

2.1 Could We Have Remained a Tech-Devoid Society? The Evolution of Technology

If one studies the course of technological development, one can see a clear evolutionary pathway. Each step of technological innovation can be traced to how it enabled man to gain mastery over an uncooperative environment.

Thus, hearing aids were devised to help the hearing-impaired person overcome his helplessness in communicating with others. When humans started moving around and were often beyond the range of normal hearing, the Morse code and the telephone were invented. Inability to communicate with people across multi-continent business led to improvements in videoconferencing technology. The above are examples from one field, namely communication technology. The ever-evolving internet, technology in the field of manufacturing and robotics, technology in banking and finance, and technology in rehabilitation are all examples of tools that have helped humans in gaining mastery over the environment. Each invention, from the wheel, to the printing press, the steam locomotive, the electric bulb, computers, spaceships, nuclear power generators, wind and solar power generators went a long way in helping man to adapt to a changing environment.

The question however is, is this need to adapt to the changing environment, a product of the present-day society, when speed has become imperative? Or going farther into the past, is this urge limited to the last few centuries (since the Industrial Revolution) or maybe to the ancient civilizations of the Egyptians, the Greeks and the Romans, or the Aryans of India, when the construction of colossal buildings, monuments, and statues necessitated the use of the wheel, the pulley, and the lever system? We would have to stretch ourselves to a much older period of time for understanding this need to control the environment. The urge to adapt dates back, in fact, to millions of years and is not restricted to man or even to primates. Adaptive behavior has been found in animals much lower down in the phylogenetic scale.

The fact is that nonhumans also employ a wide variety of tools to enhance their survival. When organisms find that their natural capabilities do not allow them to overcome problems around them, they start thinking of other ways, many of which are through the use of tools devised for a particular purpose.

2.1.1 Technology and Animal Behavior

We start with an example of animal behavior that all students of psychology are familiar with. Psychologists have shown how animals are capable of manipulating objects in a novel situation. A popular experiment on chimps conducted by Kohler (1917, 1925) required them to stack boxes to reach a banana that was hanging out of their reach. When standing on the boxes stacked one upon the other failed to get them to the enticing bananas, they acted smartly by fitting the two shorter sticks into a longer one to retrieve the banana. Essentially, Kohler was deviating from the mechanistic explanations offered by behaviorists suggesting that the cognitive skills needed to solve problems of this technological nature are far more complex than our knowledge of neurology or even the behavioral history of the animal.

A second example, that we would like to present, is the work of the famous primate researcher, Jane Goodall. In an interview with Jane Goodall, on October 12th 2015, she was asked what made her decide to work with chimps. Her reply was that it was not she who had chosen this field of study. It had been decided by Louis Leakey, the famous paleontologist and her mentor.

> "He believed there was an ape-like, human-like common ancestor six million years ago. He wanted to get a feeling as to how early man might have behaved. His reasoning was that if I would find behavior similar or the same in humans and chimps today, possibly those behaviors were also existent in the common ancestor. That's why he sent me in the forest" (Bethge & Grolle, interview with *der Spiegel*, 2015).

Once sent to the Gombe Forest of Tanzania, she started observing the chimps there, very closely, and, with the intent of discovering commonalities between chimp behavior and that of humans. What, at that time, seemed most amazing was the ability of chimps to make and use tools. Today, of course, there are many examples of how a variety of animal species use tools but this was the first time that the phenomenon was actually observed. Dame Goodall rose to become the world's most famous primate researcher and published the main body of her work in the volume entitled, *The Chimpanzees of Gombe: Patterns of Behavior*, in 1986. She describes how very early in her study she observed a chimp, whom she had named David Greybeard, experiment with a wide variety of twigs and to finally use an appropriate one to burrow through the soil and find a termite for his meal. She called this "termite fishing" and it was the first time that the traditional belief of man as the only toolmaker was challenged. Besides termite fishing, she also observed chimps taking twigs from trees and stripping off the leaves to make the twig more effective as a tool (Goodall, 1999), a behavior, one would all agree, that falls under the category of object modification and a stage often said to be the preliminary one to actual toolmaking. So revolutionary were these findings that her mentor Louis Leakey was of the view that we must either redefine man and tools or accept chimpanzees as human.

Today, it is also known that chimps use rocks to crack nuts. When unsure of the depth of water, gorillas use a stick to find out the depth of water and also use it like a walking stick to cross the water. Despite having no hands, elephants are known to tear branches from trees to fend themselves from flies or to scratch their bodies. They also have been seen to mount rocks one upon the other, to elevate their standing position. So, without doubt, there is ample proof that toolmaking is not of recent origin but dates back to our ape ancestry, being carried through the evolutionary process for the simple reason that it proved adaptive.

> ***Animal sentience****: the study of animal ability to feel, perceive, or experience subjectively*

Goodall talks of how these chimps would manifest emotions much like we do and for very much the same reasons. In the interview with *der Spiegel* (2015) mentioned above, she gives examples of how chimp communities manifested not only simple emotions such as joy, anger, and jealousy, but also more subtle ones such as awe, curiosity, and embarrassment. Other scientists have been able to induce depression in animals by keeping them in cages or by manipulating other aversive conditions experimentally. Recent research in a relatively new area of study, namely, **animal sentience**, helps scientists to understand the biological mechanisms underlying such behavior.

Controlled laboratory studies by a number of psychologists including Pavlov, Skinner, and Kohler and corroborated by observations by Goodall and other primate researchers also prove that animals are capable of learning new skills. While everyone will agree that adaptation is a type of learning, it may surprise many, that both Pavlovian and instrumental learning have been seen even in invertebrate organisms such as the mollusk (Brembs, Lorenzetti, Reyes, Baxter, & Byrne, 2002) and roundworm (Zhang, Lu, & Bargmann, 2005). In mammals and primates, on the other hand, there is clear evolutionary history revealing that with each successive stage of evolution, the brain also evolved, increasing its capabilities in manifold ways.

As the brain evolved, the organism became capable of new functions, at the same time being able to retain those that had proved adaptive in the past. Developments in the understanding of evolutionary growth of animal behavior and their comparisons led to the growth of **neuro-ethology**, a multidisciplinary branch of study involving many areas of biology and psychology. Basically, it seeks to find out how the nervous system translates relevant information from the environment into natural behavior. One example of this is the study of bats that are known to use echolocation for navigation and finding their prey. Neuro-ethology helps us understand the ways in which the acoustic system of bats is equipped with the ability to convert sound features into a sensory map of behavioral significance. It is neuro-ethology, again, that enables us to understand the cortical substrates of a phenomenon known as **Area Restricted Search** (ARS) and which may provide answers to why we could not remain a tech-devoid society.

> ***Neuro-ethology**: the study of how nervous system translates information from the environment into behavior*

> ***Area Restricted Search**: restricting the search for new foraging areas to places earlier known to be high density food supply areas*

2.1.2 *The Cognitive Shaping of Technology: Area Restricted Search*

For survival, every organism must find resources, the primary ones being food, mates, and territory. Foraging activity normally involves locomotion in search for food. Hunting activity by predators gradually causes depletion in the number of animals that are being hunted and, after a certain point of time, habitation of that area becomes unviable because of depleting dietary resources. Once such depletion has taken place, there are two choices that are available to the animal: one, that he moves to other areas in the search for food, and two, that he remains in this familiar area but suffers hunger as a result. It is at this time that the animal must start making a choice. The choice is an important one, one that is directly related to survival and concerns whether to move around and search for food, or, stay at one place, conserve energy but remain hungry. One wrong choice and the animal is doomed. Those animals that are able to make the right choice survive, while others die. Studies on foraging behavior show that most animals tend to remain in the vicinity where food had been available in abundance at an earlier point of time, or that their search for food seems restricted to known areas. Very aptly, this type of behavior has been termed Area Restricted Search (ARS). In an early study of search behavior among rats, the famous psychologist Tolman (1948) noticed that rats would move about in the vicinity of the same areas where they had earlier obtained food. He explained this through the concept of **cognitive maps** or representations of spatial directions in the brain. Later studies by Hills, Todd, and Goldstone (2008) clarify that foraging behavior among animals seems to follow a similar search strategy. When faced with the same problem, namely foraging, at a later date, the animal will use cognitive maps created

> ***Cognitive maps**: representations of spatial directions in the brain*

from earlier search behavior and will, thereby, be able to conserve energy and time by restricting its later search to locations known to be rich in food resources. Most organisms will prefer to stay in areas earlier found rich in resources and move away from low resource density areas.

It is clear that such area restricted search behavior has adaptive value. One line of evidence is that it is seen in a variety of organisms, ranging from the simple *E. coli* bacteria (Eisenbach & Lengeler, 2004) to species as diverse as houseflies, bumble bees, and even humans (Hills, Brockie, & Maricq, 2004). In every case, the choice was to move towards known high resource density areas and stay near them.

In other words, ARS is goal-directed behavior. During the course of evolution, the brain also evolved and the basal ganglia formed links with the upper centers of the brain, particularly with the frontal cortex, and it is these structures that are responsible for executing movements of the limbs. According to Hills (2004), "the evolution of goal-directed cognition emerges out of mechanisms initially in control of foraging," (p. 3). And, one result of such goal-directed cognition is the evolution of technology. The way in which tools helped was that these tools made it possible for man to make those choices that are adaptive. However, the development of these tools would not have been possible if our brain had not evolved to the point that such reward-seeking behavior became possible (Fig. 2.1 delineates the primary parts of the human brain and spinal cord).

From the literature on ARS, it seems clear that animals at various levels of the phylogenetic scale are able to make choices that enable them to not only find food

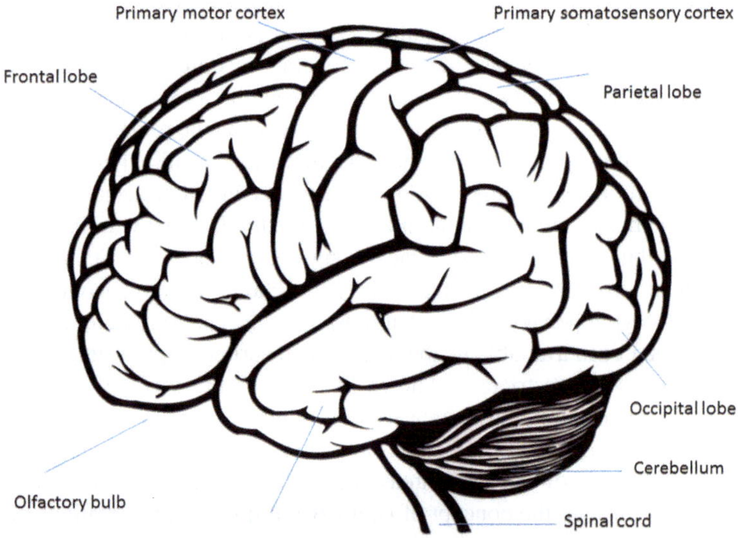

Fig 2.1 Diagram of the primary areas of the brain and the spinal cord

and thereby survive in the face of imminent death, but also to gain mastery over the environment. ARS enabled adaptive choices and offered a vehicle for a trade-off between time taken for exploration and exploitation, thus enabling the organisms to save precious time and energy.

> ***Dopamine**: a neurotransmitter which plays an important role in a variety of behavior*

Being adaptive, ARS survived through the evolution of species and neuroscientists provide many lines of evidence for this, the most important being the common denominator of the neurotransmitter **dopamine** and dopamine modulated functions across species (Hills & Dukas, 2012). Higher levels of dopamine tend to increase turning angles while selectively killing dopamine neurons reduces the capacity for ARS (Hills et al., 2004). The importance of dopamine is clear from the large number of functions that are controlled by it (Barron et al., 2013).

Recent advances in neurosciences and empirical research also clarify that at the base of this ability to gain mastery over the environment is choice making. Leotti, Iyengar, and Ochsner (2010) make this point very clear through a paper aptly entitled, *Born to Choose: the origins and value of the need for control.* According to them, not only is choice the vehicle for exercising control, but there are clear neural substrates for this ability to choose.

> "Converging evidence from animal research, clinical studies, and neuro-imaging work suggest that the need for control is a biological imperative for survival, and a corticostriatal network is implicated as the neural substrate of this adaptive behavior" (Leotti et al., 2010, p. 457).

This ability to make choices was an important adaptation and seems to date back some 530 million years ago. The vertebrate brain became similar to what is ours today, and it made possible the ability to make calculated choices or to make what can be termed a cost–benefit analysis. It was the development of this ability that made it possible for vertebrates to take a holistic view of the environment and move around in the search for rewards (Murray, Wise, & Rhodes, 2011).

It is as if we are hardwired to make choices, with those that prove to be adaptive being retained while others being lost over a period of time. By exercising choice, we are able to select adaptive behaviors and reject nonadaptive ones. This seems to act as an intrinsic motivation to repeat that same behavior, reinforcing the choice making process and inducing feelings of confidence and success, or what has been termed self-efficacy (Bandura, 1997). It is also seen that while restriction of choice is aversive to both humans and other animals (Morgan & Tromberg, 2007; Clubb & Mason, 2003), wrong choices produce negative feelings which hamper later learning and can lead to learned helplessness (Patall et al., 2008). In other words,

Behavior is chosen====⇒desired outcome====⇒perceived control====⇒ repetition of chosen behavior

As the behavior becomes more and more under the voluntary control of the species, people are able to restrict their behavior to those choices that have previously yielded desired results. Various reports conclude that when choices are provided to a person working on a task, the results are more fruitful than the ones that are obtained when no choice is provided. So powerful is the effect of choice that Brown and his coworkers refer to it as "*The lure of choice*" in their article of the same name (Brown, Read, et al., 2003).

Neurologically too, there are specific areas of the brain related to the ability to adapt and this is seen not only in animals but even in young infants long before they have had a chance to learn the value of autonomy from the society. Greater discharge is seen in areas of the pre-frontal cortex when rewards are delivered instrumentally, that is, after some form of behavior, than when they are given passively and regions of the Pre-frontal Cortex (PFC) and the striatum form a network to produce motivational states associated with choice and control (Leotti et al., 2010). Imagine two situations, one in which a child gets a reward for good behavior and the other, when he is simply given something he likes, but that something cannot be called a reward because he did not earn it. The PFC area of the brain would show much greater activation in the former situation than in the latter, motivating the child to further seek such rewards by repeating the same behavior. Also, different areas of the brain are activated depending on whether the choices are opportunities or are in the context of greater or lesser threat, thereby helping the individual to minimize threat and maximize opportunities. According to Leotti et al. (2010), the areas in the brain for choice opportunity and choice in the context of threat involve different locations, which have been uncovered through imaging of the medial sections of the left hemisphere. These are as follows:

- Choice opportunity: mainly involving the striatum for reward experience and Medial Pre-Frontal Cortex (MPFC) for increased self-relevance
- Choice in the context of threat: MPFC for control of emotion and amygdala for negative effect

It would be relevant to point out that the modulator of these anatomical structures is none other than the neurotransmitter dopamine that is crucial in directing foraging behavior and ARS besides playing an important role in a variety of functions ranging from reward seeking, to helping in adaptation to sleep deprivation to executive brain functions so important for adaptive decision-making.

The dopaminergic system, too, has evolved through phylogeny. We see it in lower organisms such as zebra-fish (Schweitzer & Driever, 2009), fruit flies (Liu et al., 2012), sea-urchins (Adams, Sewell, Angerer, & Angerer 2011), and of course, in mammals such as rodents, (Izquierdo & Jentsch, 2012; Arias-Carrion & Poppel, 2007), lamprey (an organism that diverted from the mainstream development of mammals, Murray et al., 2011; Thompson, Ménard, Pombal, & Grillner, 2008; Nieoullon & Coquerel, 2003), through the entire gamut of mammalian evolution, and finally in humans, allowing us to do a lot more in addition. Evolution has brought in its wake, human beings with a bigger, better brain (Kool & Agrawal, 2012, 2009) with clear dopaminergic and serotonin pathways found important for so many different functions of the body, including learning and reward seeking, pain and pleasure, etc (see Fig. 2.2).

Thus, we not only seek rewards but also like to talk about rewards and, going a step further, are even able to develop a mental time travel ability (the ability to

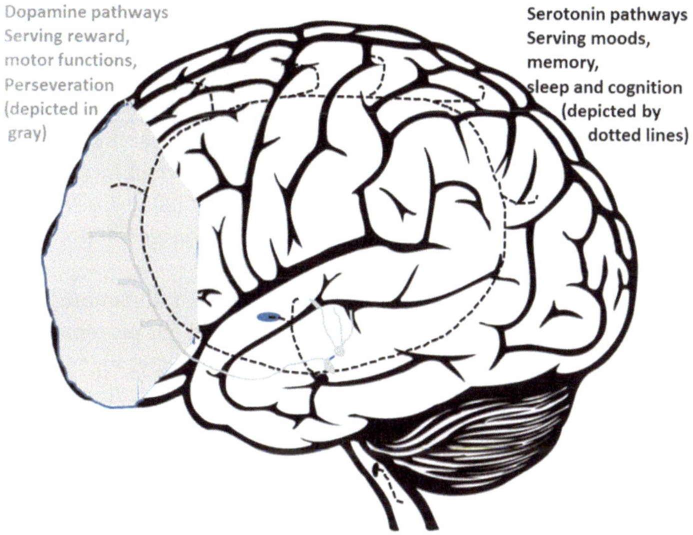

Fig 2.2 Dopamine and serotonin pathways in the human brain. Adapted from National Institutes of Health, United States Department of Health and Human Services

remember the past and predict the future (Corballis, 2009), and an ability to minimize costs (Mazzoni, Hristova, & Krakauer, 2007). Are not these two abilities, ever so important for the development of technology? Keeping in mind the needs of the future generations or even the present generation in changing times helps both the designer and the manufacturer to present technology that is not only effective but also cost-effective. Think of any realm of technology, from the lowly safety-pin to a space capsule, technological innovation does not happen in a vacuum: it builds on past knowledge. What is termed innovation is not really an innovation: it has evolved from some technology that preceded it. Such a path to innovation becomes imperative because of the very anatomical structures and physiological processes that are entailed in any cognitive processing, and beyond doubt, technology emanates from cognition.

But can this facility of choosing become maladaptive? Our own experiences probably vouch for this, that when faced with too many options, we start feeling stressed and avoid decision-making or leave it to others. Would you believe that this phenomenon, too, is supported by neurological evidence? Neuroscientists such as Fleming et al. (2010) point to the problems faced when we have far too many choices and what is the result of too many connections between the PFC and basal ganglia. The ways in which dopamine levels have changed in the course of evolution and its effects on the human mind have been described in Box 2.1.

From the above it becomes clear that the ability to make and use tools is not limited to humans or even to hominids but extends down to organisms much lower down the phylogenetic scale. Over the millions of years, the human brain evolved, becoming what it is like today, through the evolutionary processes of natural

Box 2.1: The Dopaminergic Mind Hypothesis

Recently, neuroscientists have postulated what has been called the dopaminergic mind hypothesis. It focuses on the differences between modern man and their hominid relatives in terms of changes in dopaminergic levels. It theorizes that due to the consumption of meat, there was an increase in dopamine levels some two million years ago. With further dietary, environmental, and social changes some 80,000 years ago, further changes in dopamine levels have been said to have taken place.

As a result, we modern humans are far richer in our dopamine levels than other apes and there have been consequent changes in our personality. A high dopamine personality is said to be typified by the following: high intelligence, sense of personal destiny, and a religious/cosmic preoccupation. At the same time, such an individual also shows a high risk taking and an obsession with goal achievement to the extent of becoming ruthless emotionally detached.

When all of us start showing these characteristics, it constitutes what has been nomenclatured, a dopaminergic society. Much like the constituent members, the society is extremely goal oriented, fast paced and, perhaps, even manic in its pursuits of goals. The society is seen to focus more on competition and aggression rather than on nurturance and communality.

While we do not have direct neurological evidence for such increased levels of dopamine in comparison to other members of the ape family, we do have sufficient behavioral evidence and some indirect anatomical evidence in favor of this hypothesis (Previc, 2009; Raghanti et al., 2008).

Does not the above ring a bell in our mind? Should we say that the extreme materialism and rat race seen in twenty-first century individuals, almost all around the globe, are because of this?

selection and adaptation. Evidence has also been provided to show that this toolmaking ability can be said to be based on the more primitive ability of foraging for food where making viable choices was important for the survival of the animal. In the words of Leotti and coworkers,

> "Collectively, the evidence suggests the desire to exercise control, and thus, the desire to make choices, is paramount for survival. The opportunity for choice enhances an individual's perception of control, and thus, exercising choice may serve as the primary means by which humans and animals foster this psychologically adaptive belief. Just as we respond to physiological needs (e.g. hunger) with specific behaviors (i.e. food consumption), we may fill a fundamental psychological need by exercising choice. While eating is undoubtedly necessary for survival, we argue that exercising control may be critical for an individual to thrive. Thus, we propose that exercising choice and the need for control—much like eating and hunger—are biologically motivated. We argue that while people may be biologically programmed to desire the opportunity for choice, the value of exercising specific choices likely depends on the available cognitive resources of the decision-maker in the given context, as well as the subjective value of the choice contents, influenced by personal experience and social and cultural learning" (Leotti et al., 2010, p. 459).

2.2 Beyond Darwin and Its Implications for Cognition and Behavior

Advances in modern science and technology could not have taken place if the human brain had not been of the shape and size that it is today. Paleontologists provide sufficient evidence to prove that over the millions of years of the history of mankind, the human brain has undergone considerable change in shape, size, and weight. Much of the changes that have taken place can be explained through the evolutionary processes posited by the father of evolutionary biology, Charles Darwin.

At the same time, modern advances in neurosciences and state-of-the-art neuroimaging techniques clarify that we have come a long way from the initial concepts advanced by Darwin, who explained evolutionary changes in species on the basis of the principles of natural selection and adaptation. While the former isolated structures that would help in the survival of the species, the latter fine-tuned these structures to help in the optimization of the functional aspects of the naturally selected structures. As a result, in each successive generation of species, those organisms that were able to change and adapt survived while others failed to do so. However, over the years, it has become increasingly difficult for scientists to explain the plethora of functions and structures found in any species on the simple principle of adaptation. For example, one can think of the human mouth cavity with the teeth, the mandibles, the upper palate, the lips, etc. The shape of the teeth and the jaws were predicated by the uses to which they had to be put. As man changed, from a carnivore to an omnivore, the size and the shape of the jaws and the teeth changed or became adapted to suit the changing dietary patterns. But were the teeth or the other structures of the mouth cavity basically structured for talking? This is a function being performed by structures that had been formed the way they are for some other purpose, namely, for eating and chewing (Fitch, 2011). A remnant of this function is the canine teeth still found in the human mouth. Another example is the human hand. Our predecessors had claw-like fingers and nails much akin to those of apes and chimpanzees. Look at the beautifully manicured nails and fingers of the men and women of today. As our diet changed, so did our teeth and jaws. As ways of procuring that diet changed and as lifestyles changed, so did our hands. The independence of the thumb from the other fingers was needed for holding objects firmly. But, had it been created to play games on a video console? Had it been created for hitting the space bar on the computer keyboard? How do we describe such changes? Are they adaptations, in the Darwinian sense of the word? Since they do not have a true historical genesis, in the way envisaged by Darwin-Gould and his colleagues (1982) coined another term for such changes, calling them exaptations. **Exaptations** are structures that had been adapted for one purpose but have been later co-opted for another purpose. They are initially products of adaptations that arose through natural selection for optimizing some function but are later being used for some other function.

> ***Exaptations**: evolutionary structures that have been adapted for one purpose but later co-opted for a different purpose*

Apart from such exaptations, there exist other structures that were not initially adaptations for the simple reason that they did not have adaptive value at that time. They were thrown up in the process of adaptation and can therefore be thought of as by-products of the process of adaptation, much as natural gas is a by-product in the process of crude oil refinement. Or, an even better analogy would be the spaces that are created when bridges are built. In order to support these mammoth structures, pillars are put into place. They do not serve as bridges but they are absolutely imperative for the bridge to become a bridge. In the same way, when natural selection takes place, new structures have to be put into place and in this very process of restructuring, new structures much like the pillars of the bridge have to be created. Our large sized human brain is one such example. As it became more and more complex, it threw up hundreds, no, thousands, of by-products which did not have any purpose at that time. But nature is never wasteful. These by-products were later co-opted for other purposes as and when they were needed. They were very much like the support structures of bridges and flyovers, the pillars on which the very structure of the bridge rests. These pillars create spaces that are later used for housing the homeless or for creating green spaces amidst the concrete jungle. In the same way, the restructuring of the brain created spaces that were later used to house functions such as religion and music when man invented them. And, continuing to use the analogy between bridges and brains, just as these open unused spaces under the bridges are known as spandrels, such by-products of adaptation were also termed **spandrels** (Buss, Haselton, Shackelford, Bleske, & Wakefield, 1998; Gould & Lewontin, 1979).

Spandrels*: structures thrown up as a result of adaptations for which no specific use exists at the time*

2.2.1 *Significance of Exaptations and Spandrels*

By introducing two new tools for the evolutionary scientist, Gould and Vrba (1982) were in no way refuting the Darwinian principles. It is important to understand that structures formed either through exaptation or through spandrels would not have been possible without the base process of natural selection and adaptation. Natural selection remains the primary process for creating complex biological design. It is natural selection that is responsible for creating the original adaptations which are later co-opted to become exaptations, and, again, it is this process of selection that is also responsible for producing adaptations of which spandrels are an incidental by-product. But, the importance of exaptations and spandrels is that though for a period of time, they may look like non-adaptations, they provide the flexibility that any open and resilient system needs (Kool, 2008). They form the wellspring and reservoir and the source of raw material for further selection and are probably one of the causes for the degree of complexity one sees in human functions today and which makes humans better than most other animals, including mammals. This is also the reason why our cognitive system is so much more complex than that of other animals, including primates, leading to our capability to design objects, including technology, so as to make our environmental fit better.

Today the principles of Darwin have come to be applied to a wide variety of not only psychological traits and behaviors but to areas of interest as diverse as economics, anthropology, linguistics, and even computer science, leading to what has been called *Universal Darwinism*, probably coined by notable physicist Richard Dawkins (1983) and advanced by a number of other thinkers and writers including Campbell (2009), Nelson (2007), and Hodgson (2005). Some interesting examples of how exaptations and spandrels have been used to explain changes in various areas have been described in various boxes (Box 2.2, 2.3, and 2.4) in this chapter.

Box 2.2: The Importance of Exaptations and Spandrels
The ability to appreciate and create music: evolutionary adaptation or cultural creation?

Laurel Trainor (2015) from the *McMaster Institute for Music and Mind*, Ontario, Canada attempts to explain the biological basis of music and provides an excellent example of how Darwinian principles of natural selection and adaptation cannot explain the complexities of human behavior.

While most people will insist that music is a creation of culture, there is ample evidence to show that not only is music unique to humans, it is universally seen and exhibits ontological patterns of development and should therefore have some adaptive value. Scientists have long debated on the adaptive value of music, ranging from infant management to social bonding and mating, the important question is whether musical ability is a true evolutionary adaptation or whether it is based on structures that were naturally selected for some other function but later utilized for musical ability?

In his discussion, Trainor questions whether the control and expression of musical behavior at the cortical level should be attributed to structures that can be understood to be as adaptation, exaptation, or as a spandrel. If it is controlled by a true adaptation, one should be able to isolate structures that were not there earlier but were formed to specifically serve the purpose of control of this behavior. He goes on to point out that if one looks at the capabilities needed for musical behavior, it is obvious that pitch perception, time perception, pattern recognition, and rhythm perception are of prime importance, some of which have commonality with other aspects of auditory perception such as those used in auditory scene analysis, which has an older origin and also a clear adaptive function across species. Examining evidence from human ontology along with some phylogenetic evidence, Trainor concludes that the neurological structures underlying musical behavior cannot be explained on the basis of mere evolutionary adaptations. Rather, we can understand musical behavior better if we think of it as a cultural creation which when proved adaptive, exapted structures originally selected for other aspects of auditory perception. At the same time, it is possible that some music-specific abilities rest on aspects that were actual adaptations of earlier structures.

Source: Trainor (2015)

Box 2.3: The Spandrels of Dunkin Donuts, or How the Munchkin Came to Be

Visitors to the Northeast are probably familiar with the miniature Dunkin Donuts confection, the Munchkin.

Think of a normal doughnut. What you have is a circular piece of dough with a hole in the center. You prepare a tray of doughnuts but what are you left with? You are left with as many pieces of dough, each circular in shape, only much smaller than the original doughnut, and of not much use. You have two options: you either throw away that extra dough or as most of us would do, put all the small circles together, knead it all over again and prepare some more doughnuts. But the catch is that each time you will be left with small circles that are of not any use.

Here comes in the creative baker! Rather than throwing away those circles or using more energy and time to create new doughnuts from the leftover dough, he uses them for a new purpose. In the words of Chang,

> "The Munchkin illustrates the concept of exaptation well, though it requires a baker as selector, whereas natural selection operates with no selector. A doughnut in this example is a circular piece of dough with an empty circular middle. Imagine that the shape is created by making a round of dough, and then cutting out the middle piece, leaving you with the doughnut and some extra dough. Imagine further that the baker typically throws the middle piece aside as it serves no purpose. The middle piece here is a spandrel it serves no function to the baker, but is rather a leftover portion of the functional dough the doughnut."

Chang goes on to explain how this exaptation, that is, the munchkin will also be subject to further adaptations. Maybe people will like chocolate munchkins rather than the plain ones, so gradually the baker may concentrate on the chocolate ones and the plain ones may die out.

Source: Posted in *Evolution and Psychology,* Chang (2009)

Box 2.4: Bright Idea: The First LASER

Peter Franken recalls how the laser discoveries excited physicists (interviewed by Joan Bromberg, 1985)

Franken:

> "Let me tell you about the OSA [Optical Society of America] meeting. It was held in Pittsburgh, in 1961 in Pittsburgh.... That was Panic City. The halls were packed. Normally with an invited paper at the Optical Society, you might draw a hundred people. There might be two or three click-click-clicks of cameras taking pictures of the slides. These halls were packed; the ballroom was packed, for this invited paper. I remember as a high point Charlie [Townes]—I'm sorry, Art Schawlow getting up, giving a talk: every slide he projected, there was a veritable staccato machine gun fire of Minoltas going off. It was unbelievable! Panicville. Everybody wanted to get in on it."

(continued)

Box 2.4 (continued)

What caused this excitement? Scientists were demonstrating and talking about the one invention that literally set the world on fire. This invention was the LASER and we all know of the varied uses to which it has been put, from giving sight to millions of sightless people through surgery using lasers, to communication technology and its use of the optical cable capable of transmitting many hundreds of telephone conversations via a fiber thinner than a human hair, to its uses in cosmetic surgery and other forms of more invasive yet painless surgery, and of course to its countless uses in commerce (the laser scanner used at store checkouts, the CDs and DVDs and MP3s, the laser holograms on the credit card). You could probably think of many more uses to which the laser is being put.

But did you know how the invention came to be? How scientists at various laboratories funded by both government and private corporations were trying hard to amplify radiation but failed? The theoretical know-how had been there since the early 1900s when Einstein predicted that rays could stimulate more atoms to emit rays of the same wavelength. Yet scientists at research labs such as Bell Laboratory, Westinghouse, and IBM were not successful even in the 1950s. It was finally Theodore Maiman at Hughes Laboratory who realized that all they needed was a flash lamp. He searched manufacturer catalogues and finally found a helical shaped lamp in which a ruby could be fitted.

> "A flash lamp would do. Scouring manufacturers' catalogs, he found a very bright lamp with a helical shape. Just right, he thought, for fitting a ruby inside. He assembled the components with the aid of an assistant, Irneed'Haenens, and on May 16, 1960 they observed pulses of red light. It was the world's first laser."

This is a clear example of how exaptation takes place in technology. The lamp had not been manufactured for the production of laser beams and would have soon been outsmarted by other more modern lamps and lost in oblivion. But the innovativeness of the smart scientist put to use this lowly lamp to produce something that has such an impact upon our lives.

> "Fantasies of a dreadful death ray led to the discovery of a device which helps people in a million ways"

Source: *American Institute of Physics* (2013)**;** (https://www.aip.org/history/exhibits/laser/interviews/franken_laserexcited_interview.html)

So important are these exapted structures that they even form the basis for cultural evolution. How does one explain religion or morality for that matter? We may construe religion as an adaptation that provided survival value when all else seemed beyond hope (Wade, 2015). Or we may think of religion as an exaptation of cognitive structures that had already adapted and evolved to a certain level for other purposes more directly related to survival and were later used for religion (Saad, 2012).

One can even think of cultural traits such as the ones isolated by Hofstede in his classic study of cultural differences around the globe (Hofstede, 1980). A gene-culture evolution approach has been used to explain differences on the cultural trait of individualism-collectivism and Chiao and Blizinsky (2010) have obtained a link between this trait and the frequency of a variant of the serotonin gene across 29 countries.

As our brain underwent exaptations and new spandrels were thrown into existence, our cognitive system advanced to never before known levels. Scientists attempted to understand natural phenomena while engineers used the theory so formulated to develop newer technology from the old. We can use the same concepts of adaptation, exaptation, and spandrels to understand how evolution of technology takes place.

There is however one difference. As Cattani (2008) says,

> "Intentional factors are on the contrary incorporated into an evolutionary theory of technological change. A key difference between biology and technology is that foresightful evolution can actually occur in the context of technology. As Ziman (2000) pointed out, this difference stems from the observation that "...technological change is driven by variation and selection—but these are clearly not 'blind' or 'natural.' This work is being done largely by conscious human effort, without apparently needing guidance from any 'hidden hand,' whether of Nature, the market, or God," (Cattani, 2008, p. 588).

We were able to find novel uses for obsolete technology (one example of how LASER was invented is presented in the box above). While the LASER was said to be a solution without a problem (the uses were thought of much later), some of these exaptations even led to the evolution of technology in line with customer needs and entrepreneurial activity (Dew, Sarasvathy, & Venkataraman 2004).

2.3 Could Technology Have Developed in Any Other Way: The Human Body

Although we have made significant developments in the field of biology, the understanding of the structure and functions of the human body and particularly its executive, the brain, still demands considerable investigation. With about 75 trillion cells in operation and approximately one million in attrition every day, the intricate nature of our human brain is difficult to understand (see Box 2.5).

What is not understood or sometimes not paid heed to is the fact that the things we use on a daily basis are not always compatible with the functioning of the human body. You could call it lack of scientific knowledge in our designing or manufacturing of a product, or putting it more simply, an unfortunate situation imposed upon us for commercial reasons. The keyboard of a computer or the old typewriter is one such glaring example. We have known for a long time that certain letters on the keyboard, such as the letter "a", are used more frequently than others and yet, the current layout of "a" on a QWERTY keyboard is at the far left of the middle row and as per our training in typewriting, this puts a heavy burden on the weakest and

Box 2.5: The Brain Has Been a Matter of Awe Not for Biologists or Psychologists Alone

Poet, Emily Dickinson, has put this awe into words:

"The brain is wider than the sky
For, put them side by side,
The one, the other will contain, and you beside,
The brain is deeper than the sea,
For hold them blue to blue,
The one the other will absorb,
As sponges, buckets do.
The brain is just the weight of God,
For lift them pound for pound
And they will differ, if they do
As syllable from sound."

Source: Emily Dickinson, ca 1860, published in 1921

smallest finger of our left hand. Consider what changing the layout of the design of the keyboard would mean? Well, everyone would have to *unlearn* their current skill in typewriting because of a different location of the letter "a" and then engage themselves in *relearning* a new set of movements to operate the keyboard smoothly. Ideally, this is doable, but practically, it has still not been done.

After its invention, when anything is manufactured and enters the market for consumption, the two most important considerations are profitability and ease of use. In some ways, the two issues are interrelated, but it is the latter that we, as psychologists, and even more so, as psychologists studying psychology of technology, are interested in. As far as the manufacturer is concerned, the primary issue is regarding how to get the object into the hands of the consumer such that he/she can operate it with ease. How many companies test their product, thinking about the limits of human functioning? If companies had actually engaged in such thinking, why would people be talking about the ill effects of the excessive use of the ubiquitous cell phone or the carpal tunnel syndrome resulting from continuous working on a computer?

In other words, before any technology is developed, the physical and mental limits of the individual who will use it have to be kept in mind. These limits are set by both phylogenetic and ontogenetic processes, in so far that the limits of the human body are different from those of other animals (**phylogenetic processes**), and that these limits change with the age of the person (**ontogenetic processes**). Let us attempt to understand these limits. We can start by enumerating the systems of the human body. These are as follows:

> ***Phylogenetic processes****: differences between organisms of different species*

> ***Ontogenetic processes****: differences due to age of the organism*

- The **nervous system** consisting of the brain, spinal cord, and nerves.
- The **muscular system** composed of three types of muscles: smooth, cardiac, and skeletal.

- The **respiratory system** managing the supply of oxygen in the blood through the lungs.
- The **circulatory system** consisting of heart, arteries, and veins and regulating the flow of blood.
- The **digestive system** supplying energy through the operation of esophagus, stomach, liver, and intestines.
- The **excretory system** managing the waste collected in the body through the kidneys and the bladder.
- The **reproductory** system that helps in the formation, release, and transportation of semen and the development and fertilization of the ova followed by the development of the fetus in the womb of the mother.

It was in the wake of the industrial revolution that manufacturing of a great many products was started. While the next few decades saw the arrival of many new products in the market, the focus was hardly on the comfort of the consumer. One just needs to think of the first watches and clocks, the first telephones, or the first typewriters. With time, they have not only become sleek but also more comfortable to use.

Undoubtedly, the scenario has changed but we are still struggling to create products that are best suited for human use. One reason could be the fact that individuals differ in so many ways and that the notion that "one size fits all" does not work. Consider a seat in the airplane which may be too small for an obese person, a car driver whose legs fail to reach the brakes, or left-handed persons using gadgets made for those who are right-handed. Moreover, whether we think of the dimensions of the human body as statistical means, median, or the mode, one thing is for sure, measurements have to be made to arrive at these central tendencies so that we can design products accordingly. Despite the difficulties entailed, it has become possible to measure the human body and its parts, through what has come to be known as anthropometry.

2.3.1 Measuring the Human Body: Anthropometry

In 2006 and later in 2010, famous anthropologist Zeresenay Alemseged and his team reported evidence for the use of tools used some 3.39 million years ago from Dikiki, Ethiopia (Alemseged et al., 2006; McPherron et al., 2010). He posited that though humans and chimps probably share a common ancestor, as revealed by tools used by chimps, man is probably the only primate to make sharp edged tools to hunt for food. Moreover, these primitive 3.39 million year old tools can be said to be the precursors of all the technologies today.

But, more than one question remains unanswered: why was the tool developed? Why did man need a tool more than any other animal? And, even more, what decided the nature of the tool? Purposeless action is not the rule of nature. It must have aided human adaptation in some way. The answer seems to lie in the structure, shape and size, body proportions, and movement potentialities of man vis-à-vis those of other mammals. The human species developed a much larger brain about twice the size of that of the average gorilla or chimpanzee. It reached its largest size during the Neanderthal period and was larger than that of Homo sapiens. In order to feed this brain, good quality, high energy yielding protein and fat was needed, not found in the berries and roots on which man was subsisting. Secondly, while man realized that such good quality nutrients could be obtained from the bodies of other predators, his legs were not powerful enough to chase these larger and stronger animals and his structure was not strong enough or even big enough to overpower them. The alternative was to find dead bodies of animals and eat the meat thereof. It was this primary motivation—namely, starvation and death versus adaptation to the environment that lead to the development of tools of the type discovered in Ethiopia. Man was under evolutionary pressure to make and use tools. But in order to do so, the bones of the hands also needed to evolve, such that they could grip a tool, tightly and precisely. A recent study (cf. Gibbon, 2015) has been able to clarify when and how, in our evolutionary history, this grip developed (see Box 2.6).

Box 2.6: How Human Ancestors Got a Grip

> "Squeeze a baseball or pen between your thumb and the tips of your fingers: You are using what researchers call a precision grip, a highly evolved adaptation thought to be unique to modern humans and our most recent ancestors. Chimpanzees, for example, have thumbs too short to allow them to grip objects so precisely. But a new study suggests that human ancestors in South Africa had a good grip perhaps as early as three million years ago—and so may have wielded stone tools earlier than expected."

While our ancestors have been known to be using tools at least 2.6 million years ago, the identity of the first toolmaker was not clear since a number of different species were alive at that time, including several species of Australopithecus, *A. africanus* and *A. afarensis*, along with our genus, Homo. While stone tools had been discovered, scientists had not been able to find a fossil hand gripping a tool. Some evidence is there from earlier studies that our direct ancestor, Homo erectus, used a precision grip 1.7 million years ago, and that Lucy's species, *A. afarensis*, had started using their hands more like the way in which we do some 3.1 million years ago. But it is only recent technological advances which have provided clear evidence.

As Gibbons writes in her report (*Science*, 2015), a new method of analyzing CT scans of fossils by Matthew Skinner and Tracy Kivell, a husband–wife team of paleoanthropologists has found new evidence.

According to the report,

> "When the team scanned hand bones from four members of ***A. africanus*** that lived in South Africa between two million and three million years ago, they found that the pattern of the trabecular bone was asymmetrical, as in modern humans and

(continued)

Box 2.6 (continued)

Neandertals that use tools frequently (as they also show in their study). This suggests that ***A. africanus*** was using a "human-like" precision grip "much earlier and more frequently than previously considered," the authors write. They stop short of saying that ***A. africanus*** was using and making stone tools and acknowledge that these grips could have been used for a number of different activities with tools. However, the similarity of the pattern with modern humans suggests that hominins had the capacity for stone tool use more than half a million years before such tools are securely dated in the fossil record. They also have scanned hand bones of other members of ***Australopithecus***, including Lucy's species, ***A. afarensis***, but the pattern of use was not preserved in that species' trabeculae."

Source: Gibbon (2015) and Kivell and Skinner (2015)

The second question that begs an answer is "what predicated the size and the shape of the tools?" It was the functional limits of man decided by his body structure. In fact, the anatomy, physiology, and biochemistry of the body of each creature on earth lead to not only certain strengths but also certain constraints. Homo sapiens may stand at the zenith of the animal world, yet, no matter how hard we try, we can extend our arms only so far, we can run only so fast, we can hear sounds of a limited frequency, we can see only a very small fraction of the total light spectrum. Box 2.6 explains how evolutionary processes caused changes in the human hand such that our ancestors became capable of gripping objects, so necessary for the use of any tool.

The human body occupies a limited physical space. In a fixed human position, say while standing, our range of motion is limited to the extent that we can move our limbs. Within this range of operation, we can perform better at certain angles but are vulnerable to injury in others. For example, applying pressure through your feet is not uniform for all postures.

The measurement of these body structures and its limits is the subject matter of what is called **anthropometry** (anthro = human, metry = measurement). Specifically,

> Anthropometry deals with physical characteristics of man in time and space, particularly with individual variation, ontogenesis and generic development," (Nowak, 1996).

Anthropometry*: deals with physical characteristics of man in time and space, particularly with individual variation, ontogenesis, and generic development*

Such measurements are used today, not only by anthropologists and population geographers to study migration of populations but even more so by design engineers, whether they are designing the home gadgets or a complex space station and have led to the fields of human factors engineering, ergonomics, and biomechanics, all of paramount importance for the development of any technology. With the advent of **CAD/CAM** technologies, engineers simply need to link anthropometric data to the design component and lo and behold, we have cars, airplanes, space stations that can be easily

CAD/CAM: *computer-aided design and computer-aided manufacturing*

Box 2.7: Using Anthropometrics for Garment Design

Just think about it: you go into a store looking for a shirt. Knowing that a size 40 fits you perfectly, you choose one from the size 40 shelf, pay for it and walk out. Imagine your consternation and irritation when you open the shirt to wear it and find that the sleeves are an inch short or that it is tight around the shoulder, even though it fits perfectly around the chest. How could this happen? Simple: the manufacturer did not follow the sizing chart for that region of the globe.

With an ever increasing demand for ready-to-wear clothes, sizing has become an important issue. Even in remote countries such as Albania, sizing is important even more so, now that it is becoming an important garment manufacturing center. And, with global travel on the rise, one has to cater to US sizes, European sizes, and UK sizes, just to name a few.

How is this done? While most companies were relying on one-dimensional measurements (1D) till now, 3D measurements are now becoming popular. As Spahiu and his colleagues put it,

> "Anthropometric data are a critical issue for developing products with the right fit. In the clothing industry anthropometric data are very useful for ensuring clothing fit. Using advanced technology for 3D garment design has changed the way of garment production. In Albania the garment industry is one of the most important sectors of the economy regarding to the weight of exports and number of employers. There is a growing number of garment and footwear companies working with their own brand for the home market, but with the lack of a national sizing system. The larger anthropometric studies taken years ago are not up-to-date. Actually there is a need for a national sizing system in Albania. These will help garment and footwear companies which operate in the home market to produce garment and footwear with the right fit. These anthropometric data could be used in different application.
>
> Implementation of 3D laser scanning system and advanced software for 3D data manipulation showed a fast, accurate and repeatable methodology for taking anthropometric data. Automation of the procedure for extracting anthropometric data from 3D body models will shorten the time. Now, a large group of population can be processed for extracting anthropometric data and creating Albanian sizing tables" (Spahiu, Shehi, & Piperi, 2015, p. 2141).

Source: Spahiu et al. (2015)

and conveniently manned by humans. Rehabilitation equipment designers use such data to create wheelchairs, crutches, braces, and all types of prostheses. Clothing manufacturers use them to manufacture ready-to-wear clothes (Box 2.7) and even the architect needs to know how high the door has to be. Would you believe that robotics, too, is using anthropometrics to create robots that can simulate the human being?

Unfortunately, there is a lack of sufficient anthropometric data on the general civilian working population and many a time, the only major source for safety and product development has been restricted to data drawn from studies on military

Box 2.8: Dustin Curtis' Examination of Thumb Reach for Apple Phones

Four inches is only now barely acceptable on iPhone 5 because:

iPhone 5's huge reduction in weight makes it easier to hold while contorting your hand to touch the hard-to-reach areas of the screen.

The screen's width remains narrow and only grows vertically, meaning it's still easy to reach the entire width of the device;

The device is 20 % thinner, which allows your hand to wrap around more fully and to gain slightly more reach; and

iOS's tab bars are anchored to the bottom of the screen, where your thumb more naturally rests, so it remains easy to change app sections (contrast this with Android's tab bars, which are usually located at the top of the screen, and sometimes out of reach).

Source: Curtis (2012), https://dcurt.is/4-inches

(In the Apple 6S iPhone, which is much larger, double touching the home button brings the entire screen closer to the thumb and so "reachability" is ensured).

personnel (Hsiao et al., 2002). Obviously, these data are very different from those of the average general population. With the current focus on the use of multidimensional data, as against unidimensional data, (as stated in Box 2.7), the ideal goal should be to take into account as many users as possible. There are, however, two practical considerations that have been restricting the application of new knowledge. First, many product developers are unaware of the usefulness of anthropometric data and second, the cost of implementing the use of such knowledge could be exorbitant and is therefore, sometimes, considered impractical. At the same time, the good news is that there has been a significant rise in the number of companies engaged in designing and marketing ergonomic products. For example, Apple has taken into consideration the use of the thumb dimension in developing its larger versions of mobile phones (Box 2.8).

Over the years, comprehensive anthropometric atlases have been formulated and these reveal that there are wide differences between individuals with reference to age, gender, race, ethnic groups, etc. We can, therefore, categorize anthropometry in terms of the following:

- **Population anthropometry**: body characteristics pertaining to a particular population, say, the Chinese, Caucasians, or the Germans.
- **Ontogenetic anthropometry:** changes in body proportions, etc., with age.
- **Phylogenetic anthropometry**: body proportions typical of a species, say, the Homo sapiens.

All three aspects are useful for the development of technology and have been widely used by engineers from all industrial sectors.

Molenbroek and coworkers at the Delft University of Technology describe some of the ways in which anthropometric data has been used in the development of products (Box 2.9).

Box 2.9: On Using Anthropometric Data
"There are several ways in which anthropometric data are used:

- **Ego-design**: your own body dimension as a guide.
- **Average-design**: body dimensions of the average as a guide.
- **Design for P5**: body dimensions of the smallest person as a guide.
- **Design for P95**: body dimensions of the largest person as a guide.
- **Design for P5-P95:** body dimensions of the smallest and largest person as a guide. This type is used most commonly and means that excluding 10 % is acceptable."

> "Many ergonomists are not aware that of the fact that the anthropometric data they use is mostly 1D. This does not mean it is of less value, but… it is important to realize ….how information can be extended to 2D and 3D or maybe even 4D information, which may be more appropriate and valuable for daily use in a design or evaluator's environment" (Molenbroek & Bruin, 2005, p. 289).

Source: Molenbroek and Bruin (2005)

2.3.2 *Basic Principles of Anthropometry*

We will now illustrate, very briefly, the basic principles of anthropometry in the context of psychology of technology.

1. Anthropometrics takes into account two aspects. While **static anthropometry** measures the human body parts at rest and includes indices such as skeletal dimensions, distance between joints, space taken by soft tissue and contours of the body, **dynamic anthropometry** focuses on body reach with extended arms or legs or fingers, sitting and standing heights, or clearance, as it is technically known.
2. **Anthropometric diversity**: like most other human characteristics, anthropometrics also follows a normal distribution curve, with 90 % of any population falling between the 5th and the 95th percentile. This is, however, only for populations of the same gender. When measurements are taken across gender, that is, when we take a sample comprising 50 % males and 50 % females, 95 % of the population is covered between the 5th and the 95th percentile. At the same time, just as for all other normally distributed characteristics, there are deviations, both across distributions for different groups

***Static anthropometry**: measures the human body parts at rest*

***Dynamic anthropometry**: focuses on body reach with extended arms, legs or fingers, sitting and standing heights, or clearance*

***Anthropometric diversity**: variations in body measurements for different groups of people* of people (measured by degree of skewness and kurtosis) and within distributions (individuals whose anthropometrics fall at the extremes of the distribution). The former could be due to gender, developmental stage, ethnic differences, or occupational differences, because certain parts of the body develop at differing rates depending on their usage in different occupations, while the latter is related to normal individual differences.

3. **Sitting height**: sitting height is generally 53 % of the standing height of the person. Stature differences between different groups of people whether across gender or across ethnic groups tend to get reduced while sitting down. Thus, while designing work places and when sitting down head clearance is to be considered, far more people will be covered between the 5th and 95th percentile than when these same people are standing up.

> ***Sitting height****: the height of the person when in a sitting posture*

4. **Racial differences**: different groups have different body proportions. One such difference is that native Africans have longer legs than Europeans who, in turn, have longer limbs than their Asian counterparts. The kinds of impact that this could have on technology design is clarified by considering a piece of equipment used very widely, namely, a respirator. These are artificial aids to be used either for patients who are having difficulty in breathing or for people in jobs that have potential inhalation hazards. While most of the respirators that are being used in the USA are based on anthropometric data of the USA, we must bear in mind that the workforce of the USA today is much more heterogeneous than it was earlier. Statistically significant differences have been obtained for facial measures in terms of gender, age, and ethnic groups (Zhuang, Landsittel, Benson, Roberge, & Shaffer, 2010). This is a factor that would be a mandatory consideration for designers of respiratory or even other personal protection equipment such as masks used in mining operations or even in the construction sector.

> ***Racial differences****: differences in anthropometric measurements of people of different races*

5. Height steadily declines after 20 years of age, becoming even more pronounced after 40 years of age because of shrinkage of the intervertebral disks of the spine (Box 2.10). Did you realize that your height also varies with the time of day because the spinal cord tends to get compressed by body weight and by evening, you may be 0.9 in. shorter? At the same time, over the years, the human skeleton has also become lighter (Box 2.11).

> ***Differences in body morphology****: genetic and congenital differences in body built*

6. **Differences in body morphology** have been studied over the years and we are all aware of Sheldon's somatotypes—the ectomorph, who is thin and slender; the mesomorph, who is wide and muscular; and the endomorph, who is fat and pear shaped. Just think of an obese person trying to squeeze himself into a "normal" chair. Changes in body morphology require adaptations and changes in the design of workstations and equipment to be used by them. Gordon and Bradtmiller (2012) discuss the consequences that obesity can have on office ergonomics.

Box 2.10: The Baltimore Study on Decrease in Height with Age

In a paper in the *American Journal of Epidemiology,* published in 1999, a study was reported which attempted to study the trend of height decrease among a group of individuals over a period of time. The paper was written by Dr. John D. Sorkin of the *Intramural Research Program at the National Institute on Aging*. While the study is almost two decades old, it does provide evidence for decrease in height with age.

To determine the magnitude of height loss due to aging, a longitudinal study was undertaken between the years 1958 and 1993 on a sample of 2084 men and women from the age group 17 to 94 years.

According to the author,

> "On average, men's height was measured nine times during 15 years and women's height five times during 9 years. The rate of decrease in height was greater for women than for men. For both sexes, height loss began at about age 30 years and accelerated with increasing age. Cumulative height loss from age 30 to 70 years averaged about 3 cm for men and 5 cm for women; by age 80 years, it increased to 5 cm for men and 8 cm for women. This degree of height loss would account for an "artifactual" increase in body mass index of approximately 0.7 kg/m^2 for men and 1.6 kg/m^2 for women by age 70 years that increases to 1.4 and 2.6 kg/m^2, respectively, by age 80 years. True height loss with aging must be taken into account when height (or indexes based on height) is used in physiologic or clinical studies." (*American Journal of Epidemiology,* 1999;150, 969–77).

Findings such as the above have important implications for not only design engineers but even garment manufacturers. It means that designers for the aged must use a different set of metrics in line with the decreasing height and the corresponding increase in body mass.

Source: Sorkin et al. (1999)

Box 2.11: Have Human Skeletons Become Lighter over the Years?

Chimp bones are packed with microscopic structures known as spongy bone; modern human bones aren't, increasing risk of fractures and osteoporosis. Two studies propose an explanation for this change: Chirchir et al. (2014) found that skeletons from modern chimpanzees, Australopithecus africanus, Neanderthals, and early Homo sapiens all had higher densities of spongy bone than modern humans, suggesting that our sedentary lifestyle is to blame. Ryan and Shaw (2014) also found lower spongy bone density in the hip joints of ancient farmers compared with hips from nonhuman primates and ancient hunter-gatherers, supporting the idea that a lack of rigorous exercise, not evolutionary pressure, is responsible for our weak bones.

(continued)

Box 2.11 (continued)

This too would have implications for design engineers. Since our skeletons have become lighter, we would probably not be able to exert as much pressure as our forefathers were able to do. At the same time, the amount of weight we are able to carry without hurting ourselves has also gone down. A direct consequence of this reduction in ability to carry weights is seen in the recent changes in checked-in luggage allowed while travelling to and fro from the USA. Till a decade back, this limit was two suitcases of any size and weight. It has now been reduced to two suitcases of 50 pounds each. A pound more and you must pay for extra luggage. The reason provided by airlines is that the porters who help in the transportation of such heavy luggage have often suffered injuries, as a result of which airlines have had to pay heavy insurance premiums for ensuring the personnel engaged in this type of work.

Source: Chirchir et al. (2014) and Ryan and Shaw (2014)

7. **Reach, zone of convenient reach and optimal visual field:** there is a clear difference between what one can reach and what one can conveniently reach. Just think of the times when you have had to stretch your arms in order to pick up a phone lying at the other end of your desk causing you to strain your arm muscles and compare it to the feeling of when you have to pick up a phone that is placed at a convenient distance from where you are working. The muscle pull in the former case is very uncomfortable and in extreme cases could even cause muscle strain. We can define the normal work area as the area a person is able to conveniently reach with his arms and hands with elbows flexed at 90° with a rotation potential of about 25°. However, what must also be kept in mind is the optimal visual field which is much less than the entire 180° in front of us.

***Reach**: the maximum distance one can reach when arms are outstretched*

***Zone of convenient reach**: the distance one can conveniently reach without discomfort when arms are outstretched*

***Optimal visual field**: the area over which one can see clearly without turning one's head*

2.4 Ergonomics

Advertisements constantly bombard us with products such as "ergonomic chair," "ergonomic mattress," and "ergonomically designed workstations." But what does the term really mean? Why is there so much stress on ergonomically designed products and why are they more expensive than those that are not so designed? The word ergonomics comprises two Greek words, "ergos" meaning work, and, "nomos" meaning laws. Thus, **ergonomics** refers to the laws of work and can be defined as the science of matching work or tasks to the body. This is accomplished by taking into consideration anthropometric data, physiological characteristics, biomechanical and psychological capabilities. The current emphasis on ergonomics is because it creates win-win solutions for both the organization and the employee that is user. The user, especially the employee, is much more comfortable, both physically and psychologically, and therefore manifests higher levels of job satisfaction and work morale, which in turn, makes for higher productivity, better quality of work, and reduced wastage. At the same time, there is lower fatigue, improved health and safety all leading to decrease in injuries and illnesses and thus lowering worker compensation. Which employer would not be happy, even ecstatic to have human resources of this genre?

***Ergonomics**: the science of matching work to the body*

2.4.1 How Does Ergonomics Work?

We could probably begin by attempting to understand the difference between ergonomics and anthropometrics. As Ryan (2013) puts it,

> "Anthropometrics is "the study of the human body and its movement."… Ergonomics, on the other hand, is "the study of people and their relationship with the environment around them. When anthropometric data (measurements/statistics) is applied to a product, e.g. measurements of the hand are used to design the shape and size of a handle, this is ergonomics" (www.technologystudent.com)."

TheANSI **Z94.0-1989** provides a more detailed explanation of ergonomics. It is,

> "The application of a body of knowledge (life sciences, physical science, engineering, etc.) dealing with the interactions between man and the total working environment, such as atmosphere, heat, light, and sound, as well as all tools and equipment of the workplace."

The beauty of the study of ergonomics is that it draws from a variety of disciplines, ranging from anthropometry, physiology, and psychology to kinesiology and even to human factors, industrial medicine and management. The fundamental process is fourfold:

1. Identify the physical, physiological, and psychological demands of the job.
2. Identify the physical, physiological, and psychological capabilities of the worker.
3. Identify the physical, physiological, and psychological mismatches between the demand and the capability.
4. Minimize the mismatches through education and training, and work, tool, equipment, and environmental design.

A related term is **biomechanics**, which deals with the application of principles of mechanics to living biological material, especially in terms of the level of stress that the body and its parts can withstand, as when the body is put through acceleration (be it in a car, a go-cart, or even a roller coaster). Modern technology has greatly reduced the amount of physical

***Biomechanics**: applications of principles of mechanics to living material*

stress the human undergoes, but it has not always been able to reduce the stress on small body parts. This is where the principles of biomechanics are useful, using which ergonomists attempt to design workstations and equipment whereby minimal stress is put on the human body and efficiency is improved and fatigue is reduced.

2.4.2 Ergonomics and Prevention Through Design

An important use of anthropometric data in combination with ergonomics is in the design of tools, workstations, and equipment in general so as to reduce occupational hazards, accidents, and illnesses. Governments have taken it upon themselves to introduce a variety of steps, one such being the *National Initiative for Prevention through Design, (PtD)*, undertaken by the NIOSH (*National Institute for Occupational Safety and Health*, USA) (Box 2.12). According to the *Center for Disease Control and Prevention*, it is clear that,

Box 2.12: Prevention Through Design (PtD)

The traditional method of preventing occupational accidents has been to provide safety measures and equipment such as those used for firefighting, gas masks used in mining, etc. These of course are necessary and will always be used. However, with recent awareness of psychology of technology, a new perspective based on ergonomics has emerged, called *Prevention through Design* (PtD). The idea is that worker accidents can be "designed out" through designing, redesigning, and even retrofitting and that by doing so, many an accident would be prevented and many occupational diseases would be eradicated.

It is with this in view that the *National Initiative for PtD* has been launched. PtD "is a comprehensive approach, which includes worker health and safety in all aspects of design, redesign and retrofit/will provide a vital framework for saving lives and preventing work-related injuries and illness is the roadmap."

According to CDC, "PtD encompasses all of the efforts to anticipate and design out hazards to workers in facilities, work methods and operations, processes, equipment, tools, products, new technologies, and the organization of work. The focus of PtD is on workers who execute the designs or have to work with the products of the design. The initiative has been developed to support designing out hazards, the most reliable and effective type of prevention……. The ultimate goal of the PtD initiative is to prevent or reduce occupational injuries, illnesses, and fatalities through the inclusion of prevention considerations into all designs that impact workers. Along the way, intermediate goals will be identified to provide a path toward achieving the ultimate goal. NIOSH will serve as a catalyst to establish this Initiative, but in the end, the partners and stakeholders must actively participate in addressing these goals to make PtD business as usual in the 21st century."

Source: NIOSH (2013)

> "One of the best ways to prevent and control occupational injuries, illnesses, and fatalities is to "design out" or minimize hazards and risks. NIOSH leads a national initiative called Prevention through Design (PtD). PtD's purpose is to promote this concept and highlight its importance in all business decisions" (NIOSH 2013).

While the concept of PtD is highly attractive and of great value, implementing PtD is easier said than done. One reason is the demographic changes that are taking place in the workforce in almost every industrial and business sector. As compared to even three decades back, that is, shortly before the turn of the century, the workforce, today, is far more heterogeneous. For one thing, women have entered spheres of work normally seen as male bastions. Secondly, with rapid advances in communication technology and means of travel, widespread globalization has ensued, causing the workforce to be comprised of people from a variety of ethnic groups. Thirdly, modern medicine and better health conditions have seen to it that there are more people from the age group of 60 years plus still occupied on a part-time basis if not on a full-time basis. Due to each of these, the workforce today comprises people from both genders, different age groups, and from different regions of the world, implying that if tools, equipment, and workstations have to be ergonomically designed, we would require anthropometric data for each of the above groups. The *National Institute for Occupational Safety and Health* is facing exactly this problem and the box below (Box 2.13) clarifies one such problem and the ways in which it is being handled.

Box 2.13: NIOSH Study on Truck Drivers (2015)

As mentioned by John Howard, Director, *NIOSH*, in the foreword to the report,

> "Work-related highway incidents are a leading cause of occupational deaths and injuries in the United States. Prevention of work-related highway injuries and deaths poses one of the greatest challenges for occupational safety researchers. Compared with other work settings, the work environment surrounding trucking and other transportation-related activities is fluid and dynamic. As such, it is difficult to exert direct control over a range of factors (e.g., fatigue, inattention, subpar road and weather conditions) that may negatively influence a worker's safety. Despite these challenges, *NIOSH* has been actively engaged in a range of research efforts that are likely to have the greatest impact on the reduction of work-related highway incidents. One of these research efforts is focused on reducing hazards to truck drivers by improving ergonomic design in medium-duty and heavy-duty truck cabs" (p. 176).

While American manufacturers have long realized that safe trucks cannot be designed without accurate and up-to-date anthropometric data, an earlier survey was not found to be up to the mark. In view of this *NIOSH* launched the first ever federal anthropometric study of truck drivers in the USA, in 2006, with data being collected between the years 2009 and 2010 on 1779 male and 171 female truck drivers from various states of the USA.

(continued)

Box 2.13 (continued)

The results of the study have important implications for truck designers and manufacturers. What has come to light is that there have been "widespread changes in the demographics of the American truck drivers in the last 25–30 years. In 1983, the combined category of truck drivers (heavy and light) and driver-sales workers consisted of 11.7 % African American, 5.6 % Hispanic, and 3.5 % females (BLS, 1983). In 2009, the driver-sales workers and truck drivers' category consisted of 13.4 % African American, 18.7 % Hispanic, and 5.2 % females (BLS, 2009). This new demographic reality necessitated an updating of the anthropometric data used for the design of truck cabs because anthropometric data are related to various demographic characteristics (NIOSH, 2013; Gordon et al., 1986; ISO, 2006)" (p. 1).

The results show the extent of the changes in the body dimensions of truckers. While the average male truck driver was found to be 13.5 kg heavier, the female truck drivers were 15.4 kg heavier than their counterparts from the general population. The male truck driver of today is about 12 kg heavier than those 25–30 years back. While they are not taller but they are certainly larger in body girth and body width, both important while considering seat and driving chamber design. These changes make an obvious case for revising the sizing and designing of trucks.

On the basis of the data collected on a list of 33 anthropometric dimensions, 15 body models have been worked out independently for males and females, each model representing a unique combination of body size and physique. *NIOSH* is of the view that these models will surely benefit the design of the next generation of the driver cabins of trucks.

Source: NIOSH Study on truck drivers (2015)

2.5 Applications of Anthropometry

There are a variety of ways in which anthropometry and ergonomics have been able to help man. Some of these have been detailed below. We shall start with the principles of creating a suitable workstation and go on to show the ways in which anthropometrics can make a world of a difference in a simple piece of home furniture such as a chair. From there we move to the complexities involved in the design of a manned space station. In order to illustrate how anthropometrics can help design equipment for the aged, we have included a section on anthropometrics and the aged. Another important application of anthropometrics is in the field of rehabilitation of people with impairments, thereby mandating a section on rehabilitation, and finally, we show how gender differences in body proportions can be used to create comfortable workstations for women.

2.5.1 Anthropometry and Workspace Design

A fair amount of work is done using tools that are laid out around the individual worker. In other jobs, the worker may be having a panel of switches, knobs, buttons, gears, etc., in front of him. The principle concern in each of these cases is how to lay out the material such that the worker can reach each object most conveniently and with least discomfort. Another consideration is that worker injuries have to be kept at the minimum. Thus, the workspace must be so constructed that it leads to maximal efficiency coupled with maximum effectiveness. Using anthropometric data such as given above, principles for the design of workspace layout have been delineated. Alan Hedge of Cornell University has, very lucidly, laid out certain anthropometric considerations useful for workspace design (Hedge, 2011a).According to him, workspace layout should follow five principles:

> ***Kinetosphere****: The set of all reachable points for a subject at a given position*

(a) **Importance:** items most important should be placed in the center part of the **kinetosphere**. "The dynamic reach envelope, or kinetosphere, describes the set of all reachable points for a subject at a given position. This shell of reachable space, when it is intersected with the working plane, provides what we have been referring to as the reach envelope" (Toney & Thomas, 2007).
(b) **Function**: items of similar function should be placed together.
(c) **Frequency of use**: most frequently used items have to be placed at the most convenient place.
(d) **Sequence of use**: items should be arranged according to the sequence in which they are to be used.
(e) **Normal posture**: items should be accessible from the normal posture.

A large part of work is performed in the sitting posture, with arms or maybe legs, hands or feet performing the actual task as in the case of car drivers, airplane pilots or even computer operators and other office tasks. In most such cases the apparatus on which the person is working, its knobs and handles, switches, etc. are in front of the person. Normally, everything is laid out on a table or platform in front of the person. But the crucial factor in deciding efficiency is the height of the platform and the total area to be covered by the hands and fingers. A classical study was the one conducted by renowned ergonomist Dempster and Gabel (1959). Though conducted over 50 years back, it still forms the backbone for ergonomists even today. Using very complex recording equipment and even more complex mathematical procedures, various indices were worked out. The limits of the space reached while sitting, the left and right hand overlap in this space, regions of maximum hand flexibility, mean hand positions, and degree of variability have been obtained. Using such functional anthropometrics, effective hand positions have been worked out for use in the designing of worktables, control panels, etc. But do data like the above cater for all people? It is very easy to design chairs, tables, consoles, and general workstations and most of the population would find them comfortable. But how about those whose body proportions fail to conform to the standards observed for any particular culture, age, or ethnicity? One such group would be the obese. Useful guidelines have been provided by Gordon and Bradtmiller (2012) on how obesity requires changes in office ergonomics.

2.5.2 Anthropometrics and Chair Design

Ever think about that lowly object that supports your body for the greater part of the day? The word "chair," per se, conveys little. It is only when you prefix it, arm chair, office chair, deck chair, dining chair, garden chair, and so on, that images start popping up. Make it more extravagant, and it becomes a sofa, a settee, a throne. Belittle it, and what is it now: a bench. The basic idea is the same in all, just that the dimensions vary, and some dimensions, such as the seat back taking on a value of zero, for a bench.

Four major dimensions are normally required for the design of any chair. These are seat height, seat depth, seat width, and back height. Variations occur along each of these for two reasons: firstly, the dimensions of the user and secondly, the purpose for which it is to be used. If you are speaking on the phone in office, the distance between you and the phone will decide your posture. If you are reading a novel or flipping through a travelogue, you will probably want to lean back in your chair, or even curl up; but, when you are going through an important office file, you usually tend to lean forward.

How does the maker of the chair determine the dimensions of the chair? As early as during the Roman and Greek days, various methods had been delineated to map the physical structure of man as a function of posture. Whereas the Greek used geometrical formulations and arrived at what they called the Golden Rectangle and the Golden Mean (made famous by Euclid), Leonardo used what has been termed the Vitruvian man after the Greek architect, Vitruvius, who actually took measurements of the human body. It was only much later, in the twentieth century that le Corbusier, again an architect, started looking at and measuring the human body in a more intricate manner. The greatest impetus to anthropometry, however, came during and after World War II. The landmark year is probably 1992, when a large-scale international survey was instituted. Nomenclatured **CAESAR** (the Civilian American and European Surface Anthropometry Resource), it was a collaboration between more than 35 companies and 6 different countries. The survey was based on 4431 American and European adults from both sexes aged 18–65 years (Salvendy, 2005). Each subject was scanned in 3 different poses, 59 point-to-point measurements were calculated, and 40 traditional body measurements were taken by measuring tape. The survey was the first to provide 3D human models with anthropometric data coupled with demographic data (Robinette & Daanen, 2003).

***CAESAR**: the Civilian American and European Surface Anthropometry Resource*

The chair in modern times has benefitted greatly from such surveys and has made ergonomically designed chairs possible. But this has not always been the case. Any chair will comprise three factors, function, aesthetics and material (Nelson, 1994). While some chairs focus on function, others focus on aesthetics and that is why we hear of terms such as status furniture and technical furniture (Nelson, 1994). Should there not be some sort of reconciliation between the two?

2.5.3 Anthropometry and Space Station Design

The designing of a space station takes much more than normal anthropometric data and ergonomics. There are a number of other concerns that have to be addressed. First and foremost, what anthropometric standards should be used? It is not a limited space like a workstation is and, yet, it is a workstation; it is not limited in its functional purpose such as a chair or bed is and, yet, it has chairs and beds; and lastly, astronaut postures are not fixed as on an assembly line or a computer operator, and, yet, he/she has to operate computers. Thus, the normal anthropometric parameters used for designing a workstation, a chair, or a computer are useful but with a host of other considerations factored in. Another major problem faced by design engineers of space stations is that it is extremely difficult to define the user population. Earlier, people sent into space were from a very limited population. If it was an American space ship, the users would be Americans; if the space ship was Russian, the users would also be Russians, and so on. Now, however, the population from which the selection is often made has a much wider spectrum and international anthropometrics has to be considered. Since we already know that anthropometric measurements vary across race and ethnic groups and gender, aggregation across countries is the only way out.

A detailed guide has been worked out by **NASA** from which some points are being described below. Till a few years back, the *NASA Technical Standards Program* was using a document called *NASA-STD-3000*. This was a very detailed document for manned space ship design and function and includes 14 chapters ranging from anthropometrics and biomechanics, to human performance capabilities, crew safety, workstation design, hardware and equipment and even facility management and extravehicular activity. As far as measurements for space ship and space station users are concerned, NASA has provided the outer limits. The guide provides anthropometric data for the Japanese, who are generally considered small people, and for the North American, who are generally large people. Though these parameters have been provided, NASA advises that an aggregation of different population anthropometrics be used. The detailed manual also provides guidelines for design considerations including the changes that occur in both static and dynamic anthropometric indices due to weightlessness and how these are affected by interindividual, interracial, and secular differences. NASA also advises that the same anthropometrics can be used for clothing of space ship users, that is, for the design of space suits. Of late, it has been felt that this document has become outdated and has now been replaced by a new document, entitled, *NASA-STD-3001* (2014) divided into two volumes which are as detailed, if not more, than the earlier document.

Another important facility is the *Anthropometric and Biomechanic Facility* (ABF) which is collaborating with NASA to design spacesuits keeping in mind the requirements of the future, for example, for astronauts who may be from different anthropometric pools. Rather than using specific anthropometrics, the newer concept is using whole body posture based analysis using both anthropometric data and biomechanic data (Thaxton & Rajulu, 2007).

The challenges that will face space technology in the future are very obvious. It was only a few years back that a new feat was accomplished when a spaceman skydived 24 miles from space, breaking the sound barrier and came spinning down. CBS News reported that Skydiver Felix Baumgartner made the highest skydive ever on Oct. 14, 2012. He jumped from 128,000 ft (39,000 m), or about 24 miles up, during the Red Bull Stratos mission. It was also reported at that stage that though the feat was accomplished and that too without any mishap, one serious shortcoming was that the space suit being worn by the astronaut was not suitable for this type of performance. It is in areas such as these that anthropometrists face many challenges but can provide valuable insights.

2.5.4 Anthropometrics and Designing for the Aged

Another challenge facing the twenty-first century, the world-over, with the increasing number of elderly people is in the designing of clothing, equipment, etc. for the aged.

Studies show that marked changes occur in our body proportions and ability to move around with age. According to Perry (2010), the following are the most marked changes with age:

- **Strength:** 25–30 % decrease in strength after 65 years of age.
- **Flexibility**: 18–20 % decrease with age.
- **Balance:** 1/3 of 65-year-olds fall each year.
- **Sight**: all aspects deteriorate.
- **Reaction time and speed**: both show decreases.
- **Manual dexterity and tactile feedback** undergo decrease.
- **Body fat** increases.

There is also evidence to support the observation that parameters for certain body parts are altered with aging. A box in the previous section (Box 2.9) details how height decreases with age. The most affected parts appear to be the thigh, lower leg, upper arm and forearm (Muri et al., 2008; Pavol et al., 2002), as well as the muscle mass distribution (Janssen, Heymsfield, Wang, & Ross, 2000). As a result, locomotion and mechanical work ability becomes hampered (Schuch, Balbinot, Boos, Peyré-Tartaruga, & Susta, 2011).

Apart from these physical changes there are physiological changes causing fatigue issues, problems of the respiratory and cardiovascular systems, systemic blood pressure, and ability to withstand extreme temperatures. These are further complicated when one adds the psychological changes due to aging, such as not being flexible to work schedules, requiring more structured training and education and showing greater disenfranchisement and disengagement (Perry, 2010).

In view of the above, there would be a variety of changes that would be required for elderly workers. If we think of office work, redesign of chairs and tables would be necessary to accommodate anthropometric changes with age. According to Kothiyal and Tettey (2001), even more crucial would be the placement of storage shelves for the aged. Since people tend to become shorter with age, they could have trouble reaching out for things placed at the usual heights. Another aspect that needs attention is public transport and bus seat dimensions. As larger numbers of older people set out to work, shop, or simply manage their household chores, they would need to travel. At the same time, driving often becomes tedious if not impossible, and so there would be a greater reliance on public transportation systems. Normally, trains and buses have only a few seats reserved for senior citizens. As time goes by, these will not be enough and transportation managers may have to reserve more seats for such people. Providing barrier-free environments would also be necessitated because of more people relying on wheelchairs, crutches, or braces.

2.5.5 Use of Anthropometrics in Rehabilitation

> ***Natural speed**: the speed that tends to optimize accuracy of movement*

> ***Energy cost**: total amount of energy to be expended on a particular movement*

Movement time: watching people in day-to-day living, one can easily decipher that there is what has been called a "**natural speed**" for performing any act, whether it is reaching out for an object or turning off the kitchen faucet. It has generally been said that this is because of a speed accuracy trade-off and the natural speed is that speed which tends to optimize accuracy of movement. Mazzoni, with the help of empirical research on PD (Parkinson Disease) patients, has, however, helped to analyze this further and has obtained empirical evidence regarding another factor that needs to be considered. This factor is the **energy cost**. We tend to optimize this energy cost, too (Mazzoni et al., 2007). So sometimes, if a particular act requires considerable energy or more energy than we are capable of putting in, we tend to slow down. This seems to be the case for PD patients but could also be true for those who are physically challenged, convalescents, or even the aged. Design for the aged would need to keep this in mind. Knobs and switches that require fast reaction times, say as on an elevator, a kitchen stove that heats up quickly and needs to be suddenly slowed down or even the length of the pedestrian light at crossroads need to be designed in such a manner that such people do not have problems simply because they have become slow.

Workspace design uses various anthropometric indices, one of the major ones being zone of arm reach. For many people with upper limb disability, the zone of arm reach could be different. Five anthropometric indices, namely, stature, trunk depth, arm overhead reach, arm reach forward and lateral were used by Nowak (1996) to obtain measurements for both people without disability and those with disability. There was enough evidence pointing to differences on all parameters.

Another use of anthropometric data for the disabled is in the manufacture of clothing. While general tailor measurements are used, there are someother measurements that are also necessary. These include the elongation of certain body parts while performing movements in which arms, legs, or other parts of the body are stretched out as also what has been termed **motion comfort**. Clothes for people who are ambulatory and those who use wheelchairs require different clothing-fits.

> ***Motion comfort:*** *designing of clothes so that it is comfortable to execute movements*

2.5.6 Workplaces for Women Workers

Certain occupations require overhead work. These include the automobile assembly plants and construction works. Unless designed ergonomically such tasks could lead to cardiovascular and musculoskeletal strain (Haslegrave et al., 1997). One can imagine the costs of such strain. Not only would efficiency be lower, fatigue would also be greater and injury very possible. What are some of the dynamic parameters for such work?

- The greater the distance between the body and the point of hand exertion, the less is the force exerted.
- Slight changes in foot placement also affect force exerted.
- Lying supine produces lower force exertion than sitting or standing up.
- However while supine, the maximum force exerted is increased as reach distance is reduced.
- Kneeling on two knees leads to lower force exertion than kneeling on one knee.
- Lifting and pressing lead to higher force while laterally exerted force is lower.

All of the above have been obtained for male workers and male subjects. What happens when females are put in similar situations? The last few decades have witnessed larger and larger numbers of women at the workplace and not just in office jobs and small assembly jobs but in almost all types of factories across industry sectors. Reduced efficiency vis-à-vis their male counterparts is, to a very large extent due to the fact that workplaces and tools have been designed for use by men. In view of the above, Chow and Dickerson (2009) attempted to devise indices for women. Chow and Dickerson concluded that women manifest lower shoulder strength than males, namely, that for females is 2/3 that for males. At the same time, the direction and angle of the force exerted is also important, especially when working at or above shoulder level. Maximum force could be exerted when the subjects were standing with the work handle placed at an angle of 60° from shoulder height and moved the handle downwards. In marked contrast, when this handle was placed at 90° and subjects were asked to push forwards in the horizontal plane, the force exerted was the least. In general, force was greatest in the vertical direction and least in the horizontal direction. An important recommendation made by the researchers was that since they found that different women used different postures to maximize the amount of strength exerted, workers should be allowed to alter their postures to gain maximum advantage (for example, sitting, standing, or supine).

2.6 When Anthropometrics Is Neglected

Hurtability: the degree to which it is possible to get hurt while using a product

How do we make the above happen? What happens when we do not think of the above? The answers to both of these are linked to anthropometry to a very great extent. While Norman (2008) explains how to build these features into the design of the product, Guimaraes and Antunes (2012) focus on the other issue: how bad design can cause **hurtability**, the opposite of which would be, usability. When because of commercial reasons, all we can think of is getting more and more people to buy the product, we stop thinking of how people could buy the product but also get hurt while using the product because of certain design features. Guimaraes and Antunes give us a large number of such examples of poor usability ranging from sardine cans, to ketchup packaging, bus handle supports, filing cabinets, and even things as simple as the bookbinders being used by school children. In most cases, the design is faulty because the capabilities of the human hands and fingers, the reaction times while operating certain systems, and even our mental capability in understanding system requirements are not being considered. The success of the product should not be measured by the number of pieces sold, but by user reactions concerning the achievement of her goals, with its efficiency plus with the all-important last corollary, namely, without being hurt.

2.7 Going Beyond Anthropometrics

It had long been thought that most human structures fail to show plasticity beyond the developmental period. However, recent research into expert performance clarifies that not only do cognitive structures adapt to the situations at hand, but so do anatomical structures and physiological processes. At the same time, these adaptations are easier and larger during childhood, especially during critical periods of development. Once adulthood is reached, adaptations do take place but are slower and the gains are also smaller. We have earlier said that our auditory capacities are limited not only with reference to the range of frequencies that we can sense but also with reference to our power of discrimination between sounds of different frequencies. Talent in music is based to a great degree on the power to differentiate between notes and what has been termed absolute pitch. While in the general population, only .01 % show absolute pitch, and it is generally believed that adults are unable to retain this capacity even if they manifested it in childhood, the crux of research bears on the fact that absolute pitch can, in fact, be acquired by anyone, though only a limited period of development (Takeuchi & Hulse, 1993). Even physiological and anatomical aspects show adaptation. For example, there have been observed changes in the sizes of hearts and lungs, the flexibility of joints, and the strength of bones (remember that these are all anthropometric measurements) due to appropriate training (Ericsson, Krampe, & Tesch-Romer, 1993). In fact, with a clear exception of height, a large number of anatomical structures undergo adaptation because of training (Ericsson & Charness, 1994). What is also important is that these changes have not happened through gradual refinement but by restructuring the performance and acquiring new methods and skills: clearly, an example of exaptation. That adaptation can occur in the cognitive processes is clear from empirical studies and case histories of people with exceptional memories. In most cases it has been observed that through practice the very structure of the cognitive process undergoes adaptation, in that, people learn how to circumvent the limited capacity of the Short-Term memory (STM) and directly store data in the Long-Term memory (LTM). However, even these adaptations have limits. Sustained overtraining leads to injuries and burnout and the only therapy, at that stage, is to stop practice completely. We have presented material above to show that there are normal declines with age along a number of anthropometrics. On this ground too, it has been seen that experts are found to age slower than novices, but this is limited to the particular domain in which they are experts (Salthouse, 1991).

How does all of the above relate to the development of not only technology, per se, but also to developing a coherent psychology of technology? In order for technology to develop along lines that would optimize human effort, it is clear that not only do we have to keep the human structural and functional limits in mind, but we must also not fall to the ploy of cheap commercialization. At the same time, the fact that many of our structural and functional aspects are able to undergo adaptations and even exaptations, anthropometrics should not be the ultimate limiting factor in the design of technology. Probably, Alan Hedge of Cornell University has addressed this issue in a more coherent manner. By discussing the myths of designing, he points to the many challenges facing the designer of technology (Hedge, 2011b). Designing for the average person does not solve the problem since there are no people whose body proportions are all at the 50th percentile. Neither does designing for people who fall in the 5th to 95th percentile: at the most you are covering for 90 % and not all 100 %. The best would of course be to design for an adjustable range. But would that be cost-effective?

2.8 Mapping the Body and Technology: The Unique Role of Psychology of Technology

Anthropometric data provide information about the characteristics of the body which are vital for the design and use of any equipment. When firefighting equipments, such as hoses, were heavy unlike those made of lighter materials, it was considered to be a profession mostly for men. The same can be said about heavy truck drivers who needed to lift heavy materials. Traditionally, whenever we have attempted to assess the interaction between man and machine, we have focused on the following three features:

Strength: Is the thing very heavy? How we wish the computer was in our hand or lap?
Reach: How we wished the remote control was available to us when we first used the television?
Clearance: can we avoid clutter and operate smoothly?

So, when we mentioned the designing of the Apple phone, earlier in this chapter, we noted that anthropometric data have gone far beyond the mere use of physical features, for example, the hand and its physical dimensions. While such physical dimensions continue to be relevant, we also need to evaluate the functional features of the hand in relation to the use of things. In the absence of data regarding the performance of the human body under different conditions, a considerable amount of morphological information might not be very useful or may even become counterproductive. We, therefore, need to go beyond the static and dynamic dimensions of anthropometry mentioned in the section on principles of anthropometry. The operation of most equipment is no longer based on the simple mechanical manipulation of knobs and switches, which are based on gross motor skills of the hands and feet. Instead, with most equipment becoming computerized, the operations now require the use of complex fine perceptual motor skills (for example, just think of the skills required for using a mouse to select functions) and have necessitated a deeper understanding of the mechanics of motion.

Figure 2.3 shows a force-sensing glove which is being used to collect such data regarding functional aspects of the human body for ergonomic research. Thanks to improvements in biotechnology, we now have information on several dimensions enabling us to determine how best to create conditions for human interaction with machines. Is this not a huge step forward from "human factors engineering" in which human beings were considered as simply one more factor in the design and use of machines?

The second important issue in working with tools relates to a shift from physical load to mental load. Traditionally, anthropometry dealt with physical data in terms of strength, reach, etc., as discussed above, and was considered more as an area for biology and engineering and was therefore rarely mentioned in classical books on psychology. The focus was on collecting anthropometric data useful for operating tools, appliances, or cars. Recent technological developments have changed the

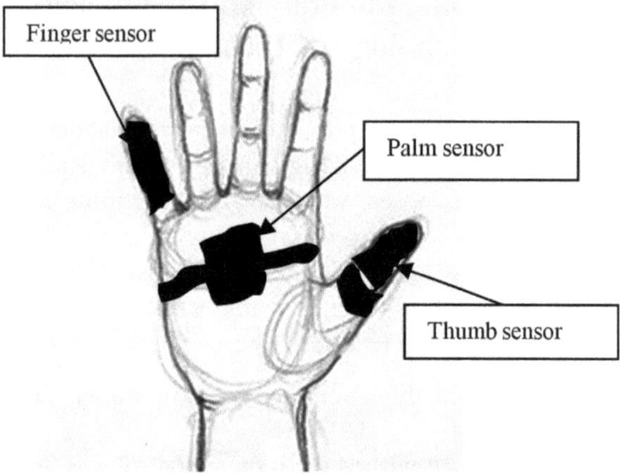

Fig 2.3 Bresslergroup's force-sensing "glove" for ergonomics. Adapted from http://www.designingforhumans.com/

scenario with a shift from manual dexterity and heavy physical operation to increased mental loads. For example, if we think of twenty-first century communication technology, the designer must consider the interface between the user's sensory and cognitive interface consisting of her associated auditory, visual and motor processes and the machine system or display. In other words, modern anthropometry looks beyond the body and gives precedence to designing for the mind. A good example is the efforts put in developing smartwatch design or Apple's mobile phones. In short, in interactive technology, the repetitive, mechanical actions of the body have now become closely linked to cognitive processes for which an understanding of the cognitive structure has become as important as that of the physical structure.

The above developments have led to the redesigning of the older workspaces. One example of how this is being done has been provided in a previous section which details how the NIOSH is attempting to improve the workspace for truck drivers of today, who are, by and large, heavier and need a workspace suited for their operations. Psychology of technology offers a platform for understanding behavioral patterns associated with a large number of unique situations that are otherwise taken for granted and thought to be "normal." In other words, it affords an opportunity to encompass the needs of different people in differing professions.

We believe that it is in this role that psychology of technology could contribute significantly at the global level. If workspaces can be improved, accidents can be prevented and anthropometric data can meet the needs of people in different cultures, the science of psychology, in general, and psychology of technology, in particular, will play a key role in the lives of people around the globe. It will help the activities of the *International Organization for Standardization* and recognize the contribution of initiatives such as PtD (Prevention through Design) mentioned earlier.

Finally, the success of any human enterprise, including the development of science and especially that of psychology which has had a weaker say in public affairs, depends on the inclusiveness of its clients. So far, anthropometry has not been able to enter the mainstream of our lives mainly because it has been seen that the principle of "one size fits all" just does not hold. Or, can we say that focusing on the majority and neglecting the minority has made a product important and personalized? We believe that the role of psychology of technology affords an opportunity for psychologists to highlight the significance of their subject in dealing with issues of global significance. As technology is proliferating at an astonishing rate, we have the momentum asking us to take a variety of initiatives. The need is to take advantage of this gathering momentum and derive psychological principles which will not only help designers of technology but also facilitate the designing out of accidents and other technology associated hazards. In this chapter, we have focused on some limited but important issues that need to be examined, clarified further, and evaluated within the classic scenario of psychology in particular and the interdisciplinary context in general. On this platform, biology and technology would find psychology a useful partner for future research.

Chapter Summary

The chapter starts by describing the use of tools and technology by animals, proving that it is not humans alone who use tools. Technological development seems to follow a similar evolutionary pathway. Neural evidence has been provided to show that we were "born to choose" and that technology does not just happen. Rather, there is a cognitive shaping of technology with each step of technological innovation enabling man to gain mastery over an uncooperative environment, thereby also proving that we could never have remained a tech-devoid society. It continues with the Darwinian process of evolution, stating that we have come a long way from the purely Darwinian analysis of evolution in terms of the laws of natural selection. Neo-Darwinism postulates that the process of evolution, including that of the human brain, throws up new structures in the form of exaptations and spandrels, which are then used for a variety of new behaviors, including new forms of cognitive abilities. The second part of the chapter asks the question that could technology have developed in some other way and answers it by providing a detailed description of anthropometrics and its application in the design of a variety of technologies, including space station design, design of equipment for the elderly and those used for rehabilitation purposes. The chapter ends by giving examples to show that though it is possible to go beyond anthropometrics, the possible consequences of neglecting anthropometrics should also be kept in mind.

Suggestions for Further Reading

Hedge, A. (2011). *Anthropometry and workspace design*. Cornell University.

Hills, T. (2006). Animal foraging and the evolution of goal-directed cognition. *Cognitive Science*, 30, 3–41.

Johnson A., & Proctor, R. (Eds.). (2013). *Neuroergonomics: A cognitive neuroscience applications*. New York: Palgrave Macmillan.

Leotti, L. S., Iyengar, S. S., & Ochsner, K. (2010). Born to choose: the origins and value of the need for control. *Trends in Cognitive Sciences, 14*, 457–463.

Chapter 3
Technology and Sensory, Perceptual, and Cognitive Processes

This chapter aims at examining the limits and expansion of our sensory, perceptual, and cognitive processes in interaction with the artifacts of technology, some specifically setting limits while others augmenting several processes. Out of our five primary senses and the enteroceptors within our body, the major focus has been laid on the visual and auditory mechanisms that are widely used in the context of technology. Beginning with how we receive, process, and store information, several topics of application based on our current understanding of sensory, perceptual, and cognitive functioning have been identified and illustrated, for example, the Gestalt laws of perception in designing things; the overloading of our sensory, perceptual, and cognitive processes with the complexity of technology; the inherent limits of our visual and auditory functioning in the development of technological products; and uses and limits of multisensory integration. All along, the theme is integrated in the context of lessons drawn, as also illustrated in Chap. 2, from evolutionary biology that while foraging has its purpose for our survival, especially in seeking and creating choices, to such human efforts that have been helping us in return to stretch beyond our own limits set for us as a biological product.

On July 2, 2012, the *New York Times* reported that several pilots, known to be elite air jockeys of the fighter plane F-22 Raptor, had been suffering from several psychological and physiological problems including cognitive loss, loss of eye–hand coordination, dizziness, fogginess, loss of concentration, and difficulty in breathing. In order to maintain superiority in the air combat with a plane at least ten times better than that of Russia and China, the contract had been given to Lockheed Martin to manufacture a unique plane with a whopping cost of over 100 million dollars for each plane (but eventually it cost the tax payers 412 million dollars per plane). In between the eagerness to keep its apex position in the air war and Lockheed Martin's acquisition of such lucrative business, the complaints of pilots were undermined until two pilots, in defiance to the restrictive orders of the Air Force, appeared on the CBS *60-Minutes* television program to describe their inability to fly this amazing piece of technology. The manufacturer claimed that the plane was built and put to operation after 5600 tests and pilot error was attributed to

V.K. Kool, R. Agrawal, *Psychology of Technology*,
DOI 10.1007/978-3-319-45333-0_3

handling the machine, including a fatal crash in Alaska in which the pilot died. Later, after substantial research it was found that the problem was caused by poor oxygen supply leading to hypoxia in the pilots.

The earlier example is just one of the many illustrations regarding how our craving for excellence can be augmented by technology but at the same it may undermine human capabilities and their functioning. Additionally, while deploying technology leads to the enrichment of our physical health and other benefits, it has its own cost. When Bill Clinton was making intemperate remarks after his bypass surgery in 2004, many, among us, wondered about his skills of public rapport. Not many knew at that time that taking advantage of sophisticated medical technology has its own cost—emergence of short-term memory loss, slow response, trouble concentrating, and emotional instability, to name just a few, these are common symptoms after surgery. Rather than addressing the issues between mechanistic explanations of cognition that restrict consciousness to internalized states of the perceiver and its replacement by broader explanations, for example, enactive perspectives that focus on broader sensory, motor, and physiological capabilities of the body in interaction with the environment (Noe, 2009), we take this opportunity to focus on the human capabilities in the context of technology.

Just as there are limits to our physical movements and the space we are capable of covering while standing or sitting or even lying down, so do our sensory and perceptual systems limit our range of input and throughput of stimulation from the external environment. In order to use any technology, or any gadget for that matter, there must be some type of input to the human body regarding this gadget. To be of any use, these inputs must be in line with our sensory and mental limits. Today, as we move forward in the twenty-first century, we seem to be living in an age of Wi-Fi. But earlier, especially in the middle of the last century, hi-fi was the big issue. What was this hi-fi and why was it important? It was important because this was the age when the radio, the transistor, the tape recorder, the video recorder, the Walkman, and the TV were invented. With all these gadgets, it was important that the sound reproduced be as close as possible to the original sound. When the degree of correspondence between the original sound and the sound reproduced through the gadget was high, the equipment was labeled as high fidelity equipment or **hi-fi**. And such high fidelity was possible only when the recorded and reproduced sound corresponded to the sensory limits of the human sense organs.

> ***Hi-fi or high fidelity***: *resulted when the degree of correspondence between the original sound and the sound reproduced through the gadget is high*

Let us start to look at these sensory and perceptual limits by attempting to understand the processes that are involved from the time we take notice of some gadget in the environment around us to when we finally use it. While basic information regarding these processes has been understood from the times of psychologists such as Woodworth (1958, 1918), it is only in the last half century or so that exactitude could be attained regarding the sequence of these processes and even the time duration of each of them. With current advances in brain imaging, especially functional

MRI (**fMRI**), our understanding of brain processes and brain areas has increased multifold.

fMRI: a brain imaging technique that detects magnetic changes in the brain's blood flow patterns.

We have five primary sense organs through which we draw information about the environment around us, namely, the eyes, ears, nose, tongue, and skin. Besides these, there are receptors within our body (**enteroceptors**) which provide information regarding our internal body environment. So when you feel a pain in the abdomen it is because there are pain receptors on that organ in the abdomen that is under pain. However, most of the time, even before these objects in the environment can stimulate any of our sense organs, we have to be attentive toward them. At any point of time, we are surrounded by all sorts of potential stimulators, but they generally activate the receptors in the concerned sense organ when we focus our mind toward that object. This focusing is undertaken by **attentional processes** and is the beginning of the stream of mental processes that ends only when the response has been started. These stimuli from the sense organs are then transmitted to specific parts of the brain through nerves dedicated for that purpose. For example, the optic nerve carries stimulation from the eyes to the brain while the auditory nerve carries sound information to a different part of the brain. We must remember that the eye is similar to a camera, and, it is only when the image on the retina goes to a specific area of the cerebral cortex that the person recognizes those wavelengths of certain colors and shapes as specific objects; it is the brain that interprets the data from the sense organ and gives it meaning according to the past experiences of the person. This process of attaching meaning to the sensory information is known as **perception/perceptual processes**. If we were to summarize the total process from the time we see, hear, feel, taste, or smell something to the time that we respond to it, it is clear that a number of processes and parts of the body are involved. These have been presented later in the form of a flowchart. Of course, one must understand that these processes do not always proceed in a serial fashion and may often be carried out simultaneously.

***Enteroceptor**: receptors within the body on muscles, tendons, and organs to signal pain*

***Attentional processes**: process of focusing our sensory apparatus toward a certain object in the environment*

***Perceptual processes**: the process by which the brain attaches meaning to the sensory information*

Objects in the environment→Focusing of our sense organ (**attention**)→Sensory data received by sense organ (**sensation**)→Data travels through the sensory nerve→Specific area of the brain is activated→Meaning added→Recognition of object (**perception**)→Further processing if necessary (**cognition**)→Processed data enervates motor nerve→**Action**.

Research shows that our sense organs are capable of receiving only a limited amount of information from the environment. Each sense organ has certain limits over which it can function. So, the human eye can receive stimulation from only a small part (called the visible spectrum) of the total light spectrum, while the human ear is sensitive to only a certain range of frequencies (20–20,000 Hz). Animals and birds are known to score much higher in their capacity to handle specific inputs from the environment, for example, dogs and bats can process sounds from a much wider range of frequencies and pigeons are capable of seeing much further into space. As far as technology is considered, the question now is, does technology keep these limits in mind? For a lay person, the implications of such limits are, at least, twofold in the context of technology:

- Things around us must be made such that they are within the limits of our sensory, perceptual, and cognitive processes.
- The earlier limits of human operation can be enhanced by technology. For example, with the help of microscopes and X-ray machines, we can see more than what we normally do and by using microphones, we can enhance the loudness of sounds so that they can be heard much farther away.

3.1 The Human Sensory System and Its Neurological Basis

While human sensory systems may lack the range shown by many animal species, whatever we are able to see, hear, touch, taste, or smell seems to have a clear evolutionary history. Tony Prescott (2007) of the University of Sheffield, UK, delineates the evolution of neural mechanisms for action selection and also discusses the formation of the brain and the central nervous system. In fact, vertebrates are the only animals with a highly developed brain although some lower animals, for example, sea urchins and star fish also show prototypical control systems. It is only at a later stage of evolution that the head and the skull became larger to house the fast increasing mass of an increasingly complex brain. The areas of the brain which are activated when different sense organs receive information from the environment are depicted in Fig. 3.1.

As Fig. 3.1 clarifies, there are clear areas in the cerebral cortex and subcortical regions that are activated for the various types of sensory data. In fact, there is a topographical distribution along the sensory areas, such that each part of the body gets representation and the amount of brain area dedicated to that part of the body determines its sensitivity. For example, areas of the face are more sensitive simply because more cortical area is devoted to it.

While we have five sense organs, and all of them are important for our survival, two of them, namely, vision and audition seem to be far more pervasive than the others. We will, therefore, focus on these sensory systems. The eye and the ear along with their pathways to the brain and the specific areas of the brain in which information from these two sense organs is encoded and processed are called the visual and the auditory system respectively. We present some more details regarding these two sensory systems.

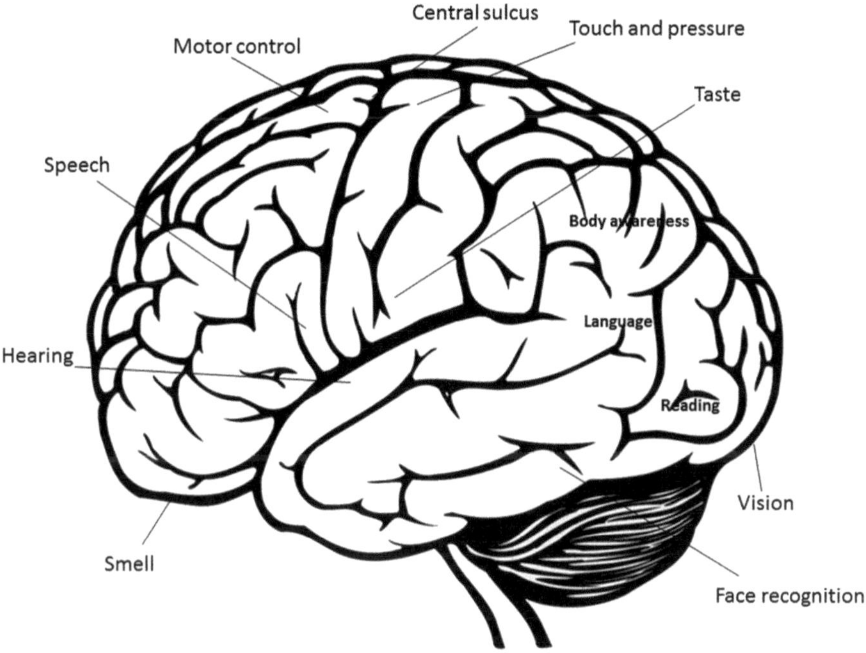

Fig. 3.1 The human brain depicting areas responsible for different senses. Adapted from www.uic.edu

3.1.1 *The Visual System and Vision*

We depend so much on our vision that almost all our activities have direct or indirect inputs arising out of it. Try to recall any event during which you have operated in the absence of light and you immediately notice that such events are extremely few. Although other senses help, we tend to rely on vision most of the time unless, of course, the sensory input is very specific such as when listening to music or tasting a new wine. Ironically, the range of our vision is very narrow, somewhere between 380 and 780 nm. It is believed that this range evolved to cope with the sunlight that we need for navigation and other activities. It is also the reason why our eyes are not stimulated by ultraviolet or infrared rays, both of them being harmful for us. Additionally, our eyes are well equipped to process colors but limited in handling ultraviolet bandwidth. If our eyes were to gain a higher range, we probably would need 1.6×10 (expo 15) bits/s or a hugely upgraded visual system. What would be the necessity of such an extraordinarily high capacity system when the existing visual system is already so complex? It appears that the current visual system provides us optimal functioning to deal with much of our environmental demands, albeit in the past few decades with the phenomenal growth of technology, one may find it of a limited value. In this latter scenario, we may, sometimes, be craving for an upgraded visual system.

The imperfections of the visual system range from a blind spot on the retina to our inability to see behind our backs. The human vision is not as smart as our ears that can cover the entire 360° around our heads. The eyes are normally able to cover 180° in the forward facing direction. Imagine if there are more than two eyes, we would have been able to cover all corners, as in hearing, but each extra eye would again contribute to increasing the size of the brain besides adding a huge burden on the visual processor. It seems that somewhere during the process of evolution, there was a trade-off between the optimal functioning of the visual system and the cost of having a better vision. Even technology has not been of much help in this context. Car manufacturers have struggled for a long time to create a side mirror that would eliminate the blind spot and make for safer driving but have not been completely successful.

3.1.2 The Human Visual Cortex

A glance at the functional areas of the brain shown in Fig. 3.1 clarifies that there are a number of areas of the brain which together comprise the **visual cortex**. Through the use of functional MRI (fMRI), neurologists have been able to decipher the workings of the visual areas of the brain. This is just one more example to show the ways in which we first develop a technology and then use that technology to understand more about ourselves, which, in turn, could be used to machine even higher realms of human functioning. Results obtained from lesions performed in the brains of rhesus monkeys provide compelling evidence that one part of the temporal cortex of the brain is engaged mainly in the act of noticing and remembering an object's qualities and not its position in space. On the other hand, some parts of the parietal cortex seem to be concerned with the perception of the spatial relations among objects and not their intrinsic qualities (Mishkin, Ungerleider, & Macko, 1983). Recent fMRI studies have confirmed the distinction between these two streams in the visual cortex. There is a ventral stream which helps in object recognition and has, therefore, been called the "**what**" **stream**. Then, there is the dorsal stream, responsible for the recognition of the spatial localization of the object and has, thus, been called the "**where**" **stream** (Grill-Spector & Malach, 2004).

> ***Visual cortex***: *areas of the brain where visual stimuli are processed*

> ***"What" and "where" streams***: *the "what" stream tells us about the object seen while the "where" stream tells about its location in the environment*

Almost a century back, famous German psychologist, Koffka, was intrigued by the question "why do things look as they do?" (Koffka, 1935). Max Wertheimer (1923) observed a series of lights flashing on and off at a railway station. To his surprise, they appeared much like the lights that seems to go round and round outside circus marquees when in reality they were just a string of lights flashing on and off. At a certain speed they appear to be moving lights. He had chanced upon the now very common phenomenon of the whole appearing other than the parts. Along with other psychologists, the phenomenon of visual perception was studied further and this helped them to enunciate several important principles, which together formed the basis of what came to be known as **Gestalt psychology**. Gestalt psychologists posited that the whole looks very different from a mere sum of the parts, or that it forms a "gestalt." Some of the key principles, which also have important implications for the design of technology and of visual displays, are as follows:

> ***Gestalt psychology****: a school of psychology which posited that the whole looks very different from the mere sum of the parts*

1. **Emergence**: the whole is identified before the parts. Is this not what you do whether you are looking at a painting, a web page, or even a new product, be it a small toy or a car? You first look at it in totality and only when the salesperson starts explaining features in detail that you notice each part in isolation.

> ***Emergence****: the whole is identified before the parts*

2. **Reification**: our mind tends to fill in the gaps. Even when a drawing is not complete, you tend to fill in the details. A good example is the IBM logo (see picture in Box 3.1). It also explains why we miss out errors while doing proofreading of a manuscript.

> ***Reification****: our minds tend to fill in the gaps*

3. **Multistability**: the mind tries to avoid uncertainty, and so when something looks unfamiliar, we link it to something more familiar and perceive it as such. This is a principle that is often used in marketing. When companies want to change consumer ideas about a certain product, they proceed with very gradual changes. Normally, manufacturers tend to make changes that are in line with the consumer's original perception of the product. Gradually, an alternative perception is created followed by efforts to strengthen the new perception and at the same time, weaken the old.

> ***Multistability****: when something looks unfamiliar, we tend to link it to something more familiar*

4. **Invariance**: as the direction of the object changes vis-à-vis our eyes, its image changes. Remember drawing a house front face. If you now want to show it sideways, you have to draw it differently. But doesn't the house seem to be the same? So, when you look at a person or an object from the side or from the front, the image projected on your retina is actually different, but you perceive it as the same object. In other words, change in size (as distance between you and the object changes) or in shape (as direction of object changes) does not change the object per se. Think how difficult the situation would be if with every minor change in distance or direction, your mother seemed to have changed, or your car appeared different!

> ***Invariance****: stability of perception despite changes in the sensory information*

Along with these, Gestalt psychologists developed other principles such as those of similarity (similar parts tend to get grouped together), proximity (parts that are closer together are more likely to be perceived as whole), closure (we tend to fill in gaps), good continuity, common fate, and figure-ground principles. A person with a talent for designing is often one who tends to use these universally seen principles intuitively. Others are taught these principles so as to use them to their advantage and get the design to convey effectively whatever they want to convey (see Box 3.1).

Box 3.1: Design Principles and the Gestalt

One technology that makes effective use of the principles of perceptual organization is web designing. The designer starts with elements that are then placed so as to make the page appear attractive but at the same time also fulfill the purpose for which it is being designed. Think of the web page of any online retailer such as Amazon. Why are you able to navigate through it so easily? It is because the arrangement of the navigation keys follows the Gestalt principles. Items have been arranged under categories and subcategories following the principles of similarity and proximity.

Similar is the case of online forms that we have to fill every so often. By requiring you to fill in information grouped in terms of personal details, academic qualification, work experience, languages known, the form designer is catering to your preference for similarity and proximity. Thus, similar information is grouped together under a common head.

We also see the use of the Gestaltian principles in the design of logos. Look at the logo used by IBM. It uses the principle of closure, and you have no problems in filling up the gaps in the letters.

Linux users will quickly recognize the old Gnome Desktop Environment logo.

It's a "G," yes indeed, but it's also a footprint and is based on using both figure and ground in an effective manner.

It is clear from the above that whether you, as a designer, use these principles intuitively or have been taught to make your design according to these principles, this is the natural way in which humans perceive.

But how is all this organization achieved and where in the visual system is it done? Today, we know that there is an executive which integrates all these diverse pieces of information in the visual cortex. The presence of such a system in the visual cortex has been obtained by Gerhardstein, Tse, Dickerson, Hipp, and Moser (2012), in that the human visual system uses a global closure mechanism by which sensory fragments of information are collated to form a holistic picture for the viewer. At the same time, depending on whether the information is to be used for perception or for action control, the level of processing differs (Janczyk & Kunde, 2012). While not all psychologists may agree to this conclusion, generally speaking, whenever a gestalt is constructed, there are some parts of the stimulus which are silenced and the human vision provides only the most useful level of abstraction to conscious awareness (Rothbart & Posner, 2015; Poljac, de Wit, & Wagenas, 2012). This, in fact, makes the system more efficient because it increases the speed at which visual information can be processed. Another strategy leading to parsimony of neural effort is that cortical information is processed in a top-down manner, the level at which processing is being done being dependent on the level of discrimination required. When easy or coarse discrimination is required, stimuli are processed at upper cortical levels only, since neurons at that level are only broadly tuned to sensory data at the upper levels. However, when we need to make fine distinctions, or when the data are difficult to interpret, the sensory data filters down to lower cortical levels where the neurons are more finely tuned to sensory data such as regarding retinal position and spatial orientation (Ahissar & Hochstein, 2004). This is the reason why **transfer of learning** does not always take place. If a high degree of precision is required, transfer of learning may not take place (Jeter, Dosher, Petrov, & Lu, 2009). Could this explain why people who are into video or computer gaming often prefer to play different and more complex levels of the same game rather than trying new games?

Transfer of learning: *when learning one thing aids the learning of something else*

3.1.3 The Auditory System and Audition

The word 'sound' has come to mean two very different but equally relevant contexts. In some cases, sound is referred to as a physical stimulus whereas, in others, it is referred to as a perceptual response to the stimulus. While the former is in terms of air pressure changes the latter refers to the experience of sound.

For each of the basic physical features of sound—its intensity, frequency, timbre—we have the correlated auditory sensations known as loudness, pitch, and characteristic sound. It is different from auditory perception which involves complex cognitive processing of a higher order such as perceived duration, anticipation, prior experience, and synthesis. The threshold for human hearing is by and large considered to be the point beyond which we can hear the lowest sound at least 50 % of the times. The range of human hearing is 20–20,000 cycles per second, with sounds beyond this range sometimes being found to be associated with negative

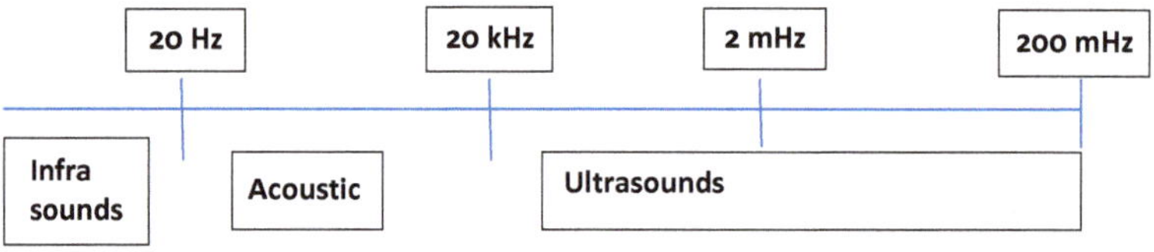

Fig. 3.2 The sound continuum

effects. We can understand these effects by categorizing sound in terms of frequency ranges. The diagram (Fig. 3.2) depicts these ranges.

Ultrasounds, that is, sounds beyond 20 kHz and upto 200 mHz are being used for a wide variety of medical diagnostic purposes for both humans and animals. It has also been put to wide industrial use. However, it is also known that sounds beyond 20 kHz may be harmful for humans. The exact degree of harm is actually a function of both sound frequency and sound amplitude (or loudness). Thus, high frequency sounds of low amplitude may not cause as much harm as low frequency sounds of very high amplitude. At the same time, continuous exposure to ultrasounds of loudness greater than 120 dB has been shown to lead to hearing loss, while ultrasounds beyond 180 dB may lead to death. A report by an independent advisory group of UK (*AGNIR*) recommends an exposure limit of ultrasounds of 70 dB at 20 kHz and 100 dB at 25 kHz and above (Advisory Group on Non-Ionising Radiation (AGNIR), 2010).

Research on human hearing shows significant individual differences. It is estimated that between the ages of 65 and 74 years, approximately one-third of Americans develop hearing problems mostly due to sensory-neural causes, called presbycusis. It affects their lifestyle and their adjustment to technology, ranging from the design of hearing aids to especially designed telephones, with louder sounds needed for the aged.

As our society is growing increasingly technological, the world we live in has also grown noisier. From music concerts to leaf blowers, the intensity of sound levels takes an enormous toll on our inner ear culminating in rise in heart rate and blood pressure. In acute cases, acoustic trauma leading to temporary hearing loss may occur or may even result in a permanent change in the threshold for hearing. Those who work around jet engines are commonly known to suffer from a hearing loss of 10 dB or more. So much so that Gary Evans (2006), of Cornell University, has pointed out the catastrophic effects of high noise levels on developmental outcomes of cognitive, emotional, and social significance. Whereas, compared to short-term memory losses, long-term losses were significantly higher under high noise levels, disregarding or ignoring auditory input was also consistently observed and has been considered a major risk factor for the development of children living under noisy conditions. It has often been noticed that children who live near airports have significantly higher reading delays than those who live under less noisy conditions.

But our senses also have a remarkable capacity to adjust to changes in the environment. Our eyes adapt to changing levels of illumination through the adjustment of the pupil. People living close to a factory get so used to hearing the loud sounds that they realize that there were sounds only when the factory stops operating. Most of us get used to the whirring of the fan. Both are examples of auditory adaptation. A recent study at *Rotman Research Institute* in Toronto, Canada showed that as compared to nonmusicians, who have never played an instrument, amateur and professional musicians were significantly superior at picking out speech against noise. The researchers Zendel, Tremblay, Belleville, and Peretz (2014) concluded that being a lifelong musician could be an advantage as continuous use of the auditory system tends to mitigate or attenuate changes in the brain due to aging.

3.1.4 The Human Auditory Cortex

As compared to the detailed mapping of the visual cortex, knowledge about the auditory pathways to the brain and the **auditory cortex** is less complete. We do, however, know that sound is interpreted in the temporal lobe of the cortex and the exact areas have been depicted in the diagram presented earlier. At the same time, we do not have enough information regarding the neural correlates of auditory distance perception, though there are some results which suggest that neurons in, or near, areas processing other aspects of the spatial features of the sound could also be sensitive to cues to auditory distance perception (Kopko et al., 2012).

> ***Auditory cortex****: areas of the brain that are responsible for processing of sound*

The sensations received at the local ear level are transmitted to the brain via two pathways. One is neural and the sound impulses from the inner ear travel through the auditory nerve to the auditory cortex of the brain. This is the normal neural pathway. There is, in the case of audition, another pathway for the transmission of sounds reaching the individual. Since sound is generated whenever there is vibration of anything in an air media, the sound waves cause the very bones of the ear and the skull to vibrate, and these vibrations also travel to the brain. While the former is known as **air conduction**, the latter is called **bone conduction**.

> ***Air conduction****: the neural pathway for audition in which sound impulses travel through the auditory nerve to the auditory cortex*

> ***Bone conduction****: when the vibrations travel through the bones of the skull to the inner ear*

Have you ever noticed that your voice as it normally sounds appears unrecognizable when you hear a recording of it? The difference is that as we speak, vibrations from the sounds we produce reach our brain through both air conduction and bone conduction. Contrary to this, a sound recording of our voice consists of only vibrations that reach the microphone and the recording system through normal air conduction. This is the reason why our voice normally sounds much richer than when we hear a recording of the same. It also explains why bathroom singers often feel that they are “good” singers!

One advantage of bone conduction is that as compared to neural conduction of sound, bone conduction signals require smaller increases in stimulus to produce equivalent changes in sound loudness and thereby serve an important purpose of helping us to make fine discriminations between sounds (Patrick, Kring, McBride, & Letowski, 2012).

There are other advantages to bone conduction, too. For people who suffer loss of hearing because of deformities of the external and middle ear, bone conduction provides an alternate route for hearing. Where the traditional cochlear implant fails to serve any purpose, **BAHA** (bone anchored hearing device) units have been designed, which work by passing sound through microphones to a magnet implanted below the patient's skin, which convert the sound to vibrations in the skull and thereby reach the inner ear. One such hearing aid company manufactures hearing aids called *Cochlear* and it is reported that the sounds reaching through these BAHA units do not suffer any distortion.

Bone anchored hearing device: *auditory devices which use bone conduction*

Also important is the fact that bone conduction-based headphones leave the normal sound passage free. Thus, when you use normal headphones, it cancels all other sounds and all you hear is the sound flowing through the headphone. BAHA units, on the other hand, leave the normal sound passage free for ambient sound. You can therefore listen to music on your iPhone and at the same time, keep your ears open for a car honking at you as you walk, your kids yelling for more food, or, as far as military personnel are concerned, they can get their orders through BAHA units leaving the normal sound passage open for sounds of approaching danger. More details have been provided in Box 3.2.

Box 3.2: Hearing Through Your Bones: Bone Conduction of Sound

Bone conduction-based headphones have proved their worth in various facets of life, ranging from music listening to parenting and even defending the country.

Bone conduction (BC) is not new. Are you aware that the famous German composer Ludwig Von Beethoven was deaf? How did he hear the music he composed? He would bite on an iron rod attached to his piano and as he played the music on his piano, he could hear the piece only because the vibrations travelled through the bones of his face, skull, and ear and thereby reached the brain.

During the early 1900s, BC was being used in telephones and hearing aids. Today, in the twenty-first century, the same technique is being used by the entertainment industry and it seems to be the future of headphones. You may have heard of what are called bone phones, such as *Madcatz, Audio Bone*, and *Aftershokz*. They are all BC headphones. The reason for their popularity is that they allow the wearer to hear ambient sound or sounds from the environment

(continued)

Box 3.2 (continued)
while at the same time listen to music through their BC units. One of the more recent entrants in the field is Panasonic, which in 2013 announced the debut of their BC headphone, namely, *RP-BTGS 10*. In fact, even the controversial Google Glass uses bone conduction for its speakers systems. But as *Aftershokz* CEO Bruce Borenstein puts it, "the difficulty has been in transmitting vibrations through bone with enough power to be musical" (as reported for CNN by Monks, 2014).

Parents, too, would find BC units useful. Denmead, best known for his *New York Times* bestseller *GeekDad* series, has shown how they can be useful for parents who want to enjoy music but also keep an eye on their kids in the living room or those playing outside (Denmead, 2015).

Probably the most innovative use is that by ad agency BBDO of Germany which has tested a BC-based unit attached to windows of metro trains of *Sky Deutschland*. As tired or bored commuters lean against the window, advertisements are streamed through special transmitters named *Audiva* attached to the window and from there they reach the brain. It is as if you were suddenly hearing voices in your head (Kelion, 2013).

3.1.5 Psychoacoustics

As soon as the sound passes through the ear, it stops being a physical phenomenon and becomes a matter of perception. What we hear is almost always different from the actual sound, due to the peculiarities and limitations of our hearing. At the same time, what we hear can largely differ from what we think we are hearing, due to the many tricks that perception plays on our awareness, as shown in Box 3.3 on phantom voices.

It is examples such as that described in Box 3.3 and many others from our daily lives which go to show that the perceived sound may often be very different from the actual sound. To arrive at explanations for this difference between what is veridical and what is perceived, we have the upcoming area of **psychoacoustics**, a juncture of the psychology of sound and the physics of sound. The current digital age with its emphasis of audio outputs in a variety of settings for both work and entertainment has brought psychoacoustics to the forefront of audio research and development. On the basis of the research, certain principles have been enunciated. These have been briefly described as follows:

Psychoacoustics*: a multidisciplinary field dealing with the physical, physiological, and perceptual correlates of sound production, transmission, and reception.*

Box 3.3: Phantom Voices

"The appearance of a phantom voice can be traced to the Eastern ritual of the Cuncordo de Castelsardo, a brotherhood on the Island of Sardinia. A choir of four brothers evokes a fifth voice, the quintina. This higher female voice is said to be the spirit of the Holy Virgin Mary."

So writes vocalist and ethnomusicologist, Mark van Tongeren in his PhD thesis (Tongeren, 2013), on hearing a group of only four men singing but with sounds of a fifth female voice also emerging alongside. Tongeren has not only tried reproducing the quintina with singers from his own vocal ensemble but has also travelled to several other similar brotherhoods on Sardinia to find the quintina. However, he has been unsuccessful in all his attempts and has never been able to find the phenomena anywhere else, except for this specific place with this specific brotherhood.

Various people have tried to explain the appearance of the quintina. One explanation is that the emergent female voice resides in the very way in which the brothers sing combined with the specific acoustical features of their church. It seems that there is a direct connection between the spoken vowels and particular resonances in the voice that correspond to bands in the overtone scale. Under the right acoustical conditions, a strong emphasis on those vowels while singing could evoke certain higher tones of the voice to become more clearly perceived. It is these tones that are then perceived as a female voice.

Source: Tongeren (2013)

Some principles of psychoacoustics:

- As sound travels from the outer to the middle ear and then to the inner ear, the sound gets attenuated and this attenuation is maximum at the midrange frequencies, that is, in the range of 3000–4000 Hz.
- Thereon, as the sound progresses inside the ear, noise caused by blood flow is added. This noise, known as the internal noise spectrum, is maximal at low frequencies, basically for frequencies between 1000 and 3000 Hz.
- The absolute threshold of hearing is, therefore, the sum total of the transfer function and the inner ear noise plus some other minor effects.
- Further attenuation of the sound takes place in the cochlea (a part of the inner ear), called frequency smearing. This is what produces **masking** effects such that a weaker adjacent tone becomes inaudible if it is too close in time to a stronger tone.
- Temporal smearing also takes place because the hair cells of the cochlea need some time to adjust to the incoming sound.
- Added to all of this is delay due to the time it takes to respond to the sound or **reaction time** delay.

***Masking**: when one sound prevents another sound from being heard*

***Reaction time**: lapse of time between the original stimulus and the final response to the stimulus*

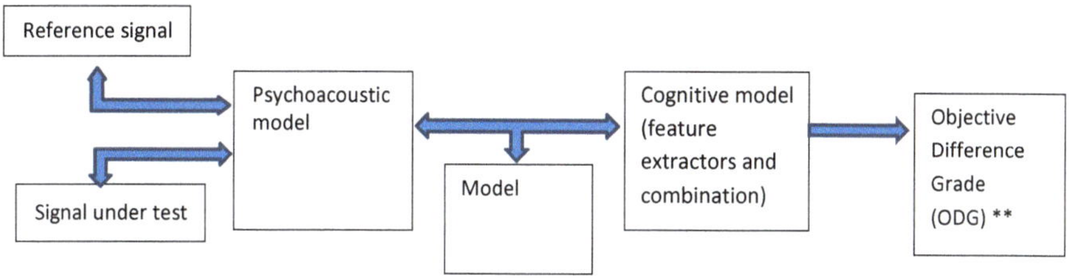

** ODG: single index to rate perceptual audio quality of signal under test compared to reference signal

Fig. 3.3 Basic PEAQ model (adapted from Temme et al., 2009). *Double asterisk* ODG: single index to rate perceptual audio quality of signal under test compared to reference signal

3.1.5.1 Using Psychoacoustics in the Design of Loudspeakers

The twenty-first century has seen a flood of audio–video gadgets as also various audio formats such as the compressed disc (CD) and the MP3 format, just to name a couple. Whatever the format being used, a major task is the audio coding and its inherent sound quality. Till about 15 years back, the only way by which the audio quality of systems could be tested was by elaborate listening tests using experienced human subjects. Today, however, these subjective tests have been replaced by objective computer-based methods, resulting in what has been nomenclatured PEAQ (Perceptual Evaluation of Audio Quality) which has been recommended even by the International Telecom Union (ITU, 1998). The PEAQ uses a number of psycho-acoustic tests combined together to yield a measure (ODG) of the difference between a reference signal and the signal to be tested.

A basic PEAQ model has been elaborated by Temme, Brunet, and Keele (2009) and has been depicted in Fig. 3.3.

Taken together, all these physiological effects of the ear create an inner representation of the sound that is further sent to the brain for it to cognize the sound in terms of various cultural and personal preferences. You can well imagine the degree of subjectivity that can result regarding what has been heard and how it is interpreted. Remember playing the game of Irish Whispers as a child, in which the first person whispers a phrase or small sentence into the next person's ear. The second person whispers whatever he has heard to the next person and so on down the line. The last person is told to speak out whatever he had heard. The last person's version is almost always nothing akin to what had been whispered by the first person! The almost inaudible whisper added to make the effect even more pronounced and led to much more fun.

Findings from the realm of psychoacoustics have important implications for the designing of headsets of the type worn by military personnel. Army personnel in the field have to maintain constant vigil over not only wild animals but also human enemies and infiltrators. It requires them to be always on the alert and the equipment used in the standard settings includes vision and audition augmenting gear. While a binocular is used to augment vision to distances otherwise not seen by the naked eye, headsets help the soldier to focus only on certain types of signals by not allowing other sounds, for example, those of animals and birds to interfere with the relevant sounds. An interesting application of psychoacoustics in the realm of music has been detailed in Box 3.4.

Box 3.4: Applying Psychoacoustics to Music

Human hearing is restricted within a very small range of frequencies, namely, between 20 and 20,000 Hz (the range becoming even narrower with age). Therefore, the musical instruments are made so as to produce frequencies within our normal range of hearing. In 1958, a landmark was created by Stockhausen in his composition *Kontakte*. For the first time in the history of music, he was able to find ways to bring all the properties of sound (namely, intensity, pitch, timbre, and duration) under a single control.

A potent illustration of these connections was that when he used a high, bright tone which descended in several waves, becoming louder as it gradually acquired a snarling timbre, and finally passed below the point where it could no longer be heard as a tone. It was clear that once the tone passes below the threshold of 16 Hz we stop perceiving tone, and, start to hear beats.

The range of hearing was never explored in this way before for the simple reason that there was no instrument that could perform this frequency range. Up until that time, beats and tones were considered separate musical properties, with beats belonging to the realm of rhythm and tempo, and tones to melody and harmony. With *Kontakte*, Stockhausen showed how beats and tones form a continuum, with absolutely no distinction between them. Whether we perceive sound as beats or tone depends solely on whether the sound is above or below our hearing threshold.

Source: Paul Oomen, Psychoacoustics: An Introduction (2014)

3.1.5.2 The 13 dB Miracle

There are various ways in which technology has attempted to override the physiological and psychological limits of our senses. Audio compact disc development especially that of MP3 players and disks is a clear example of the ways in which the knowledge of psychoacoustics has been exploited by technology. A concept, aptly called the 13 dB miracle, has been used to minimize the effects of noise. According to this concept, if any relevant signal is 13 dB louder than other noise signals then the noise will not be heard. This concept is being used in MP3 players, in which there is a possibility of noise to exist but if the sound of the music is 13 dB louder than the noise then the noise is eliminated and the user can enjoy the music.

The basic principle being used by the engineer is that of auditory masking which delineates that a louder sound can stop a softer one from being heard. With rising levels of sound pollution, we are all victims of auditory masking. Our conversation may sometimes become inaudible due to the sounds of a passing truck or an overhead airplane, or, the melodious voice of a singer may become drowned in the raucous noise of a rock band. Is it not a technological feat that while in our daily life we witness negative effects of auditory masking, technology has been able to

Box 3.5: The Making of MP3 Discs

Psychoacoustics theory describes the human hearing system as a set of band-pass filters, each one covering an area of the audio spectrum known as a critical band. Each of these critical bands corresponds to a separate area on the basilar membrane in the inner ear. Two independent tones that lie within the same critical band will often be perceived as a single sound; two tones that are not within the same critical band are perceived as two separate tones. When the two are combined our brains will hear them as a louder overall signal.

The key to the MP3 format is to split the audio signal into multiple frequency bands, work out how many bits to allocate to each up to a limit set by the user and then use more conventional data compression techniques to further squeeze the bit stream.

Another interesting fact is the way in which MP3 got its name. The file-naming limitations of the Microsoft DOS operating system helped name the prevalent digital music format of our time. "At that time, file extensions were limited to three letters," Brandenburg recalled at the Audio Engineering Society Convention in London: "On 14 July 1995, we started to use the extension '.mp3' for our software. If there is an official birthday for MP3 that is it."

use this same phenomenon to its advantage in the making of compressed discs of various kinds?

What is the basic difference between an MP3 and a CD? Remember that an MP3 disc contains much more information than can be put on ordinary compressed disks (CDs). How is this done? The problem was to devise a method that would greatly reduce the amount of data required to represent the audio recording and yet sound like a faithful recording of the original. In fact, it called for compressing information to one-tenth of the original bit rate and normally information theory tells us there will be a considerable loss of information if such a compression is made. Thus, a way had to be found for dealing with this problem and ensuring that we do not perceive the loss of information. The solution came from the psychoacoustic principle of auditory masking (see Box 3.5).

The 13 dB miracle was first demonstrated by two scientists Johnston and Brandenberg in the AT & T Bell Laboratory in 1990. In what they termed the 13 dB miracle, a short selection of linearly coded music was played as a reference and then played again with its Signal/Noise (S/N) ratio degraded to only 13.6 dB. The sound was grossly distorted. The musical section was played a third time, again with measured S/N of 13.6 dB, but this time the noise was virtually inaudible.

The surprising finding was the way in which the sound distortion vanished. It is important to remember that though the signal/noise ratio in the two conditions was exactly the same, namely, 13.6 dB, the two conditions led to vastly different perceptions. What caused the difference? It was the way in which the frequency and amplitude of the noise was distributed. The noise constantly shifted in frequency in relation to the music's spectral balance and amplitude so that it remained hidden by the music. The poor-sounding signal had exactly the same amount of noise, but the noise was flat in amplitude and static in level. Thus, though having the same physical properties, the two sounds were perceived as being vastly different.

Brandenburg played samples of masking effects and pointed to what he called the "13 dB miracle." The masking threshold is, on average, 13.6 dB below the louder sound.

As Christensen writes (Christensen, 2008) "ever since the demonstration of this "13 dB Miracle" at AT&T Bell Labs in 1990 by K. Brandenburg and J. J. Johnston laid the ground for what we now know as MPEG-1 Layer 3 (MP3) and MPEG-2/4 AAC, much effort has been devoted to developing models of the human auditory system (also referred to as perceptual models) for signal processing purposes" (p. 1).

3.2 Cognition

3.2.1 Improving Our Ability to Collectively Improve Ourselves

Over a period of millions of years our cognitive capacity has not only helped us to develop tools but it has also helped us to improve ourselves. We have progressed tremendously: starting from the use of crude stone tools sharpened at one end to dig out marrow from the bones of dead animals to feed the ever increasing demands of a brain that was increasing in size (due to the exaptations and adaptations mentioned in the last chapter) to the use of robotic tools today. From stone tools to iron tools and then to steel and plastic, and now to the ever evolving new substances that are being used to create highly advanced technological wonders that often seem to astound even the creators of that technology, the progress is unbelievable. But what is at the base of it all? A simple but highly effective phenomenon called choice and the ability to make choices that prove to be adaptive. It is through such choices that we are able to master the environment and develop a sense of control over the environment. Though this desire for control often seems to be grounded in the societal values for personal autonomy and self-determinism there is enough evidence which suggests that this could be an innate ability found not only in animals but also in children not old enough to understand the rewards of such personal autonomy (Leotti, Iyengar, & Ochsner, 2010). Once even 4-month-old children learn to feed themselves, they do not like being fed by adults, showing that the restriction of choice is aversive (Kochanska & Aksan, 2004). Findings such as these clarify that we are indeed hard wired for making choices and that there are specific areas of the brain that are called into play while making choices (Samejima, Ueda, & Kimura, 2005), so much so that certain regions within this brain network explicitly code for actions that are most adaptive in a given context. This suggests that there is a biological basis for organisms to be causal agents rather than to be passive observers in their interactions with the environment. In other words, the ability to make adaptive choices is the vehicle for the development of feelings of control over the environment. It is, probably, this aspect that is responsible for why we keep on creating tools to help us cope with the exigencies of, first nature, and then of man-made environments, all made possible through our highly specialized and complex cognitive structure.

3.2.2 *Cognitive Foraging: The Evolution of Cognitive Structures and Functions*

Studies in comparative neurobiology have helped us to understand more about the human cognitive architecture. Much of this insight has been obtained from studies of foraging behavior and the area restricted search strategies used by various species of animals (described in Chap. 2). Recent literature clarify that just as Area Restricted Search (ARS) helps in preserving scarce energy during movements in search of food, so, too, does man conserve limited cognitive resources while searching for solutions to problems. So pervasive is this manner of search that in humans, this perseveration of strategies used earlier holds not just for spatial foraging but also for more abstract cognitive search. This is the reason why this process of cognitive search has often been called **cognitive foraging**.

> ***Cognitive foraging***: *a process of cognitive search much akin to ARS foraging strategies seen in animals*

There are several lines of evidence which suggest that the goal direction strategies we see in human and other primate cognition is born out of the same evolutionary processes that led to efficient foraging strategies among lower animals. First, when faced with novel situations, individuals, typically, go back to well-tried solutions reached earlier, mediated by the prefrontal cortex in the brain. This is analogous to what animals do when faced by food shortage (Eichenbaum & Cohen, 2001). In fact, Hills (2004, p. 3) notes that,

> "In this way, cognition tends to wander down well-worn paths in our day to day behavior presumably represented by specific network activation patters, which may range in complexity from motor habits to verbal rants."

Think of the last time you misplaced your cell phone and tried to remember where you could have placed it. You would have probably mentally tracked the same places that you are most likely to have placed it. Or, think of the time when you felt that a particular document is very important and that it needed to be placed safely. In all probability, the next time you searched for it, you are not able to remember where you had kept it mainly because it had not been kept in its usual place. So, we often say that whenever we are extra careful in keeping something, we are not able to find it. Let it lie among the mess on your table and you would probably know exactly where in that mess the document is!! Much of experimental data on cognitive processes such as **cognitive priming** (think of the word

> ***Cognitive priming***: *a nonconscious process by which exposure to one stimulus influences response to another*

yellow and your next immediate response might be banana!) and **perceptual masking** (the ways in which one perception can interfere with the perception of something else) are reminiscent of cognitive foraging. Similarly, if we ask you to focus on the words in the center of the page, you may not be able to 'see' those at the side. The explanation is again the same. Even the well-known phenomenon of the use of **mnemonic aids** for enhancing our limited short-term memory has, at its base, ARS mechanisms (see Box 3.6 for what are mnemonic aids).

Perceptual masking: *the interference of one perception over another*

Mnemonic aids: *memory devices to help learners memorize large lists of information*

Box 3.6: Mnemonic Aids

Going shopping: trying to remember your shopping list consisting of **eggs, bread, milk, cheese, butter, bagel, olive oil, and fruit?**

Well, it won't be too difficult if you form a sentence like the following:

Emily brought many children to her big blue and orange farmhouse.

And, it did not take me more than two minutes to make the sentence!

And how much easier it is to remember the sentence than to remember the list of things to be bought. Most lists can be similarly converted into jingles, sentences, rhymes, or even abbreviations and made easy to remember.

What have you used? These tricks which convert hard to remember lists into interesting and often queer sentences are, what are known as, mnemonic aids that have been devised to help you learn. The term mnemonic has been derived from the ancient Greek word meaning "of memory or relating to memory" and had been used even by philosophers such as Plato and Aristotle. The idea is to relate the list to something more familiar so that you are able to memorize it easily. Have you noticed that most customer care telephone numbers do not ask you to remember the number? Instead, you are asked to simply remember the name of the product: for example, call 1800-CAT-Help might be the call center number for a company which provides help regarding cats. Is it not easier to remember than the number, 1800-228-4357?

Besides its utility in daily life, mnemonics have great use for teaching and learning and have been found useful for subjects as diverse as psychology (McCabe, 2011) to elementary accounting (Liang, 2010).

Are there different types of mnemonics?

Well, we may classify them into nine types (Music mnemonics, Name mnemonics, Expression Mnemonics, Model mnemonics, Ode/Rhyme mnemonics, Note organization mnemonics, Image mnemonics, Connection mnemonics, and Spelling mnemonics, Congos, 2005) and if you take the first letter of each type of mnemonics from the list earlier, lo and behold! What do you get?

The world of mnemonics is truly unbelievable!

Source: adapted from Congos (2005)

Second, Hills and his colleagues (Hills & Butterfill, 2015; Hills, 2011) also provide results proving that those individuals who tend to perseverate in spatial foraging also do so for cognitive foraging and vice versa. This tendency to perseverate, or not to do so, was pointed out by psychologists such as Guilford (1967) some 50 years back in his distinction between **convergent** and **divergent** thinking and how these styles of thinking are more or less permanent tendencies (Box 3.7). Those who indulge more in cognitive foraging strategies are less likely to exhibit creativity. At the same time, people who are able to stray from normal cognitive foraging show high creativity and, therefore, greater out-of-the-box thinking.

Convergent thinking: *using all possible facts to arrive at a single solution*

Divergent thinking: *using creativity to arrive at unusual solutions*

Box 3.7: Convergent and Divergent Thinking

These two terms were coined by noted psychologist J. P. Guilford (1967) to differentiate between two styles of thinking. It is important to remember that both styles are important and necessary, though, for different kinds of tasks.

Convergent thinking refers to a kind of thinking that emphasizes speed, accuracy, and logic. As such, it relies on recognizing the familiar aspects of the situation, on reapplying strategies used in the past, and on using stored knowledge and the manipulating of this knowledge to solve a given problem (Cropley, 2006). A good example of convergent thinking is when a person is faced with a problem, say, when her car is having starting problems. What does she do? She starts to think of how and what she did the last time this happened. She may also think of what she had read in the manual as far as troubleshooting is concerned. She may even ask her friends as to what to do. So, in other words, she is using all her old knowledge to solve the problem. She is using convergent thinking.

In contrast, divergent thinking relies on the spontaneous, free flow of ideas, which may not even be logical. In fact, divergent thinking is often termed out-of-the-box thinking. The more unexpected the solution, the better it is. The importance of divergent thinking is that it leads to creative solutions.

How do we test for divergent thinking? If a group of people are asked to think of as many uses to a common object such as a newspaper, the person who is able to generate the largest number of uses is the one who is said to be high on divergent thinking. Probably the most common strategy for generating divergent thinking in people is to do brainstorming: a strategy which focuses on obtaining the maximum number of solutions, with evaluation of each solution just not being considered.

Recently, research has been undertaken to isolate the neural correlates of convergent and divergent thinking. According to one cognitive scientist from the Center of Neural Basis of Cognition at Carnegie-Mellon, Mark Smith, convergent thinking is related to left hemisphere activity while divergent thinking is related to right hemisphere activity (Smith, 2012a). This is the reason why you hear of people calling themselves left brained or right brained.

The third line of evidence comes from patients who show cognitive pathologies. Nieoullon (2002) provides a useful review of the link between these pathologies and the evolutionary history of foraging. Pathologies such as those seen in Parkinson's Disease, schizophrenia, Attention Deficit Hyperactive Disease, Obsessive Compulsive Disorders, autism, or even in drug addicts are apparently more than mere cognitive pathologies. They often range from motor pathologies to even ritualized behavior. One can thus think of obsessions and compulsion along the lines of cognitive foraging. In this case, the person simply keeps searching in the same brain area and therefore performs the same act or has the same thoughts over and over again, illogical though they may seem. Schizophrenic persistence of ideas, withdrawn though they are from reality, is also reminiscent of ARS. And, just as ARS is dependent upon dopaminergic action, so are many such cases which show dopaminergic defects and have been successfully treated with dopaminergic drugs.

In our discussion on the psychology of technology, we have been focusing mostly on the user and her physical and mental limits and the extent to which technology does, or does not, match these limits and the problems thereby. Our understanding of the psychology of the techie designer becomes much deeper with our recent understanding of the curious ways in which cognitive foraging operates. Why did the Wright brothers take so long in developing the technology which would enable them to fulfill their ambition of flying? Why did Edison have to cast away so many ideas that led to naught while trying to design ways to transmit electric current and be able to light an incandescent bulb? Why do we normally hear of technological evolution with revolutions in technology occurring only very rarely? Engineers often struggle for years before the 'Eureka' experience startles even the very person who had been struggling. Why did Steve Jobs and his partner have to spend so many years in their garage shack trying to design the Mac? Even in the sphere of marketing, whether it is the marketing of cars, cell phones, operating systems for computers, each new model has just so much of change in it that the customer can be made to sense it. By doing so, the manufacturer ensures that the image of the product remains in the same cognitive area and, thereby, allowing the advantage gained from cognitive foraging to take place. The rest is marketing hype. It would not be incorrect if all of the earlier is traced back to this tendency of using ARS in cognitive foraging. That the vast majority of individuals show this tendency also vouches for the fact that people who are able to do out-of-the-box thinking, or move out of cognitive foraging, are really looked up to and are in high demand in any industrial or commercial sector.

3.2.3 Schumpeter's Gale: Creative Destruction and Leapfrog Technology

The effect that technological innovations can have on society was very succinctly brought to the fore by the Austrian-American economist Joseph Schumpeter. One section of his book, *Capitalism, Socialism and Democracy,* first published in 1942, starts with some very provocative sentences. He writes, "Can capitalism survive? No. I do not think it can." He predicted that capitalism would be destroyed by its own successes. According to him, capitalism sparks entrepreneurship, the hallmark of which is opportunity encashment. Further, such innovations would lead to gales of destruction. He named this process destructive creativity since it was creativity that was destroying older technology. We have many examples from the modern world where companies that have failed to innovate have been guzzled down by newer and smaller organizations. Think of Xerox! Though the process of photocopying is still often referred to as xeroxing, what happened to the company itself (Box 3.8)?

Another example is Polaroid. With digital cameras the need for instant photography became a thing of the past. And sometimes, an innovation takes the world by storm: think of Apple. Such storms do not occur on a daily basis, they are the result of years of dedicated skunk-work. But once there, these Schumpeter gales destroy everything in its wake, much like a literal storm. It is these gales of technological innovation that produce continuous progress and help the company to leapfrog to heights, hitherto unknown, to older and more established companies.

Box 3.8: From Haloid to XeroX to Fuji-Xerox

What we know as the *Xerox Company* today can trace its roots to 1906 when a photography paper company was established in Rochester, New York, USA. The company, named, *The Haloid Company* was doing well and in 1947 it entered into an agreement with *The Battelle Memorial Institute* to produce a machine based on a new process called xerography derived from the Greek words for 'dry' and 'writing.' In 1949, the first XeroX copier made its debut (please note the capital X at the end of the name!), with the popular XeroX 914 copier being introduced in 1960. It was an instant hit, so much so, that the *Fortune* magazine was to later call it "the most successful product ever marketed in America." In fact, the entire period of the 1960s was a time for tremendous success of the company.

Unfortunately, this did not last long. In 1970, IBM introduced its first office copying machine followed by Kodak coming in with a more sophisticated machine and the Japanese company Ricoh introducing smaller and less expensive machines. Xerox started losing its worldwide market share by 1985, dropping to 40 %, from the 85 % it had commanded in 1974. The next two decades were followed by periods of ups and downs for the company.

A long lasting change in fortune required much more than an organizational restructuring. It needed to reinvent itself to overcome the effects of the Schumpeter gale that threatened to destroy it. This occurred in 1994. The company decided to shed its image as a copier company and take on a new image as *The Documents Company*, offering a wide variety of document processing solutions, microprocessor technology, and laser technology. Today it continues to innovate, with over $1.4 billion being spent by the Xerox group on R & D and two patents being awarded to it each day.

It is when the innovator, rather than engaging in standard cognitive foraging, moves ahead to a higher level of development, without going through the intermediate phases, that he is said to create **leapfrog technology**. Today, leapfrogging is standard usage in developing countries. Rather than starting out with traditional technology, developing countries are benefitting tremendously by leapfrogging to newer technology. One such example is that of developing countries, who have been without electricity. Yet, they are able to move directly to solar energy rather than going through the process of first using fossil fuels and then making a move toward other unconventional means of producing energy. One may also think of communication technology and the ways in which mobile telephony has penetrated to remote areas of Africa and Asia. There are developing countries where the cell phone is the first phone that its citizens have vis-à-vis those countries where the cell phone actually replaced the wired phone or was an additional phone for greater mobility. Farmers in India are able to get help from the agriculture department via SMS, while runners in Kenya compare running speeds through SMS. We even have examples of reverse innovation, whereby a product is developed for a developing nation and then moves on to the developed countries. A good example is of how *General Electric* developed a $15,000 portable ultrasound machine for use in remote areas of India and China and is now being used in the US. So great is the demand for such leapfrog technology, that these Cinderellas of the global economy are becoming the newest and largest markets for companies dealing with all kinds of innovative technologies. Small time entrepreneurs like those mentioned in the example later (Box 3.9) are

***Leapfrog technology**: a technological development that is revolutionary and which does not go through the intermediate evolutionary stages of development*

Box 3.9: Leapfrogging

Andy Rabagliati would find it difficult to explain his business to people in the US or in Europe. The vast majority of the hinterland in South Africa lacks internet connections or even when internet access is available it is too slow to be of any practical use. Andy overcomes this problem through a very handy technology called *Wizzy Digital Courier*. It is based on a simple USB stick of the kind that we all use to store data on. Andy has designed an open-source software program through which a school with no internet connection can compose email messages and specify topics for internet search. These are stored on the USB stick and a messenger, even a gardener, is sent on a bicycle to another school which has a fast internet connection. There, the information is collected through the internet and sent back stored on the USB stick. One doesn't need a phone line; one doesn't need costly internet access. Unlike other technological innovations, this has been designed keeping in mind Africa's needs. It is just one of those many businesses that are helping Africa leapfrog to a new era for people who lack access to banks, credit cards, or even phone lines.

Source: Lindow (2004)

surely creating Schumpeter gales of creative destruction and, as predicted by Schumpeter himself, form the basis for the progress of these nations.

How has it been possible for man to develop technology of the type described earlier? What appeared as science fiction half a decade back is now reality. Is this not because of our amazing cognitive capabilities?

With modern imaging techniques such as Positron Emission Technique (PET) and especially fMRI, our understanding of the human cognitive architecture has advanced greatly. The ways in which this knowledge has been acquired is also indicative of the fact that technology is interactive in nature. Man developed technology and technology helped man to understand himself better, the end result being better and more adaptive choices.

3.3 The Cognitive Architecture

> ***Information processing approach**: an approach which likens man to a computer, both capable of taking in inputs, doing internal processing and producing outputs*

Long dominated by Behaviorists such as B. F. Skinner, the cognitive revolution was brought about not by people in the field of psychology but by three external influences, namely, communications research, computer modeling, and linguistics. The conjoint effect of advances in these three fields led to what is known as the **information processing approach**, based on an analogy between man and computer, both capable of taking in inputs, internal processing, and producing outputs. The 1950s and 1960s saw a flush of models of human processing, producing yet another Schumpeter gale, leading to a paradigmatic shift in almost all areas of psychology. These models included the primary and secondary memory model developed by Broadbent (1958), the two-store memory model of Waugh and Norman (1965) and the three-stage memory model of Atkinson and Shiffrin (1968), to name just a few. The paradigmatic change was even more apparent in the flood of new experimental paradigms being developed and used leading to a deeper understanding of our cognitive structure and its functioning. The concepts of chunking (Miller, 1956), attentional filters (Broadbent, 1958), the ways in which we go beyond the information given (Bruner, 1974), and the sensory register (Sperling, 1960) trod the unbeaten path and provided new insights. These models of human cognitive performance were given a sound experimental base when psychologists such as Michael Posner (1986) used various reaction time techniques in the psychology laboratory to explore the workings of the mind and have been discussed in a volume entitled, *The Chronometric Explorations of the Mind.* A physiological base was added to the insights gained from information processing when new technologies of brain imaging came into existence and cognitive neuroscientists started using PET and later fMRI to delve into the physiological basis of cognition and we have upcoming fields such as cognitive neuroscience coming into existence. Describing the field of information processing and cognitive neuroscience would be beyond the scope of this book, but an attempt will be made to collate most of our available knowledge in the form of a model.

In short, cognition or the **cognitive process** refers to a system of interrelated processes that start when the nervous impulse from our sense organs reaches the brain and undergoes processing at various levels. However, it does not end when the response is decided upon and the brain commands the appropriate part of the body to make a response. Evidence from brain imaging studies seems to favor an interactional approach in which higher level processes amplify and reduce sensory input and thus modify the input.

***Cognitive process**: system of interrelated processes that start with the nervous impulse from our sense organs reaching our brain and then undergoing processing thereafter.*

The pervasiveness of cognition had been brought to the fore by Neisser (1967), who was of the view that cognition seems to be involved in everything that a human being might possibly do. According to him,

> "The term "cognition" refers to all processes by which the sensory input is transformed, reduced, elaborated, stored, recovered, and used. It is concerned with these processes even when they operate in the absence of relevant stimulation, as in images and hallucinations…." (Neisser, 1967, p. 4).

These processes may be further categorized as follows:

- **Primary functions**: sensory stimulation, attentional filters, sensory gating
- **Secondary functions**: perception and learning, assimilation of new information
- **Executive functions**: analyzing, problem solving, creating

***Primary functions**: sensory stimulation, attentional filters, and sensory gating*

***Secondary functions**: perception, learning, and assimilation of new information*

***Executive functions**: analyzing, problem solving, creating*

3.3.1 The Cognitive Pyramid

What powers this high powered function of the human body? Empirical research has enabled us to understand the architecture of the cognitive system in the form of a pyramid consisting of four interactive hierarchical aspects (see Fig. 3.4).

(a) **Cognitive requirements**: Let's start at the base of the pyramid. The biological basis of the cognitive system may be understood in terms of two aspects, namely, the anatomical basis or the areas of the brain and the physiological basis, that is, the neurotransmitters used by these areas. Current cognitive neuroscience posits that various parts of the brain are called into play, including the **striatum**; the **basal ganglia**; the prefrontal cortex; and large parts of the frontal, occipital, temporal, and parietal lobes of the cortex. A diagram (Fig. 3.5) clarifies this further.

***Striatum**: The striatum is a subcortical part of the forebrain.*

***Basal ganglia**: The basal ganglia comprise multiple subcortical nuclei in the brains of vertebrates and situated at the base of the forebrain.*

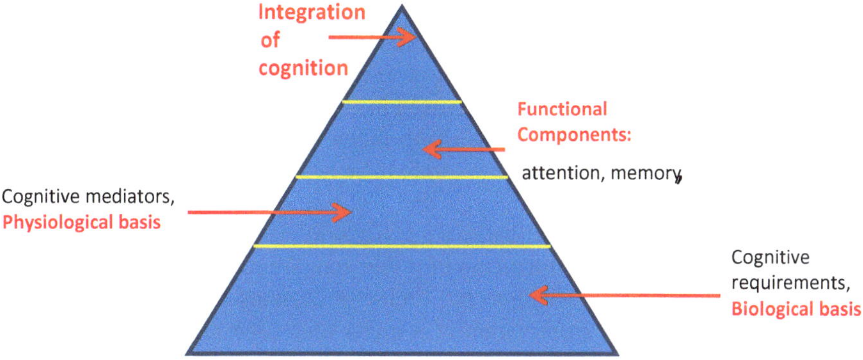

Fig. 3.4 The Cognitive Pyramid, four hierarchical interactive components of cognition

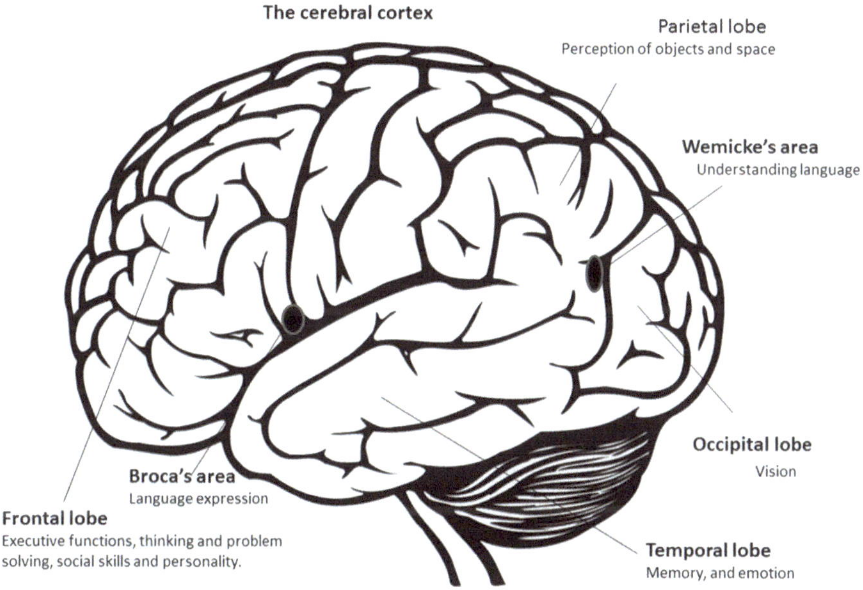

Fig. 3.5 Lobes of the cerebral cortex and cognitive functions. Adapted from www.headway.org.uk

Cognitive mediators: the second level delineates the cognitive mediators or the physiological basis of cognition. Through recent imaging techniques we have come to know that dopamine (which we have mentioned in earlier chapters also) is an important **neurotransmitter** through which our choices are regulated, reward-seeking behavior is modulated, and cognitive foraging is made possible. This dopaminergic system served our evolutionary ancestors and continues to serve us even today. The world has changed,

> ***Neurotransmitter****: a substance in the body which carries a signal from one nerve cell to another.*

we have moved from natural jungles to concrete jungles, but in the true Darwinian sense, natural selection processes have helped us retain this highly effective dopaminergic system along with the pathways for another important neurotransmitter, serotonin, as discussed in Chap. 2.

(b) **Functional components**: **the working memory**: If each and every piece of data that our sense organs captured were to be retained by us and processed, the capacity of the required processor and the size of the store, that would be required to cope with this burgeoning mass of data, would be gargantuan. Evolution has worked out a method for this, too. We receive a large amount of input all the time, but owing to our limited processing capacity, we can hold only a small part of information at a given time. The rest fades away very quickly. For visual input, it is estimated that a large chunk of information would be lost in roughly 2 s.

The structure that holds information for this extremely short period has been called **working memory**, a term coined by Miller, Galanter, and Pribram (1960), but used by a number of theorists including Baddeley and Hitch (1974) and Atkinson and Shiffrin (1968). It has also been called the **short-term store** though it should be clear that there are differences between the two: the former is a structure that not only holds the information but also manipulates it to make further processing easy. The latter, on the other hand, acts only as a storage system and does not have much to do with the manipulation or organization of information. Although the concept of working memory was proposed more than 30 years back, it continues to stimulate empirical research even today (Baddeley, 2003). In the year 2000, further extensions were made to the model by Baddeley (2000) in which he proposed an **episodic buffer** in addition to the two buffers proposed earlier. This new buffer, as the very name episodic buffer suggests, is a place where short-term memories for events and temporally extended episodes can be stored and manipulated. Empirical research has also been able to clarify the role of the neurotransmitter, dopamine, in the working of the working memory. According to Sawaguchi (2001), increase in the level of dopamine leads to improved working memory capacity.

Working memory*: a part of our cognitive system that holds information for very short periods of time and also manipulates it to make further processing possible*

Short-term store*: a part of the cognitive system that holds information for very short durations*

Episodic buffer*: a part of the cognitive system where short-term memories for episodes and events are stored*

Working memory also plays a role in the goal-directed activities of individuals. One would normally expect that if you can keep goal relevant information highly activated, goal attainment would become easier. Well, it has been seen that people with high working memory capacities are better able to keep goal relevant information highly activated and ready for use whenever needed (Gray, Chabris, & Braver, 2003).

More details about the sensory register and its operation have been described in Box 3.10. Ways in which we manage to increase the size of our sensory register and short-term store have been explained by George Miller through his famous paper

Box 3.10: Sperling's Experiment on the Sensory Register

In a very clever experiment, George Sperling (1960), at the *Bell Laboratory*, conducted a series of studies in which he showed randomly arranged alphabets in three rows for a fraction of a second. When subjects were asked to recall the entire set of letters, say, in all 12 letters, equally distributed in three rows, he found that the total number of letters recalled was fewer than in another condition in which, instead of recalling all 12 letters, the subjects were told to recall merely the items of top, middle, or lower row only (partial recall condition). The significance of this experiment is that it demonstrates our inability to hold a large chunk of visual information for more than a few seconds. By the time we start processing the entire input, much of it is lost, much like the bursting of soap bubbles, in a very small fraction of a second. When our optic nerve receives visual information, it relays the messages to the brain, but nerve fibers carrying this information cannot transmit all the information simultaneously; only a given set of fibers transmit information; and they, too, fire within their limits, say approximately 10 megabits per second. Our source of sensory information is limited right at the very inception. Even after the input from the optic nerve reaches our brain, the question that arises is the degree to which accuracy and precision of the inputs can be maintained in the visual working memory. Recent results show that though the precision of the working memory can be voluntarily controlled and that it does show a certain degree of flexibility, both of these are possible only when the number of items to be retained is low (Machizawa, Goh, & Driver, 2012). This could also explain why Sperling was able to obtain greater accuracy of reports when he asked subjects to recall only a part of the total stimulus array.

entitled *Seven, plus or minus two* (Miller, 1956), wherein he showed that while this is the ultimate number of bits that can be retained by us, we can increase it by the process of **chunking**, or reorganizing the material to form groups such that these groups should then be of the order of 7 ± 2 (Box 3.11 gives an example).

> ***Chunking****: grouping of information so as to reduce the number of bits of information*

One striking difference between animals and human beings is this remarkable ability of being able to combine new information with the old and thereby building an immensely rich model of the world. Maria Popova (2012) wrote that the human "working memory is hardly different from that of a monkey, even though the monkey's brain is roughly one-fifth the size of ours. Experiment after experiment has shown that on an average, the human brain can hold four different items in its working memory, compared to 3–4 for thc monkey." What makes the difference? In his book, *The Ravenous Brain: how the new science of consciousness explains our insatiable search for meaning,* Bor (2012) argues that it is through this ability of

Box 3.11: Trying to Memorize aapdaaapdaaapdaaapdaaapdaaapdaaapda

Try to memorize the above: there are 35 characters in all, that is, 35 bits of information, much too large for our limited capacity system to handle. But, we are also sure that you would have realized that these 35 characters can be organized without changing the order of the letters? Break it up as shown as follows.

aapda/aapda/aapda/aapda/aapda/aapda/aapda

The result: you have organized it into 7 bits of 5 letters each. Now, all you have to remember is the 5 letter series repeated 7 times. Was it not easy? This is what Miller (1956) called chunking. It is one of the mnemonics (tricks to help us memorize) that we so often use whenever we have to memorize long lists, whether it consists of random letters, telephone numbers, lists of people, or any other type of list. Miller also posited that our processer is most efficient when the number of bits is of the order of 7, plus/minus 2, or in the range 5–9. Telephone departments have used this fundamental fact in the formation of telephone numbers. They are normally segregated into groups, the country code, followed by the city code, the area code, and then the number. An example would be 044-0214-351-5776. Most people will be able to remember to affix the country and city codes, all that they have to remember is the 7 or 8 digit number after that.

chunking that we are able to expand the limits of our working memory, so much so that it plays a key role in gaining expert skills such as those required for language acquisition and even chess (Moore, Cohen, & Ranganath, 2006; Olesen, Westerberg, & Klingberg, 2004; Gobet et al., 2001).

Because chunking is known to aid memory improvement in healthy adults (Bor, Duncan, Wiseman, & Owen, 2003), training in the use of chunking leads to significant increases in working memory capacity and has become a major strategy underlying memory training programs (Mcnab et al., 2009; Olesen et al., 2004). In his book referred to earlier, Bor (2012) illustrates the immense benefits that can be accrued through training, and describes a person who could increase his working memory capacity from the normal 4 items to 80 items. Today, chunking has also been seen to bring relief to elderly people with cognitive deficits and those with mild Alzheimer's disease (Huntley, Bor, Hampshire, Owen, & Howard, 2011; Belleville, 2008; Bier et al., 2008). It is also useful for cognitive skill training for elderly people. One such intervention is the *Advanced Cognitive Training Program for Independent and Vital Elderly (ACTIVE)*. In a report, the results of this intervention on 2832 people have been released and the study shows that the effects last for a period of 10 years (Rebok et al., 2014).

The ways in which modern technology has enabled the finding of neural correlates of chunking is a good example of the nexus between psychology and technology. Through studies on both animals and humans, functional MRI (fMRI) has revealed that encoding, storage, and retrieval from working memory are associated with parts of the prefrontal cortex and parts of the parietal cortex. A network generally referred to as PPN (prefrontal parietal network) enables this process of chunking (Bor & Seth, 2012; Bor & Owen, 2007).

The implications of chunking for technology are brought to the fore in the realm of eLearning, where one does not have an instructor to guide oneself through the course (see Box 3.12). The ways in which chunking can be used in computer-assisted language learning have been empirically established by Li and Swanson (2014). In a report of the *Suntrust Equitable Securities*, on companies involved in technology-aided learning, it has been pointed out that "ultimately we believe the improvements in how we learn will be the single greatest change that the Internet has on the society" (cited by Willis, 2004). Willis (2004) goes on to explain that the

Box 3.12: Chunking in the Age of e-Education

With more and more people relying on e-education or what has generally called online learning programs, the importance of chunking is being increasingly recognized. The reason is that many people who opt for online courses are people who are not able to attend regular college for various reasons and are therefore forced to devise self-help techniques for learning.

One of the drawbacks of eLearning is that you do not have an instructor who can guide you through the course. As such the course material often appears unwieldy because of which the dropout rate from such courses is often very high.

Chunking provides an easy way to help the learner go through and digest the material. One method is to divide the process into four steps.

Step 1: start at the highest level by determining the hierarchy of the course content, namely, the number of modules.

Step 2: divide the modules into smaller parts, and each of these can then become a lesson.

Step 3: you can now start chunking at the screen level, that is, chunk all the material into related parts. One can devise a metaphor for that, or, one can construct a sentence with the first letter of each word being used to denote one aspect of the material. In other words, one starts using memory mnemonics.

Step 4: for each page on the screen you could do a working memory check: keeping the material down to four items. If there is more than that, you can try rechunking the information.

One trick is to remove all that is of no use or even all the repetitions. By doing that, one is often able to cut down on the total amount of information.

Source: based on Malamed (2013)

challenges thrown up by the internet are even more so as far as older people are concerned. Tech-aided learning for the elderly brings up two issues: not only must they learn the substantive material but before they can go on to that, they must learn how to operate the technology that is carrying this material and on which the course is being run. While learning how to chunk can help in the former, a different learning regimen is required for the latter.

With higher and higher order chunking, there is, in fact, no limit to the number of bits that an expert is able to retain. Most technical gadgets around us, though, do not require the monitoring and holding of information from such an unusually large load of sensory input. Gadgets today have their own memories and these help us to obfuscate our own memory limitations. Of course, for tasks which require exceptionally high degree of vigilance coupled with a synchronized response such as in radar detection, there is a speed/accuracy trade-off, the degree of which can be analyzed using **signal detection theory**, initially proposed by Green and Swets (1974) but valid even today.

> ***Signal detection theory****: a theory which enables the measurement of the ability to distinguish signals from noise*

3.3.2 *Visual and Auditory Perceptual Processes*

A very important aspect of cognition is the process of perception. While perception takes place in all our senses, we will limit our discussion to visual and auditory aspects in relation to use of things. For example, our ability to determine the location of sound is based on the time difference that occurs in the reception of sound by our two ears, that is, the ear closer to the source receives it earlier. This **inter-aural time difference** (ITD) has a lot of implications for things we use in everyday life and in critical conditions.

> ***Inter-aural time difference****: the time difference between the reception of the sound by the two ears*

3.3.3 *Visual and Auditory Displays: Using the Laws of Perceptual organization*

A cardinal principle in designing a console involves using the laws of perceptual organization, which were, as stated earlier, first enunciated by the Gestalt psychologists over 90 years back (Wertheimer, 1923). According to these laws, while the eye merely captures the various parts of the visual stimulus it is the brain that organizes these into a perceptual whole, not randomly, but according to certain principles. These principles include the laws of similarity, contiguity, closure, and good figure. Readers interested in the ways in which these laws of perceptual organization operate would do well to consult any introductory text of psychology for a detailed account (we have described them briefly earlier in this chapter). That these laws serve an important adaptive function in the life of human beings is clarified by Bhatt and Quinn (2011) which shows that even among infants of about 3–4 months old, at least two principles of perceptual organization do operate: the principles of common motion and good continuation. Such perceptual organization enables infants to transfer knowledge from one source to another, a process known as scaffolding or bootstrapping. While technological gadgets obviously stimulate transfer of information, does it help on all occasions? It seems that though children are smart enough to relate events but at the same time they find themselves limited in overexpanding the organized information (Smith, 2012a).

> ***Visual attention guidance****: whatever is directly in front of the eyes has greater potential to be focused upon than objects in other parts of the environment*

The world around us contains innumerable visual objects within the visible spectrum of light waves and which therefore have the physical potential to stimulate our eyes. But do you notice each and every one of these objects? Certainly not. A common experience is that when you are working with your head down and you lift your head suddenly, whatever comes in front of the eyes in the same direction is usually what our eyes are attracted to. This is known as **visual attention guidance**. Technology has taken notice of this phenomenon and has used it to its advantage. On the other hand, when this phenomenon is disregarded by the designer, problems ensue. Just think of the problems created by Windows 8 or 8.1 wherein the start button doesn't appear at the center of the computer monitor but at the side.

Some examples of the effective use of the visual attention guidance are advertisements that catch the attention of the viewers. Certain areas of the advertisement have a greater potential for attracting viewers and a good advertisement is one in which the designers concentrate on that part of the advertisement to make their advertisements more appealing to the consumers and to tempt them to buy their products. This concept is also used by storeowners. The way in which the items are arranged in the store is a core concept in store styling such that the consumers are attracted toward the very items that the owner wants them to be attracted toward. For example, if the storeowner puts the cosmetics section after the vegetable section, then there is a high probability that this arrangement will attract more women toward both of these items creating higher sales and more profit for the store. Have you ever thought of how many products you buy from those kept near the checkout counters and to which your attention is attracted as you wait for your turn at the cash register? If you introspect, you will find that most of this shopping has been impulse shopping!

3.3.4 Visual Displays

Also known as monitors, visual displays can be seen all around us: on our computers and tablets, cell phones, television, vehicles, or even the mundane household gadget such as the microwave oven. They range from simple displays such as that seen on a thermometer to extremely complicated ones as the displays in a cockpit.

Over the years, the technology being used by such displays has changed, with the earliest being the cathode ray tube. Today, of course, we have moved to the LCD and now to the LED. In order to be effective, many of the aspects of visual perception discussed earlier need to be kept in mind while designing the console. Three basic ones would be the principles of visibility, comprehension, and compatibility. Thus, if the symbols are two small, or if they are using light rays from beyond the visible human spectrum, our eyes may not be able to see them. Similarly, if we do not know what the icon stands for or if the icon is not compatible with the object for which it is being used, the visual display would hardly be effective (two examples have been presented in Box 3.13).

Visual displays are not the only type of displays being used. With the realization that the visual system is already overburdened and requires constant focusing of the eye toward the display, auditory displays are being created for a variety of settings. Displays using other sensory systems are also used, such as tactile displays and combinations of tactile and auditory displays for people with visual impairments.

Box 3.13: Have You Ever Noticed?

The logo on the Hershey's Kisses product has a hidden logo: an extra Kiss. Turn your head to the left and you will see that between the 'K' and the 'I' there is a Hershey Kiss baked into the logo!

Another clever logo is that of Toyota! According to Toyota, the three overlapping ovals on American vehicles "symbolize the unification of the hearts of our customers and the hearts of Toyota products. The background space represents Toyota's technological advancements and the boundless opportunities ahead." And even more impressive, if you look at the overlapping circles you will see Toyota spelled out.

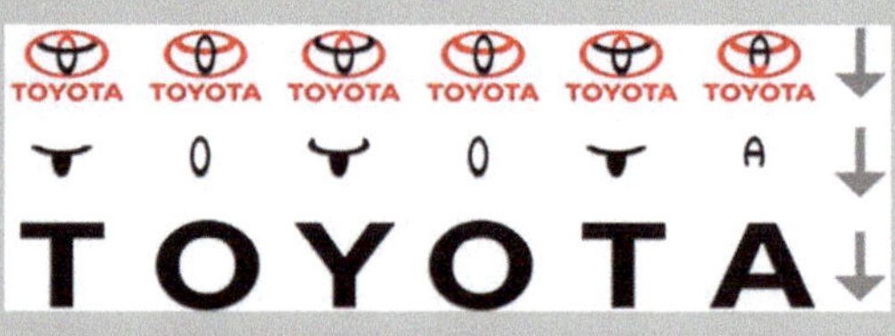

(Source: http://a.msn.com/00/en-us/BBjnc1G?ocid=se)

3.3.5 Auditory Displays

Various types of auditory signals and displays are being used today, including the use of streams of sound (vis-à-vis, simple signals, which are actually alarms) to represent objects which do not naturally make those sounds or to even communicate streams of data.

> ***Properties of sound waves****: amplitude, wavelength, and timbre*

> ***Perceptual attributes of sound****: loudness, pitch, and characteristic tone*

The physics of sound has shown that **sound waves typically have three properties**, amplitude, wavelength, and timbre depending on the source of the sound. The corresponding **perception of these three properties** is in the form of loudness, pitch, and characteristic tone. By varying one or more of these attributes, different tonal qualities can be created. This aspect of sound perception has been used to create technology that uses audio signals to add perceptual data to the normal visual input. Some very common examples are the ring tones on your cell phone or the audio signal telling you that you have received an SMS. Many household gadgets employ this technique, for example, the alarm on your oven, or the musical indicator on your water purification appliance. The purpose is to give you an auditory signal which can be heard from wherever you are and thus help you to move to some other spot or to some other task, without having to maintain a constant vigil as would have been the case had there been only a visual signal. But these are alarms. There are a variety of uses to auditory displays. Some of them have been described later. But, in order to serve their purpose, they must show ecological validity, without which they may prove to be disadvantageous. Ways in which this can happen has been explained in Box 3.14.

- The alerting function, as in alarms, warning systems
- Status, process, and monitoring function, as displays used in factories to tell about which stage the factory process is in at a particular point of time, or even in hospitals to monitor heart rate

Box 3.14: Acoustic Ecology

With the increasing use of various acoustic devices, acoustic ecology has emerged as an interdisciplinary branch of study involving mainly sounds, living beings, and their environment. For example, too much noise and its variety may cause desensitization to the critical source of information. Imagine you driving on a noisy road and also attending to the siren from an approaching emergency vehicle compared to listening to the same siren on the same, but quiet, road. Addressing the importance of this issue in automobile driving, Nees and Walker (2011) showed how auditory displays could be useful for vehicle technologies.

At the same time, a report from the *Boston Globe* (Kowalczyk, 2011) had analyzed the downside of auditory displays, by reporting how alarms may be neglected by citing a case from the Massachusetts General Hospital, Boston.

(continued)

Box 3.14 (continued)

Calling this phenomenon Alarm Fatigue, a number of suggestions have been made to control the potentially harmful effects of overlapping sounds in the Emergency Room of a hospital. Some of these have been detailed as follows:

Monitors with similar frequencies should be spread out

- Monitors must be decentralized, that is, onset time should be different for different equipments
- Auditory signals should be supported by visual signals
- Dynamic displays are better comprehended in the auditory mode
- Issues of auditory fatigue must be addressed in a larger context of Emergency Room operations

Source: Nees and Walker (2011); Kowalczyk (2011)

- Data exploration as in sonification, where auditory graphs may be used instead of visual graphs
- Art and entertainment.

Advantages of auditory displays:

1. Auditory displays are sometimes more effective when a person is unable to look at a visual display (as in the case of a nurse who has to attend to several patients and it is not feasible to constantly look at all the visual displays constantly or simultaneously) or is unable to see the display because she is visually impaired; it is also useful when the visual system is overtaxed, as in the case of constant radar monitoring, such that any further signal would not be noticed.
2. Auditory displays are sometimes more effective because of the ability of human beings to process multiple sounds simultaneously (i.e., parallel processing is possible), unlike the visual system which senses objects only in the line of the eye, and is therefore capable of only serial processing.
3. With mobile systems becoming increasingly smaller, visual displays are also decreasing in size, and so there is only a certain amount of information that can be put on the screen and still be visible. It is here that auditory displays provide additional space for presenting information.
4. Where data are complex, have inherent complex patterns, or where there are changes in the data over time, or there are warnings that require immediate action, auditory displays provide a much stronger stimulus than that provided by visual signals or even visual displays. Just think of what you are more likely to pay attention to: a warning siren for a passing ambulance or the visual sign of the ambulance.

3.3.6 Sonification

> ***Sonification**: the use of continuous nonspeech sound patterns to communicate data and especially changes in data.*

One form of auditory display is what is known as **sonification**. Sonification is much more than the presentation of an auditory alarm and to be effective it must use an interdisciplinary approach drawing from fields as diverse as audio engineering, audiology, computer science, informatics, linguistics, mathematics, telecommunication, music and not the least, psychology.

While an alarm is sounded when a particular object rises above a certain threshold, say for example a smoke alarm, sonification uses continuous sound patterns to signify changes in whatever aspect it is being used for. One use of sonification is in hospitals to signal changes in physiological parameters for nursing staff where rather than nurses having to constantly monitor heart and respiration rates for critical patients, sonification has been used to present a continuous sound pattern with rise and fall in pitch designating a rise and fall in heart rate or rate of respiration. The earliest successful use of sonification was probably the **Geiger counter** which is used even today to measure the amount of radiation in a room. The number of audio clicks indicates the amount of radiation. Today many other uses have been found for sonification. An important use is to help people with visual defects to understand visual experiences. The *Sonification Laboratory* at the School of Psychology at the *Georgia Institute of Technology*, USA, is making innovative use of the principles of sonification to aid people with visual impairments. One such project which will enable them to get a more complete and richer experience of museums in which live exhibits are kept is called the *Accessible Aquarium Project.* The scientists have used sonification to convey information about the number of animals, their movement, and location. Other equipments that have been developed by the same lab include advanced auditory menus for use in mobile phones and even vehicles and a wearable auditory navigation system called *SWAN* (System for Wearable Audio Navigation). The *NASA Goddard Space Flight Center* is also in the process of developing a system which can display numerical data as sound and can therefore be used by people with visual impairments. Another use is for training in dancing or swimming. If only movement could make sounds, telling the person what kind of movement he has made would have been very easy. The day is not far when this will become a reality. It is already been tested in laboratories and would soon be released on a commercial basis.

> ***Geiger counter**: first equipment using sonification and which measures amount of radiation in a room*

3.3.7 Ecological Validity and Auditory Displays

> ***Ecological validity****: degree of correspondence between artificial settings and the real environment*

Auditory displays are not difficult to design, but the problem lies in creating effective designs. Auditory displays are being used in a variety of settings, some of which are very crucial, for example, in the flight cockpit or even in skydiving. However, such displays will be effective only if there is a correspondence between sounds in real life and the auditory signals, or when **ecological validity** is maintained. This correspondence can be said to vary along a continuum, ranging from complete correspondence at one end to a complete lack of correspondence, at the other end. For example, one may want to present an auditory display for clinical monitoring. The high correspondence end would consist of the actual parameter, say heart rate, being presented as an amplification. Another high correspondence cue could be the actual word itself, say, oxygen, when the patient needs it.

3.3.8 Acousmatization of Sound

> ***Phenomenology****: the study of conscious experience, specifically related to perceptions, judgments, and emotions*

Both visual and auditory modes of presenting information and its processing by the brain have deep implications for psychology of technology. Unlike vision, audition is not intrinsically spatial. When we see, the object is found localized in space, but in audition it is extended in time. In other words, while vision is considered robustly spatial, making us aware of space and its characteristics, audition is, by and large, nonspatial and in most cases may *not* be judged in terms of spatial features. This issue is obviously concerned with the nature of our experience of inputs from the environment and refers to the **phenomenology** of vision and audition.

We normally experience sounds such as the ripping of paper or the sound of a loose fender as events outside our body (O'Callaghan, 2010). The same is true for another sense, namely, olfaction, which also gives us an aspatial experience. Therefore, comparing senses becomes a very interesting albeit challenging subject for investigation from the point of view of psychology of technology. As discussed earlier, as far as the direction of sounds is considered, it is the ITD (Inter-aural Time Difference) that provides us with a sense of direction and some spatial cues. But otherwise, it is quite clear that audition at the phenomenological level remains less spatial in its content than vision. Are there any implications of such experiences with auditory stimuli in the designing and using of products? Does the elimination of spatial direction in audition make our experience different, for example, does it contribute to an alteration in what we hear? Can we hear sounds from different sources simultaneously and still find the experience enriching?

> ***Acousmatization:*** *hearing sound without being able to see what is causing it*

This interesting phenomenon of **acousmatization**, or the hearing of sound without being able to see its source, dates back to the days of the great geometrician Pythagoras who, during his discourses would draw a veil between himself and his students, so as to stop distractions because of his physical appearance. So, all that students heard was his discourse, without seeing him in person delivering it. It was clear that when vision does not distract, the effect of the sound becomes more intense. Called acousmatic experience, it has become a significant trend to highlight sounds detached from the source of their production and thereby induce esthetics in musical listening. This experience has been used by modern audio technology in the form of the radio, records, telephone, and the tape recorder. We are seeing more and more of such experiences being exploited in the art and science of music production (Boxes 3.15 and 3.16). For example, to create suspense in horror movies, acousmatic experiences provide an important adjunct to the repertoire of devices used by the director.

Box 3.15: Using Human Auditory Constraints to Gain Advantage

The marvel of modern technology is that it has not only been able to override many human constraints but has also been able to use some of these constraints to its advantage. One example is the use of the nonspatial characteristic of human hearing in the production of films to create mystery and suspense artificially. French film sound theorist Michel Chion (1994) explains how this is done. In cinema, sound is created in two ways: one, where the source of the sound is shown and then the sound comes on, or the source becomes "acousmatized." This is what has been called visualized sound or direct sound. The second modality is to reveal the sound first and then subsequently show the source of the sound. This veiling of sound creates suspense because we do not know who is speaking or from where the sound is emanating. This has been termed deacousmatization of sound. Together these two types of sounds are said to create a notion of off-screen space (Chion, 1994).

Box 3.16: The Technological Wonder of Stereophonic Sound

Just imagine the amount of physical space a full sized orchestra would require and that with the strings at one end, the horns at another and the bass at a third, it is but natural that the orchestra provides a sense of spatiality to the music created. As space has become more expensive and musicians are hard to come by, we have started relying more and more on technology to create orchestra-like music. Thus, one of the principal requirements would be to keep the ITD intact and use it in a way that creates the atmosphere normally possible only

(continued)

Box 3.16 (continued)
by a large orchestra. This is all that stereophonic sound is. Think back to your home cinema systems. It normally consists of multiple loud speakers placed at a distance from each other, such that you get to hear different frequencies from different directions. One set of speakers may pick up the string sounds, while another may pick up the trumpet and the saxophone. In a movie, you may get the sounds of people talking from one loud speaker, and the birds chirping from another, and the sound of a car starting from a third. The effect has greater ecological validity because in real life these three sets of sounds would emanate from different directions. The sound productionist has thus created spatialization of sound through the use of what has been called acousmatic music, in which not only is the music score provided but a diffusion score is also provided which would detail the placement of different loud speakers of the multiple speaker system being used (Emmerson, 2007). Thus, an acousmatic performance with prerecorded music and without performers in concert halls would provide the same richness of experience that an actual recital with live musicians would.

3.4 Multisensory Integration in Sensation and Perception

3.4.1 Integration of Cognition

After this brief description of the human cognitive architecture, including visual and auditory perception and its use in technology, we will now discuss some other aspects of perception and especially multisensory integration with special reference to how technology uses or exploits our cognitive processes because of the limits of our cognitive architecture.

In real life, sensory data are not partitioned with reference to its source. We generally tend to "see" the environment as a unified whole with all kinds of sensory data adding up to create a multidimensional and rich experience. This is possible because of the capacity of our cognitive system to integrate all the information and to, sometimes, even add its own flavor, as posited by Bruner (1956) in his famous article: "*Going beyond the information given.*" This executive tends to fill in the gaps, create pieces of information from the sometimes scanty or insufficient data received by the brain. This process of unification is what is known as multisensory integration in sensation and perception.

There are many aspects of daily life in which we simultaneously receive sensory data from two or more senses. One such example is speech. Imagine a person speaking to you standing in such a position that you are not only able to hear the sound but also able to see her face and body while she is speaking. If this same person walks behind a wall but continues speaking, you are well aware of the loss of sensory data and the problems associated with only hearing sounds and not having the visual image in front of you. This becomes even more obvious if we are trying to hear what the person is saying in a scenario of considerable noise. We experience the same difficulty while watching a program on TV in which the sounds that we are hearing do not match the lip movements of the speaker. In other words, the incongruence between the sounds that we hear and the images that we are seeing creates problems in our understanding of what the person is saying.

3.4.2 *Technology and Multisensory Integration*

> ***Multisensory integration:*** *integration of information obtained from different senses so as to form a coherent whole*

A number of studies using a variety of imaging methods help us to understand that audiovisual **multisensory integration** does take place (Mercier et al., 2013) though audiovisual speech perception seems to be following a mechanism different from that used in other situations involving multisensory integration. In a very interesting study, Tuomainen, Anderson, Tiippana and Sams (2005), provided subjects with sound data in the form of sine waves under two differing conditions. When the subjects were not aware that the stimuli were speech data, they failed to show any audiovisual integration. However, as soon as they got to know that this was speech data, they integrated information from both the auditory and the visual part of the signal much as they would do in the case of normal speech perception.

> ***Audiovisual asynchrony:*** *when the visual and sound emanating from the same object are received by our sense organs at varying points of time*

One important feature of audiovisual integration arises out of the fact that the speed of light is faster than that for sound. Thus, when we see an object in the environment and are also able to hear sounds made by it, it seems very likely that the two pieces of information would fail to reach us simultaneously, even though the period of time intervening between the two would be very small. We are all aware of the phenomenon of first seeing the lightning to be followed by the thunder only later and this is clearly perceptible. This is what has been termed **audiovisual asynchrony** which is due to a natural property related to the varying speeds of light and sound waves, respectively, but could create problems if not attended to. The human perceptual system is apparently able to deal with such asynchronies. Laboratory studies clarify that audiovisual integration of speech is highly tolerant of such asynchronies but only when the visual stimulus precedes the auditory stimulus. When the auditory stimulus is preceded by the visual stimulus, asynchrony across modality is readily perceived (Grant, VanWassenhove, & Poeppel, 2004). Is this not amazing? It seems that the human brain is hard wired for the maintenance of ecological validity. Since light waves would precede sound waves simply because of their inherent properties, our neural system has been so created that it can tolerate this temporal difference.

The results of the study also show that as soon as audiovisual asynchrony is detected, multisensory integration is affected. At the same time, the range of auditory-visual temporal asynchrony is fairly broad (roughly 20–200+ ms) and is also quite asymmetrical.

3.4.3 Technology and the Audiovisual Asynchrony

> ***Anticipatory coarticulation****: rounding of the lips of the speaker for the word about to be said even before the word is actually spoken*

A common problem in speech recognition is what has been called **anticipatory coarticulation**. For example, when you are going to say "sound" there is a rounding of the lips before the word is actually heard by the listener and this adds considerably to the recognition of the word, becoming even more pronounced under noise conditions. While speech understanding is not just about hearing and involves the sight of the person speaking and sometimes even the tactual feel of the person, an important function of the brain is to integrate all these differing pieces of sensory information so as to create a coherent whole.

Technology faces a similar problem. In today's era with the use of live audio–video streaming, as in video conferencing, even a tiny lag between the visual and its auditory component could be a cause for concern. So we often hear terms such as audio–video sync, AV sync, lip-sync and the opposite, that is, lip sync error or lip flap. These aspects are normally in the realm of the audio mixer or the sound producer and the positioning of the camera and the microphone can cause considerable harm. While some people are of the opinion that people who watch movies or live streaming will normally not be able to notice the lag, and even if they do will probably discount it as just one of those things in using technology, others fail to agree (e.g., Reeves & Voelker, 1993). Whether people like it or not, and even whether they notice it or not, asynchronies tend to have disastrous effects on the image of the person whose conversation is being live streamed. Since this problem exists because of the fact that the technology is still evolving, people have started to look for temporary solutions. At IBM, at important board meetings or when senior management is communicating with other employees of the company, instead of using live streaming of data as in video conferencing, they use a still photo accompanied by a self-paced audio track. They say that while we do lose some information in the form of body language and facial expressions, it, at least, does not cause negative images being formed of senior management and even CEOs because of the lack of synchronization.

The McGurk Effect: Sometimes called the McGurk–McDonald effect, it was first described in a paper entitled, *Hearing lips and seeing voices*. It is a phenomenon that was discovered by accident when the two scientists, McGurk and MacDonald were working on a dubbing experiment (McGurk & MacDonald, 1976). They noted that sometimes, when the visual for the lip movement for one phoneme is coupled with the audio signal for a different phoneme, the resultant perception is of a third phoneme (so 'ga' and 'ba' may sound like 'da'). In reality, an illusion, it is caused by the difference between the visual information from the speaker and the sound information that accompanies it.

Imagine a scenario in which the person is experiencing a poor audio and a good visual. In such a scenario, the chances for the McGurk effect to take place are very high and it might lead to the obtaining of wrong information from the experience. If there is a proper synchronization between the audio and the video then the person will not experience the effect, but if the synchronization is poor then the person is definitely going to experience this effect. Ironically, it is the people good at synchronizing the sensory inputs who are the ones who are most prone to this effect. At the same time, one has to remember that there could be many other reasons that could be responsible for this illusion. For example, a person suffering from some types of brain damage may also feel the McGurk effect, that is, they may be seeing one thing and hearing another (Schmid, Thielmann, & Ziegler, 2009).

The McGurk effect is an important phenomenon for people who are trying to dub a movie, because if the technician is not careful with the placement of the microphones and the camera, there are chances of a strong McGurk effect. There is, of course, also the probability that people who are used to watching dubbed movies may not experience this illusion since they are apt to ignore information obtained by looking at the mouth of the speaker. Most of the time they are busy reading the subtitles placed at the bottom of the screen.

3.4.4 Everyday Listening and Acoustic Ecology: Soundscape Mapping

> ***Soundscapes**: patterning of sounds in the environment so as to give a unified whole*

Much as we see the world as landscapes so also there are **soundscapes** created based on the spatial localization of the sounds around us. As you stand at a bus stop—the picture around you can be broken up into two parts. One is the visuospatial map consisting of visual images of objects and persons around us (i.e., the landscape) and the other is the audiospatial map consisting of sound patterns emitted by these objects or persons. We do not simply hear these sounds but are also able to localize them in space and this can be called the audiospatial map or a soundscape. A typical soundscape can be constructed on the basis of sonic characteristics such as magnitude, frequency spectrum, temporal dynamics vis-à-vis that of the listener. Organizing soundscape classifications according to perceived sound quality, esthetic or emotional content, and spatial characteristics has helped to create a number of what are called soundscape ontologies which can then be used for community design.

Acoustic ecology: Using data from five villages of Europe, Droumeva and Mcgregor (2012) have found that there are strong connections between the aural world local culture and everyday life. They found that we can categorize communities into acoustic communities according to the typical soundscape map because characteristic sounds and sound patterns vary from one community to another. They have also differentiated between three types of sounds: signals, soundmarks, and keynote sounds. Among these, the most important are the **soundmarks** (similar to landmarks), or important sounds in the vicinity (church bells, songs of typical birds, factory siren, etc). These acoustic communities are dynamic, changing and shifting with time, much as the landscape changes whenever a new building, a new road, or a new field comes up.

> ***Soundmarks****: auditory landmarks in a community*

In real time, acoustic communication positions the listener to the sound and the soundscape in a dynamic two-way flow of interaction, communication and interdependence. Both our listening and our sound making are functions of the context in which we listen and sound, not only culturally but also literally. In fact, our ears pick up only relevant cues. This explains many well-known phenomena such as acoustic masking and the **cocktail party phenomenon** (Truax, 2001). Everyday listening proceeds in a set of sequential steps. We first attend to the listening context, trying to situate the listening experience (i.e., we try to focus on where the sound is coming from). Once we have obtained the spatial location or at least have been able to identify the direction of the sound, we start focusing on the sound itself, alternating between the foreground and background sounds, trying to decipher and identify the sound. Combining what is heard and what is known about the context and memories of similar experiences, we attempt to make a coherent narrative of the whole, that is, we try to associate the sensory experiences with our memories for similar experiences or add meaning to our sensory sound data or obtain a percept of the same. We are not the only ones affected by what is known as the cocktail party phenomenon. It has been observed in animals such as frogs, song birds, bank swallows, cliff swallows, and even king penguins. Attempts have been made to analyze the nature of processing behind this. It seems to be based on both sequential and simultaneous integration of information arriving through top-down as well as bottom-up processing (Bee & Micheyl, 2008). More recently, it has been pointed out that both high- and low-level processes may be involved in auditory perception and cognition and that it is the interaction between these processes that allows us to create the rich experience of soundscapes (Snyder, Gregg, Weintraub, & Alain, 2012). The above helps us to understand what is being prioritized, what the aspects that are being focused upon are, and what is being lost. Designers of auditory displays would definitely find such information very useful. A number of studies have been undertaken in order to help designers of auditory displays (e.g., Droumeva & Wakkary, 2010; McGregor, Lepltre, Turner, & Flint, 2010; Coleman, Macaulay, & Newell, 2008). One such study conducted by Droumeva and McGregor (2012) and described above attempts to build an ecological model that can be used for the designing of auditory displays. According to them, auditory displays would be most successful when they are able to parallel everyday listening which seems to entail three levels outlined above: moving from contextualizing to identifying and then to

> ***Cocktail party phenomenon****: special attention to one's name being used even though there is considerable noise*

building up a coherent whole. Designers of auditory displays should see that the display is in line with the earlier stages and would thereby be able to ensure ecological validity.

3.5 Augmenting Cognition

3.5.1 Is Augmentation Possible? Synaptic Plasticity

It has already been pointed out that as we evolved from other hominids, our brains also evolved, with changes in both size and structure. Even when we compare man to other primates, the size of the primary structures, that is, those which deal with the sensory motor functions, is similar. However, once we enter the realm of the secondary and tertiary structures of the brain, that is, those that deal with the higher mental functions, there are massive differences (Garcia-Rill, 2002). Our brain continues to evolve even during the lifetime of a person. While older neurological theories were of the view that we are born with all the neurons we can ever have, it is now becoming increasingly clear that not only do we continue to add neurons but also that neurons can take over functions performed by other neurons (McKhann, 2002).

The new concept of neurogenesis is becoming increasingly established through both empirical research and practice, causing Kolb and Gibb (2011) to state that,

> "The development of the brain reflects more than the single unfolding of a genetic blueprint but rather reflects a constant dance of genetic and experiential factors that shape the emerging brain. Understanding the dance provides insights into both normal and abnormal development" (p. 265).

Connections between synapses also become stronger or weaker, through what has come to be known as Hebbian plasticity (Abbott & Nelson, 2000). In fact, a recent report in the *American Scientist* provides empirical evidence of how experiences can change the rewiring of the brain in mice (Kaneko & Stryker, 2014). The way in which these connections undergo change has been very aptly expressed,

> "Neurons that fire together, wire together—and neurons that fire apart, wire apart" (Mercola, 2015).

However, not all plasticity is good and may sometimes lead to abnormalities in behavior. A good example is drug-induced changes in the drug addict. Much of the behavioral abnormalities are due to drug-related changes in the prefrontal lobe of the person (Robinson & Kolb, 2004).

Taken together, it is clear that while we are born with certain pathways and brain structures responsible for sensory, perceptual, and other higher mental processes, they are malleable, modifiable, and have the necessary plasticity such that they can be increased.

Augmentation of cognition can be brought about through various ways. As discussed in the previous chapter (Chap. 2), in the section on *Going beyond Anthropometrics*, we have already provided evidence of how through intensive and continued training for at least 10 years, we can cause structural and functional changes in our body. At the same time, it was also pointed out that while this is possible, there is a caveat: we must start early in life, while development is still ongoing, or else, modifications are few and far between.

Before we start to explain ways in which this can be done, it is important to note that when the term cognitive enhancement or augmentation is used, it refers to improvements in core cognitive capabilities and not to domain specific skills. Over the years there have been a variety of measures through which such augmentation has been attempted. While philosophers such as Socrates and Confucius spoke of methods of learning how to learn and effective pedagogy centuries back, the twentieth century brought in its wake experimental psychology findings proving that the functioning of our cognitive system can be enhanced. We have already discussed the ways in which our memory capacity can be expanded using chunking, while short-term memory limitations can be overcome by getting the information to be transferred to the long-term store through rehearsal. The noted Nobel Prize winner, Kahneman (1973, 2011) showed the ways in which we use heuristics to save our limited resources and allocate them only to those mental processes that cannot be automatically processed. Cattell (1963) similarly provided data that there is much in our intellect that is genetic but there is also a portion that is built up during the ontogenetic development of the individual. Metamemory, the use of mnemonic aids, and training in divergent thinking are all methods that are used to enhance our cognitive system. Of course, we cannot forget schooling and education (does not education improve our cognitive abilities?), drugs, mental training such as yoga, martial arts, genetic modification, etc. However, for the twenty-first century, probably the most pervasive method would be through information and computer technology (ICT).

3.6 Brain–Computer Interfaces

We are all familiar with the use of a variety of external aids ranging from the simple pen and pencil with their well-sharpened pointed edges allowing us a degree of accuracy and calligraphic beauty that we could never hope for by using our fingers as in finger painting, to all the gadgets and applications through which we are not only able to encode a large amount of material but also store and process it. Just try to imagine life without your cell phone, your iPad, your laptop computer, or even your desktop computer. What do these help us to achieve? It is clearly a case of augmentation of our limited cognitive capacity to encode, store, and process the increasingly large amounts of data we are being bombarded with. So much a part of us have they become, that we can often use them with our eyes closed or looking elsewhere. They have become part of what we automatically process without using our limited attentional resources. But are they seamless in the true sense of the word? One might tend to think so, until the day one loses one's gadget and starts thinking of buying a new one. No matter how tech-savvy you may be, it takes time to learn how to use all the features of the new phone or a new laptop computer. But supposing you could have a system implanted in your brain, which naturally stays with you, which you cannot lose or get fed up with, even if you wanted to? This may

> ***Cyborg**: seamless man–machine systems*

sound like science fiction, but so did a trip to the moon sound a mere 60 years back and travelling by air, a century or two earlier. We are on the brink of using such seamless man–machine systems, or becoming **cyborgs**, as Clark (2003) calls them. In his captivating volume, entitled *Natural-Born Cyborgs Minds, Technologies, and the Future of Human Intelligence,* he discusses how there is nothing unusual about cyborgs, because it is in this ability to engage in complex relationships with nonbiological objects that we are different from other animals. It is human nature to use and develop tools and it hardly matters whether those tools are physically attached to us (as writing with a pen on paper, or using the computer keyboard to key in data) or whether they are embedded within us. In fact Gray (2012) provides empirical evidence to show that the human mind does not differentiate based on from where the information is coming. It could come from our sense organs or it may come through wires leading to an especially implanted system. It would be processed in the same way if it reaches the areas of the brain that process information. An interesting example of this is that of Professor Warwick (Box 3.17), who had an implant

Box 3.17: Prof Warwick and His Brain Implant: What Was Implanted and What It Allowed Him to Do

Kevin Warwick, an engineer and distinguished professor at various universities including Oxford University was nicknamed Captain Cyborg. The reason: he voluntarily participated in Project Cyborg at the University of Reading, UK, to become one of the first people to become a real cyborg. The experiments started in 1998 when he had a simple transmitter implanted underneath his skin on his arm. Through this transmitter he was able to control various objects in his vicinity including doors, lights, and heaters. The next stage was more complex and involved the implantation of an electronic electrode array that was implanted directly into his nervous system and was connected to an external glove which contained the electronic support system. Through this implant he was able to control a robot arm further away to mimic the actions of his arm. Another interesting aspect was when this implant was connected to the internet at Columbia University, USA and from there he was able to control the robot arm at the University of Reading in UK. Perhaps what drew greatest publicity was when a similar though simpler array was implanted in the arm of his wife. The two electronic systems were able to communicate with each other, the first direct and purely electronic communication between two humans.

You might be apprehensive that such implants would cause damage to his nervous system or even his arm. So were scientists at the University. So they measured the impact on his arm and nervous system. Surprisingly there was no damage or interference of any kind nor was there any kind of rejection process at work. Indeed, nerve tissue was seen to grow around the electrode array enclosing the sensor. This just provides evidence for the plasticity of the human nervous system as discussed by Guy McKhann (2002) in his very interesting article, *Then, Now and the Future.*

embedded in his brain which in turn afforded him the capacity to perform simple direct communication with another implant embedded in his wife's brain (Warwick et al., 2003).

The last decade has seen augmented cognition research striking new frontiers. The *Defense Advanced Research Projects Agency (DARPA),* USA is working on a series of research projects with the aim of creating real-time computer interfaces for augmenting cognitive processes. Going a step further from the graphical user interfaces being used today, it is using complex psychophysiological data to determine the cognitive state of the individual and then adapting the computer interface to help the individual deal with the task on hand by augmenting the bottlenecks and glitches seen in the areas of attention, working memory, and executive functions of the brain. The beauty of the whole project lies in the fact that it is a multidisciplinary team consisting of not only researchers and academia but also people from the industry from fields as diverse as human factors and experimental psychology, neurobiology and neuroscience cognitive neuroscience, mathematics, and electrical engineering. Various companies such as Lockheed Martin, Boeing, Honeywell Labs, and Daimler-Chrysler are taking an active part in the project, the main aim of which is to develop interface systems that can be used by the military, even on the war front. Details of another project, the Google self-driving car is presented in Box 3.18.

Apart from the efforts of DARPA, there are other agencies such as the University Hospitals Eye Institute, USA which has already developed the first ever bionic eye

Box 3.18: Driverless Cars No Longer Science Fiction

The *Google self-driving car* is a project by Google that is involved in the development of autonomous electric cars. The software of the car is called *Google Chauffeur* and it uses detailed laser generated maps of the surrounding conditions to help it navigate itself.

By 2012, the project team had fitted a large number of cars including the Lexus RH450h and the Toyota Prius with the driverless equipment and most of them have been tested across various states in the USA (Lavrinc, 2012). They have also traversed the hair pin bends of the streets of San Francisco, gone across the Golden Gate Bridge and around Lake Tahoe. The best test was when a visually impaired person, Steve Mahan, was driven around in a driverless Toyota Prius car. He was taken to a restaurant, then to a dry cleaning shop, and finally home. And guess what, he is 95 % blind (Mahan, 2013)! You can watch this drive on Youtube!

In May 2014, Google unveiled a new model which was without steering wheels and pedals; a car that was autonomous in the true sense of the word. The team is very, very optimistic and in an article in the *International Business Times,* the team head claims that it will be ready for widespread use by the year 2020 (Halleck, 2015).

to be used to help people with Retinitis Pigmentosa, a condition characterized by the gradual loss of the ability to see. It works by converting video images captured by a miniature camera in the person's glasses into electrical impulses which are then transmitted to the remaining cells on the surface of the retina. These activations create patterns of light in the brain and the patient learns how to interpret these light patterns. The gadget called *Argus II Retinal Prosthesis System* is due for commercial release soon and has been approved by various countries. In fact, a keratoprosthesis consisting of a customized cylinder has been designed and even inserted into the right eye of a person who has not been able to see for the last 18 years. The prosthesis is named *AuroKpro* and the surgery was performed at the world famous facility, Aravind Eye Hospital at Madurai, India (Shastry, 2013).

Cognitive enhancement is also being targeted by scientists in the realm of **nanotechnology**. Nanotechnology, or the use of micro particles to perform more efficiently, functions that are hampered mainly because of the size of equipment (just think of how the chip is getting smaller and smaller, and what would be the advantages if it became miniscule), is currently being explored for a variety of purposes, hitherto not even thought of, one of them being ways by which to augment human cognition. Studies are already being conducted on the use of nanoparticles for therapeutic purposes. One such study has been able to stop cancer from reappearing by using nanoparticles to stream the anti-cancer drug directly to the tumor. Experiments on mice and even humans with breast cancer clarify that the nanoparticles killed the cancer stem cells and destroyed the tumor (Wei et al., 2015). But just imagine what would happen if the gray goo that can replicate itself unendingly were to be implanted into the brain of a person. If the person was a creative person, well and good, we would get more of Schumpeter's gales of creative destruction. But if he were a despot or if he were a schizophrenic, just think of the types of cognitive processes nanotechnology could become capable of creating?

***Nanotechnology**: the process by which micro particles are used to perform functions earlier performed by bigger gadgets*

3.6.1 Convergent Cognitive Enhancement

The twenty-first century has brought in its wake many new technologies that have been used to enhance a variety of cognitive processes. Pharmaceutical biotechnology has provided many ways of improving working memory and has even produced genetic changes in rats and mice (e.g., Tan, Liu, Koshiya, Gu, & Alkon, 2006). Research using the new technology of Transcranial Magnetic Stimulation (TMS) provides an easy yet effective method for improving performance on a wide variety of tasks in both people with cognitive deficits and healthy adults. TMS has been found to enhance performance of healthy adults on perceptual and motor tasks and even improve the executive function (Luber & Lisanby, 2014), on episodic memory tasks (Gagnon, Schneider, Grondin, & Blanchet, 2011) and on tasks involving motor skills (Boyd & Linsdell, 2009). People with Alzheimer's disease have also been seen to show improvements (Cotelli et al., 2011) as also those who have suffered from brain strokes (Emara et al., 2010). But, many of these are at the research stage and though they offer huge prospects, the risks can also be great. However, what has been achieved with considerable success is in the treatment of various mental disorders, especially for people with migraine, major depressive disorders, and even some cases of schizophrenia (Lefaucheur et al., 2014; FDA, 2013; Melkerson, 2008).

Even prior to the twenty-first century, in fact, all along the history of modern man, there have been in use many mental training methods and modern pedagogy relies on many such methods. The positive effect of collective enhancement has been brought to the fore through methods of group discussion and team work. Added to these would be the idea of PDAs, now fairly common and wearable computers such as the *Google Glass*, though it failed to click the way its developers had thought it would. One must, of course, mention the potential offered by nanotechnology, even though its use for cognitive enhancement is still in the nascent stage.

What is being visualized now is to go a step further and combine many of these methods such that the disadvantages of one can be offset by another method. Or, it may be possible to fine-tune methods by using a unified technology, whereby we combine pharmaceutical biotechnology, information technology, with current advances in cognitive neuroscience, medicine, and general health. To this we can add the effects of education and mental training, better diets, and enriched environments. The end result would be what can be called **convergent cognitive enhancement**. What Douglas Engelbart had written about in 1962 in his volume entitled *Augmenting Human Intellect* would be closer to culmination. He had defined the goal of augmented cognition as "increasing the capability of a man to approach a complex problem situation." One such attempt is detailed in Box 3.19.

> ***Convergent cognitive enhancement****: increasing cognitive ability by using a variety of processes developed by diverse disciplines*

It is, indeed, true that today's complex world and lifestyle require not only speedy comprehension, but also better comprehension and probably a new type of comprehension that can deal with the twenty-first century global village. This would be possible only if core competencies of the human being are enhanced and not by simply developing better and better computer hardware and software. We require enhanced psychological capabilities and this is not just a wild dream of an armchair philosopher but something that is very possible, seeing the current status of technological and neuroscientific know-how. While it is easier to increase specialized abilities than to increase general cognition, the rewards of the latter are much greater. As way back as in 1994, Herrnstein and Murray (1994) had estimated that a 3 % general increase in overall IQ could lead to the following: reduction in poverty rate by 25 %, males in jails by 25 %, high school dropout rate by 28 %, parentless children by 20 %, and welfare recipients by 18 %. Twenty years later, this increase in cognitive capabilities seems possible: technology has taken

Box 3.19: America's Next Moon Shot!

When an American political scientist Robert Horn conceived of the *Human Cognome Project* and suggested it to the *National Science Foundation*, USA in 2002, the scientific world took its first steps toward attempts at understanding the human brain. Akin to the *Human Genome Project* which tried to unravel the mysteries of the genetic structure of humans, this project envisaged the reverse engineering of the human brain in an effort to replicate the brain and thereby understand more about its structure and functioning, with special reference to cognitive functioning. While the project is still to be completed, a landmark is the release of a new brain atlas, entitled the *Allen Mouse Brain Connectivity Atlas* (2011).

An even more ambitious project and one that has been likened to John F Kennedy's Moon Mission of 1969, is *BRAIN*. Here is what NIH Director, Dr. Francis Collins writes:

> "Some have called it America's next moon shot. Indeed, like the historic effort that culminated with the first moon landing in 1969, the *Brain Research through Advancing Innovative Neurotechnologies (BRAIN) Initiative* is a bold, ambitious endeavor that will require the energy of thousands of our nation's most creative minds working together over the long haul. Our goal? To produce the first dynamic view of the human brain in action, revealing how it's roughly 86 billion neurons and its trillions of connections interact in real time. This new view will revolutionize our understanding of how we think, feel, learn, remember, and move, transforming efforts to help the more than one billion people worldwide who suffer from autism, depression, schizophrenia, epilepsy, traumatic brain injury, Parkinson's disease, Alzheimer's disease, and other devastating brain disorders."

Through this initiative, the NIH awarded $46 million to more than 100 researchers in 15 states and three nations and will involve four federal agencies, the NIH, the National Science Foundation, the FDA, and DARPA.

And he ends with the following words,

> "So, just as NASA kept the nation informed about the many successes (and a few notable failures) of the space flights that paved the way for the Apollo 11 moon landing, you can expect to hear a lot more from NIH and its partners in coming years as we gear up for our big goal: capturing a dynamic picture of the human brain—and enabling the use of such knowledge to advance human health" (Collins, 2014).

Source: https://directorsblog.nih.gov/2014/9/30

gargantuan steps and our understanding of the human brain and the ways in which our cognitive capacity can be enhanced has grown by leaps and bounds. All of this could be achievable through cognitive enhancement, especially by using, not single techniques, but by using convergent technology. There are, however, people who have reservations as far as the ethicality of cognitive enhancement is concerned

(Bostrom & Roache, 2008). Others have differing opinions. It seems that this step toward cognitive enhancement is just one more choice that we are wired to make. As Anderson (2012) puts it in an article in *The Atlantic,*

> "Using technology to enhance our brains sounds terrifying, but using tools to make ourselves smarter may be part of humans' nature."

Probably no one can frame it better than Buchanan, scientist, consultant, *Presidential Commission for the Study of Bioethical Issues*, and author of the volume, "*Better than Human: the promise and perils of enhancing ourselves*" (Buchanan, 2011). He claims that twenty-first century cognitive enhancement is not something that is new or out of the blue. In an interview to Ross Anderson (2012), he outlines his take on the issue. According to him,

> "Humans have done enhancement in the past by developing literacy and numeracy, and the institutions of science, and more recently we've done it with computers and the Internet" (from interview given to Anderson, 2012).

We are sure that few would disagree with Buchanan's statement.

Chapter Summary

The chapter starts with the human sensory and perceptual processes to clarify how our use of technology is restricted by these systems and the complex ways in which these have important implications for the development of man–machine interfaces in particular and use of things in general. For this purpose, in this chapter, we have focused on the visual and auditory systems, primarily owing to their dominance in our daily life. Issues related to the ever increasing complexity of technology on human information processing along with its consequential cognitive load have been addressed in the context of cognitive foraging, cognitive enhancement, and dysfunction. In continuation with Chap. 2, we have further elaborated in this chapter, the concept of Area Restricted Search as human beings seek and build choices not just biologically but also in the context of social preferences as illustrated through what has been come to be known as Schumpeter's gale. In other words, this chapter differs from traditional chapters presented in books on human factors engineering and on man–machine systems, albeit both deal with human artifacts. Technology, in this chapter, is viewed in a broader perspective, that is, in terms of both technical and social factors acting not only simultaneously but also interactively, whether it is in the design of the simple IBM logo or the complex Amazon website. The chapter closes with more details of the auditory system.

Suggestions for Further Reading

Baldwin, C. L. (2012). *Auditory cognition and human performance: Research and applications.* Boca Raton, FL: CRC Press.

Bor, D. (2012). *The ravenous brain: How the new science of consciousness explains our insatiable search for meaning.* New York: Basic Books.

Hills, T.T., & Butterfill, S. (2015). From foraging to autonoetic consciousness: The primal self as a consequence of embodied prospective foraging. *Current Zoology* 61, 368-381.

Poljac, E., de-Wit, L., & Wagenas, J. (2012). Perceptual wholes can reduce the conscious accessibility of the parts. *Cognition*, 123, 308-312.

Rothbart, M. K., & Posner, M. I. (2015). The developing brain in a multitasking world. *Developmental Review*, 1, 42-63.

Chapter 4
Technology and Motor Behavior: The Cinderella of Modern Psychology

> "When you're taking a walk around the block, your body is mostly on autopilot—you don't have to consciously think about alternating which leg you step with or which muscles it takes to lift a foot and put it back down. That's thanks to a set of cells in your spinal cord that help translate messages between your brain and your motor neurons, which control muscles" says Samuel Pfaff, a professor in Salk's Gene Expression Laboratory.
>
> He writes further,
>
> "Using optical methods to be able to watch neuron activity has been a dream over the past decade," "Now, it's one of those rare times when the technology is actually coming together to show you things you hadn't been able to see before."
>
> **Source: Salk Institute for Biological Studies. (2015, September 2). Scientists see motor neurons 'walking' in real time.**

"Scientists see motor neurons 'walking' in real time," unbelievable does it not sound? But it is true! The report above is based on a recent article in a 2015 issue of the journal *Neuron* (Hinckley et al., 2015) and describes just what these scientists have been able to do, a perfect example of the interplay between technology, neuroscience, and psychology.

4.1 The Importance of Motor Behavior

By using a completely new method, (again, thanks to technological advances), scientists at the *Salk Institute of Biological Sciences* and at other places have been, finally, able to decipher how complex movements are carried out. They knew that there are cells in the spinal cord collectively known as the locomotor Central Pattern Generator (CPG). But what was not clear was how these cells managed to coordinate with other motor neurons to execute even a simple task such as walking. When your cognitive system gives instructions to walk away to avoid someone or to remove your hand from a hot stove, these are very simple signals. Not only must the body walk or the hands move, but the muscles must move at a certain speed and use

V.K. Kool, R. Agrawal, *Psychology of Technology*,
DOI 10.1007/978-3-319-45333-0_4

just the right amount of force. For the movement to be accurate, computations have to be made by the CPG. Specific neurons have to be moved at a definite speed and muscles contracted using a specific force. How is all this accomplished? The manner in which the connections are forged by the CPG was far from clear. But, this has now become possible and "scientists can see motor neurons walking in real time."

Exciting it is but, while on the one hand, technology helps in the understanding of motor behavior, it may actually be hindering motor development in children of this tech-powered era. Here is what occupational therapist and biologist Chris Rowan has to say,

> "Reminiscing about the good old days when we were growing up is a memory trip well worth taking when trying to understand the issues facing the children of today. A mere 20 years ago, children used to play outside all day, riding bikes, playing sports and building forts. Masters of imaginary games, children of the past created their own form of play that didn't require costly equipment or parental supervision. Children of the past moved… a lot, and their sensory world was nature based and simple. In the past, family time was often spent doing chores, and children had expectations to meet on a daily basis. The dining room table was a central place where families came together to eat and talk about their day, and after dinner became the center for baking, crafts and homework" (Rowan, 2009, p. 1; 2013).

She goes on to narrate how technology has changed all of this. The fast paced life of the nuclear family with dual bread earners may bring in comforts, gizmos, and gadgets, but at the cost of disintegrating the very fabric of the family. No longer do families spend time together; entertainment from infants to grandparents depends on the iPad, internet, TV, or video games.

Would this not impact the growth and healthy all-round development of the child? Neither the sensory-motor nor the emotive systems of the child have evolved for such a sedentary lifestyle. As a result, behavior problems among children are on the rise, child obesity and diabetes, once unheard off, is becoming increasingly common, and syndromes such as ADHD (Attention Deficit Hyperactive Disease) are becoming rampant.

It seems to be clear that not only is motor behavior necessary for the all-round balanced development of the child, but also that technology could become a devil in disguise: helping mothers to find more time for themselves but at the same time, keeping children sedentary and hindering their motor development. Yet, the irony is that we cannot and should not attempt to stop the evolution of technology. Instead, we need to become the "master" and not the "slave" of technology.

This is where psychology of technology comes in, and, in this chapter, we focus on the psychology of motor behavior and what it has to offer to psychology of technology. As one goes through psychology books, research articles, conference proceedings, and even college and university courses, it becomes increasingly clear that cognition and the cognitive system have been disproportionately represented while motor behavior has remained on the back burner, much akin to the fairy tale character, Cinderella, ignored and abused as she was. This is the reason why the psychology of motor behavior is often referred to as the Cinderella of modern psychology (Rosenbaum, 2005). While mainstream psychology may have ignored motor behavior, psychology of technology would be very much incomplete without clarity about the psychology of movement.

"A thing of beauty is a joy forever." So goes the adage and is probably true for a wide range of objects in our environment. A Van Gogh painting, an immaculately woven Persian carpet, a Roman sculpture, or even a rose are objects that one longs to own only to be able to gaze at it. But think of your latest Lexus, your iPhone 6 Plus, or any other gizmo, you would probably want to do more than own it, simply to be able to gaze at it. Even before you buy the Lexus, you would want to test-drive it. Before you buy the latest iPhone, you would want to not only know about its latest features but also be able to hold it in your hand and operate it. A wheelchair may be a thing of beauty for its designer, but would be liked by a person with motor impairments only if she finds it useful and easy to operate. This is true for all technology, for that matter. Technology is meant to be used; it is not developed simply as a thing of beauty.

Any tool or any technology can be used only if we humans are able to hold it, manipulate it, and navigate with it. Technology necessitates the user to move, whether it is her fingers, her eyes, her hands and feet or even the whole body. Be it the traditional hammer, the horse-drawn plow, a bicycle, a supersonic jet plane or any of the computers, tablets and smart phones that we are seeing today, they would all be useless if they were not devised keeping our human body and musculature in mind. Our sense organs may be able to sense them and our brain may be able to cognize them but in the end these would be possible only if they fit into our hands or whatever part of the body is to operate them. As mobile screens become more cramped with multitudinous apps and their icons, extremely fine finger movements have become necessary. As man takes strides in outer space, the ability to maneuver in an environment of weightlessness needs to be enhanced and its after-effects countered. The ways in which lessons from psychology of motor behavior could help train astronauts has been detailed in Box 4.1.

Many a time, we do something so automatically that we do not even think of how we are doing it. If you are asked to randomly locate the position of letters on the keyboard, you will take more time than in typing those letters organized in words. Try to think where the letters "t" "h" and "e" are located on your keyboard and now type "the" and see the difference. Similarly, while playing a musical instrument, as soon as we know that we have made an error, we try to correct it by stopping the movement of our finger, but it's too late. It is because the practiced movement runs faster than the speed of correcting the movement. In our everyday life, we pay a lot of attention to "how we will do it," that is, regarding the thinking and the planning aspects, but not so much about "what will happen when we do it," that is, the damage to the apparatus, namely, the body, that is engaged in the performance of the act. So, we keep on typing the pages of this book without realizing that it would damage our fingers in the long run.

While ease of operation is a consideration that is exceptionally important for business purposes and manufacturers spend a lot of money and time to market such user-friendly products, the body mechanisms involved in the use of things have, hitherto, not been very well explored by psychologists. They have been intensely preoccupied with the ways in which we acquire knowledge and understanding about

Box 4.1: Motor Imagery: Applying Lessons Learnt from the Psychology of Movement to Space Science

One of the major problems besetting astronauts is adapting to weightlessness while in space and readapting to gravity once they come back to earth. The former is normally managed through extensive training including underwater training and parabolic flight training, helping them to adapt to symptoms such as sensorimotor disturbances and postural instability while in space. However, what happens when they are to return to earth and once again adapt themselves to gravity? Can such training be given in-flight?

A group of scientists from sports institutes in Germany and France have suggested a novel method: that of using mental imagery and mental practice, much as is already being used, and, very successfully in sports training. Outlining the utility of the training program in a paper in a 2015 issue of the journal, *Frontiers in Systems Neuroscience*, Bock, Schott, and Papaxanthis (2015) write,

> "Mental practice could be applied to prepare astronauts not only for the sudden onset of gravity upon landing, but also to the sudden absence of gravity after launch. ... Mental practice could offer an expedient alternative or supplement to those established methods. Likewise, mental practice could be applied to prepare astronauts for the landing on and the launching from celestial bodies other than the Earth. Finally, MP could be used as countermeasure for decrements not only of motor skill, but also of muscular force" (Bock, ibid, p. 3).

The model that they propose to use is PETTLEP, the letters of which stand for the seven critical aspects on which not just resemblance but complete congruence is required. These aspects are the **Physical**, **Environment**, **Task**, **Timing**, **Learning**, **Emotion,** and **Perspective** aspects.

The training program would be undertaken in two stages, the first taking place on earth, in which they would carry out their normal activities and would then be asked to practice them mentally. The second phase would start several days before landing and would consist of a number of sessions in which the astronaut listens to certain scripts based on their normal activities on earth and mentally imagines the sensations, observations, and even emotions accompanying these activities. However, they do not actually perform the tasks. It is expected that such mental imagery of tasks which had been mentally practiced would reduce their problems post-landing. "For example, practice of the "TUG" script could enhance astronauts' ability for standing up and walking without assistance after landing, and the benefit could generalize to other postural tasks as well" (Bock et al., 2015, p. 3). The authors of the plan envisage that such mental practice could be used to prepare astronauts for a variety of tasks ranging from adapting to sudden absence of gravity after launch of the spaceship to landing on or launching from other celestial bodies.

Source: Bock et al. (2015)

things, either through self-report or other forms of observed behavior, but have failed to pay equal attention to what happens between the executive orders of the brain and the observable behavior. Commenting on the significance of the application of our knowledge of motor control in technology, Schmidt and Donald (2005) wrote,

> "Many industrial skills, such as using a lathe, typing, woodcarving and handwriting, are of critical importance to this field of study" (p. 6).

In a very stimulating article published several years ago in the *American Psychologist*, David Rosenbaum (2005) expressed concern over the bias initiated by prominent psychologist B. F. Skinner and other behaviorists in contemporary psychology. For a behaviorist, it is enough to know the connection between the origin of any behavior (the stimulus) and its termination (the response). Rosenbaum noted that although from the founding father of modern psychology, William James, to contemporary British psychologist, Donald Broadbent, many psychologists had felt the need to study the topic of motor behavior, not much has been accomplished in this area as compared to the rapid growth of psychology in other fields.

4.2 Technology and the Nonexecutive Functions of Our Body

The muscles of our body bear the load of all the hard work involved in using gadgets of a wide variety—be it lighting a cigarette to playing a computer game to operating a heavy machine. Muscles are part of the motor system and along with our joints and tendons have an important role to play, particularly in positioning our body to perform tasks effectively and at the same time, efficiently. Basically, the motor system consists of three major parts: the muscle fibers, the motor units and areas in the brain and spinal cord which control such movements.

4.2.1 *The Muscle Fibers*

> ***Muscle fibers****: tube-like structures which join together to form a muscle*

There are about 600 muscles within our body silently performing tasks that range from pumping blood throughout the body to overt activities such as washing our car. Basically, tiny fibers join together to form a muscle that may be compared to elastic tubes very similar to the common rubber band. Muscles are spread all over our body and are of three types: s**mooth muscles** that conduct involuntary functions of our body, for example, muscles of the stomach, bladder, and digestive system. They function without our actively being conscious about it; **cardiac muscles**, as the name suggests, manage the flow of blood in our heart; and **skeletal muscles**, varying in size and shape, perform the voluntary work of the brain and enable us to move different parts of our body. Attached to the bones, muscles provide power and strength to our movements. These bones and muscles are further connected to each other by tendons, made of tough tissue.

4.2.2 The Motor Unit

***Motor unit**: consisting of three parts, an axon, a synapse, and an end plate*

***Muscle recruitment**: the number of muscle fibers called into play to perform a particular movement*

This consists of a synaptic junction at the ventral root of the spinal cord, a motor axon, and a motor end plate in the muscle fiber. The number of muscle fibers under each motor unit varies and may range from as few as three to as many as 2000 fibers depending on how fine a movement is required (did you know that even the simple smile involves the operation of an extremely large number of muscles?). Different movements call into play a different number of muscle fibers and this phenomenon is known as **muscle recruitment**.

4.2.3 Areas of the Spinal Cord and the Brain

Important areas in the brain that help in the execution of movements and its control are depicted in Fig. 4.1.

- **The cerebrum** in which the frontal lobe contains the motor cortex. A feature of the motor cortex is that finer movements such as those of the hands and the tongue have a proportionally larger representation in the brain than gross movements such as those of the legs.
- **The basal ganglia**, which is a group of neural centers located in the cerebrum and regulates voluntary movements. As mentioned earlier, it plays a vital part in

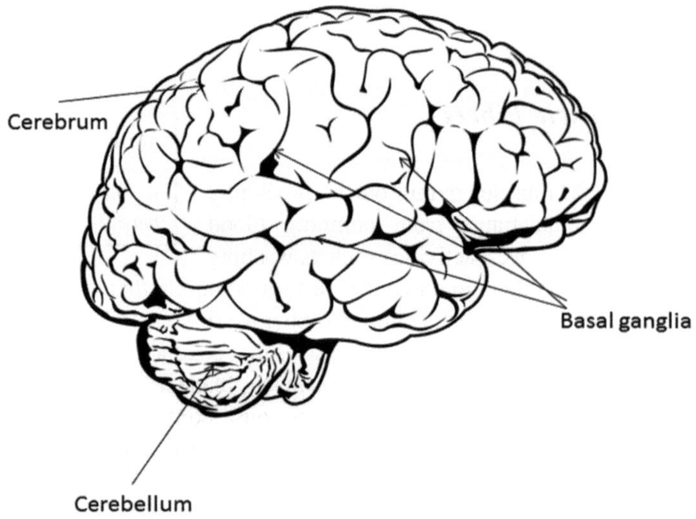

Fig. 4.1 Parts of the brain involved in motor behavior. Adapted from How StuffWorks (2005)

perceptual-motor integration, organization of movements, and in regulating its amplitude.
- **The cerebellum**: Consisting of 30 billion neurons, this second largest part of the brain receives inputs from sensory neurons connected to muscles, joints, and tendons and plays a key role in coordinating movements in communication with the basal ganglia and the motor cortex.

Thus, the physical machinery underlying every act of movement is possible only when all of the above three aspects are intact. Dysfunction of this system of the body is caused whenever there is damage to any of these three interrelated aspects.

So important is movement for life and survival that the vertebrate brain has evolved accordingly. The brain has evolved in order to govern motor activity with the basic function of transforming sensory inputs into patterns of coordinated movement. This is clear from the ways in which a perception-action coupling is seen, such that perception is the means to an action and action is the means to perception.

Perception ←=====→ Action

4.3 Role of Dopamine in Motor Behavior

In the last two chapters, we had focused on the role of dopamine in various aspects of cognitive behavior and especially in reward seeking and making choices as in the case of Area Restricted Search. Recent studies in neurosciences show that dopamine is important in motor behavior too, and surprisingly, this is so for both vertebrates and invertebrates. Animals with certain mutant genes leading to defective dopamine synthesis show differences in motor behavior. They show large fluctuations in their locomotion because they are not able to make due adjustments in their speed based on the distance between themselves and the target. That dopamine is responsible is clear because this abnormality is rectified with dopamine treatment (Omura, Clark, Samuel, & Horvitz, 2012). The results also suggest that animals and humans might share a common mechanism by which dopamine restricts activity level and coordinates movement. Dopamine in the basal ganglia is also responsible for calculating the cost of reaching movements, while its projection to the striatum provides a signal for what has come to be known as "**motor motivation**" (Mazzoni, Hristova, & Krakauer, 2007). In a study on people with Parkinson's Disease (PD), who generally show a slowness of movement known as bradykinesia, Mazzoni and his colleagues noted that even though the PD patients were as accurate in their movement as those without PD, they did not seem to have the motivation to perform them. This reluctance to move faster could be because of the high energy cost of making those movements. These PD patients manifested a shift in balance between perceived reward of arriving at the target quickly and the amount of effort or energy required to achieve the speed of the movement.

> ***Motor motivation****: the drive or urge to perform a movement*

It seems that the motor system has its own motivational circuit that operates much in the same way as explicit motivation, though independently. Even more interesting is the fact that this motor motivation works outside the realm of conscious awareness of the person and has therefore been named **implicit motor motivation** governing spontaneous and automatic speed of reaching behavior. It is almost as if there was a neuro-economic scheme which studies the reward and cost of speed of movement and regulates it accordingly. Thus, speed of movement depends not only on distance and size of the target but also upon dopamine levels.

> ***Implicit motor motivation****: an urge to perform a movement about which a person is not conscious*

4.4 The Psychology of Movement

4.4.1 Motor Development

The child at the time of birth is born with the above motor apparatus and for the first few weeks, though seemingly still, is able to make those movements that are necessary for survival. He may not be able to move his legs, leave alone walk, he may not be able to raise his hand when you call out to it, but he is able to move his eyes, he is able to blink his eyes; if something is put into its mouth, it starts sucking on it. It is able to breathe, and his heart muscles function to pump blood throughout the body. So, while skeletal muscles may not be able to move, cardiac and other smooth muscles in all the internal organs are definitely operational. Because such movements are normally not under voluntary control of the infant and they are called **reflex movements**. As the child grows physically, the skeletal muscles also grow stronger and gradually all types of hand and leg movements become possible, at first randomly but later, very much in the control of the child. In fact, indices of both **kinetics** and **kinematics** change with age and are a clear sign of the motor development of the child. While some of these developmental changes are maturational in nature, a large part of them are learnt patterns.

> ***Reflex movements****: movements that we are capable of performing from birth though not under the voluntary control of the person*

> ***Kinetics****: study of movements based on the force applied at the source of the movement*

> ***Kinematics****: study of the mechanics of movement*

4.4.2 Motor Learning

That motor learning does take place is evidenced by the fact that with practice, a small child becomes proficient at various types of movement. It is well known that

> ***Motor learning****: learning how to move different parts of the body*

keeping a child from moving his limbs delays the onset of walking. Talking of locomotor skills, we see a clear progression from turning on one's back to crawling and then to walking, running, jumping, and so on. The same is the case with talking. Sounds that were due to the random movement of air in the vocal chords of the 6-month-old soon give way to clear language patterns. The fact that all this is due to learning is clear. You enable the child to rehearse the movement and he soon becomes skilled at it.

Almost all motor learning (except for those seen in young infants and toddlers) appears to progress through three stages.

The first stage is the cognitive phase in which the individual thinks about how best to make the movement. Motor cognition involves a number of cognitive processes including the process of motor imagery or the mental rehearsal of a movement before it is actually made and motor priming or learning by watching, much like the vicarious learning posited by Bandura (1977).

Motor learning takes place in three stages: *cognitive phase, associative phase, and autonomous phase*

The associative stage is the second stage during which the child becomes capable of making those movements in a more effective manner. While she fumbled in the beginning and could press that button on the cell phone, only accidently, she could later control her movements such that she hit the right button at the first go.

This brings in the third stage, or **the autonomous stage**, in which movements that once required conscious effort become automatic. Gradually these movements will not only become more effective (that is, enabling the child to do what it wants to do) but also more efficient (that is by using fewer muscles, or smaller muscle recruitment). This is when we can say that a **muscle synergy** has developed. Some optimal synergies are task independent and arise irrespective of the context of a task; others are task dependent and arise in the context of a task. We use both of them to adapt to the environment (Chhabra & Jacobs, 2006).

Muscle synergy: *performing movements by using fewer muscles*

Remember your first driving lessons! Comparing your driving during those days to your driving today would enable you to envision each of these phases in learning how to drive.

4.4.3 Motor Control

Motor control: *set of processes that enable creatures to stabilize or move parts of the body*

The psychology of motor control is the study of how living beings are able to control their movements (Schmidt & Lee, 2005), or, as Rosenbaum (2005) puts it, it is a "set of processes that enables creatures (living or artificial) to stabilize or move the body or physical extensions of the body" (p. 308). Empirical work on special populations, for example, those with visual impairments (Kool, 1980; Kool & Singh, 1988) provides valuable insights regarding motor control.

Discrete movements: *specific movements that have a beginning and an end*

Continuous movements: *movements that are ongoing*

Two types of movements have an important bearing on our understanding of the psychology of technology: **discrete movements** and **continuous movements**. The former are generally involved in motor behavior such as reaching for targets and require precision. Consider moving a pencil toward the tip of your nose as fast as possible and bring it as close as possible without touching the nose. Considerable motor control is needed as your hand reaches near the tip of nose; otherwise you might hurt yourself. You may notice that such movements require a posture orientation with initiation and termination of a response. **Continuous types of movements** involve activities that are ongoing, for example, control of a moving object using a device. In their research on monkeys involving a brain–machine interface (BMI), Hatsopoulos, Joshi, and O'Leary (2004) noted that different brain mechanisms are involved in regulating these two types of motor behavior: the primary motor areas of the brain being more active in managing continuous movements while the dorsal premotor area is more effective in regulating discrete movements. They further argued that:

> "Motor control can be hierarchically characterized as the selection and planning of discrete movement classes and/or postures followed by the execution of continuous limb trajectories" (p. 1165).

A number of highly skilled tasks such as the operation of gadgets require both discrete and continuous movements. For example, our taking notes (handwriting) or moving the cursor to track a reference requires a combination of both discrete and continuous movements.

Research in the area of motor control shows that a lot of what the brain does is based on the feedback provided by the muscles in our body. When we trip, we quickly balance ourselves or if something slips out of our hand, we try our best to retrieve it.

The question before us is: did the manufacturer think of these muscles and their limits when designing a gadget? Did we, as customers and consumers of that technology, think of the effect the gadget would have on our hands, shoulders, or our feet? We may love our iPhone, but how much do we like its keyboard? Is it easy to operate? The idea that I have an iPhone (material possession) with a keyboard, albeit of any size, is a good enough reason to underscore the threat caused to our "mindless" muscles. Are muscles simply subservient to the executive brain and perform mechanically without alerting the brain about its correct or incorrect decisions?

If this were so, we would never be able to excel in any sport. When a high jumper fails to clear the bar, she immediately tries to think back as to what was the reason for this. She also tries to plan for her next jump and adjusts her posture accordingly in a way that would increase the chances of clearing the bar. What could be behind such adjustments? This is the much debated and yet little understood aspect of motor control.

4.5 Systems of Motor Control

4.5.1 Open Loop Motor Control System

> ***Open loop hypothesis**: the idea that attention is required only for the initiation of the action*

One of the earliest descriptions of motor control was provided by William James as far back as in 1890 (James, 1890). He postulated an **open loop hypothesis** according to which attention is required only for the initiation of the action. Once the action is started, there is nothing that can change either the direction or the speed of the movement. So, if there are unexpected changes in the environment the person will not be able to change the course of his action. Of course, many an action is explained by this hypothesis. For example, the cause of many a road accident is that the driver was not able to change his action when a pedestrian or even a stray deer appeared suddenly on the road.

> ***Motor program**: the program through which our brain controls our movements*

In other words, William James was pointing to an abstract **motor program** controlled by the executive functions of the central nervous system and allowing the brain to anticipate, plan, and guide movement.

There are various lines of evidence supporting the idea of motor programs. First and foremost, ongoing modulation of rapid movements is near impossible seeing the speed of afferent processes. Secondly, the positive correlation between reaction time (RT) and movement complexity clarifies that even before the movement is initiated, some forward planning has taken place. And thirdly, movement is possible even without feedback from the moving limb (for example, when visual feedback is denied).

At the same time, the incompleteness of this hypothesis is clear if you ask a batsman in cricket as to how a "late cut" is executed.

4.5.2 Closed Loop Motor Control System

Near about the same time that James was formulating his open loop theory of motor control, Woodworth (1899) started studying rapid arm movements by measuring the time taken and the degree of precision achieved. He noted that there were two components to each movement:

1. An initial impulse phase (similar to RT)
2. A current control phase which relied on visual feedback to make corrections

He also found that movements with a stronger initial phase are faster but less precise than movements which had a weaker initial impulse phase. At the same time, faster movements resulted in increased errors.

This hypothesis that includes a current control phase is what came to be later known as the **closed loop theory** (Adams, 1971) that focused on slow, graded tasks wherein error detection and correction to meet environmental demands is possible. Adams was of the view that the motor program consists of two states of memory: one, which is akin to the verbal recall stage and determines the early stages of action execution. This memory state is strengthened by practice and feedback about movement outcome in a Hebbian manner based on the reverberation of neuronal circuitry. The second state is perceptual in nature and is based on sensory feedback from past performance regarding the position of the limb. Incorrect positions are corrected and each time a perceptual trace is collected and retained for further use in an iterative fashion. While this was an important step in the understanding of motor behavior, the drawback is that it presupposes a multitude of programs (one for each movement executed) to be stored in our limited capacity memory store. It also fails to explain the performance of novel movements or motor equivalence (whereby we may perform the same act using a different coordination of muscles). Consequently, the notion of a **Generalized Motor Program** (GMP) was developed which was said to be constant for a class of movements with smaller ones being added for specific movements. Further analyses by Richard Schmidt (1975) revealed that the stored motor program comprises four aspects, namely,

Closed loop theory*: attention is required at both initiation stage and during the ongoing process with feedback loops enabling correction of action*

Generalized Motor Program*: a brain program constant for a class of movements*

1. The **initial conditions** of the movement (postural information regarding the body as a whole and the limbs in particular).
2. The **response specifications** for the motor programs, such as speed and force of movement required.
3. The **sensory consequences** of the movement response, that is, how the movement felt and looked, or even sounded.
4. The **outcome** of the movement along with the knowledge of results.

In other words, motor learning and control is a continuous process that has schemas updated each time a certain movement is performed (Fig 4.2).

4.5.3 *Helping People in the Use of Prostheses*

One prime example of closed loop motor control systems is the human grasp. Even infants are seen to grasp anything coming near them. Most mothers will remember their amazement at watching the infant grasp the mother's finger and pull it to its

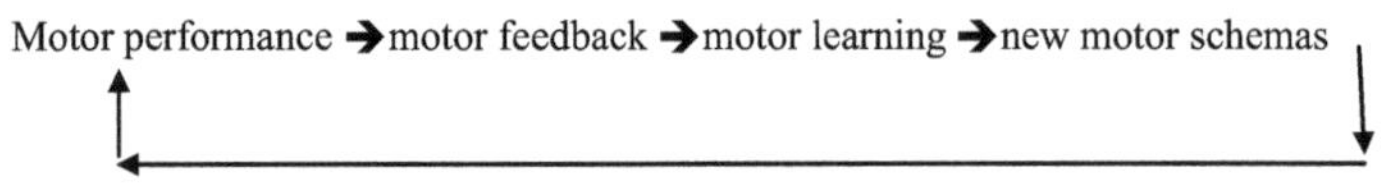

Fig. 4.2 Flowchart for closed loop theory of movements

mouth. Or, watch toddlers grasping anything and everything, whether it is a tiny insect or the odd shoe lying around. The psychology of motor behavior has established empirically that infants move from global to more localized grasping: from a **palmer grip** in which they use their entire palm to grasp even a small object and so sometimes failing in their attempts to the finer **pincer grip** which enables the fine muscles of the fingers to move in coordination to pick up tiny objects, as tiny as an ant.

***Palmer grip**: initial attempts by the infant in grasping objects by using the entire palm*

***Pincer grip**: enables grasping using the fine muscles of the fingers*

Grasping is one of the most remarkable of human capabilities. We are not only able to grasp objects but also lift them and manipulate objects of various textures, weights, and sizes. How do we manage to do it? Do you find yourself trying to lift something and then noticing that you have to close your fingers in tighter as the object is smaller? Yes, if the object is slipping from your hand, you do try to grasp it in such a way that it does not slip. Imagine how difficult it would be if we had to adjust hand strength each time we picked up something. This is obviously possible through closed loop motor control based on both feed-forward and feedback mechanisms. The former leads to economical grasping through which we are able to maximize force expenditure while ensuring that the object does not slip away, while the latter is saved as a motor program to serve us on subsequent events.

A seemingly simple task performed instantaneously and what seems effortlessly becomes extremely difficult for those using artificial limbs. A variety of empirical findings support the use of closed loop motor control systems in the use of artificial limbs and is clear evidence of how psychological findings and knowledge regarding movement control have helped in the design of prosthetic aids. Some examples include studies such as those by Pylatiuk and his colleagues (2006) regarding the design and evaluation of a low cost force feedback system for prosthetic hands; the testing of a prosthetic haptic feedback stimulator (Chatterjee, Chaubey, Martin, & Thakor, 2008); studies on the role of feedback and feed-forward closed loop prosthetic control (Saunders & Vijayakumar, 2011); and even for the training of people using prosthetic aids so that they are able to grasp better (Stepp, An, & Matsuoka, 2012). Today, such a feed-forward pathway between the brain and the limb is normally delivered through mechanical devices such as vibrator motors or pressure cuffs, commercially available in prosthetic aids such as *the Touch Bionic i-Limb* (2013) or even the *Otto Bock Sensor Hand Speed* (2013). Another example is presented in Box 4.2. A recent collaboration between scientists from Serbia and Germany (Jorgovanovic, Dosen, Djozic, Krajoski, & Farina, 2014) has attempted to develop a system using electrotactile feedback and the results of their study have been published in the journal, *Computational and Mathematical Methods in Medicine*. In the words of the authors of the study,

> "Closing the control loop by providing somatosensory feedback to the user of a prosthesis is a well-known, long standing challenge in the field of prosthetics. Various approaches have been investigated for feedback restoration, ranging from direct neural stimulation to noninvasive sensory substitution methods. Although there are many studies presenting

closed-loop systems, only a few of them objectively evaluated the closed-loop performance, mostly using vibrotactile stimulation. Importantly, the conclusions about the utility of the feedback were partly contradictory. .....We have developed a realistic experimental setup for (using electrotactile) virtual grasping, which operated in real time, included a set of real life objects, as well as a graphical and dynamical model of the prosthesis. We have used the setup to test 10 healthy, able bodied subjects to investigate the role of training, feedback and feedforward control, robustness of the closed loop, and the ability of the human subjects to generalize the control to previously "unseen" objects. Overall, the outcomes of this study are very optimistic with regard to the benefits of feedback and reveal various, practically relevant, aspects of closed-loop control" (Jorgovanovic et al., 2014, p. 1).

Box 4.2: Training Upper Limb Amputees in the Use of Prostheses

"We wanted to see if there was something we could improve in therapy that helps amputees—something to refresh the rehab," said Lewis Wheaton (2015), an associate professor who led the study and directs Georgia Tech's Cognitive Motor Control Lab. "If people with a prosthesis can't figure it out in the first 3 days, they tend to give up."

Earlier research had clarified that 75 % of amputees consider their prosthesis to be primarily aesthetic and 33 % reject the device, mainly because they feel that it is impractical. In fact, if the first experience is dissatisfactory, most amputees give up on the prosthetic aid.

So what did Wheaton and his colleagues find? They conducted a study with people wearing elbow-to-hand prostheses with movement sensors embedded onto the elbow and an EEG cap. On the first day, they practiced various movements using the device followed by 3 days of watching videos with people who looked like them but were either wearing the same device or had nothing on their arms. On the fifth day, they were again asked to reproduce the same movements. The findings, published in the journal *Neurorehabilitation and Neural Repair*, show that

"Those who watched a matched-limb participant did significantly better after 3 days of training," said Wheaton, a faculty member in the School of Applied Physiology. "Their arm movements were more consistent and fluid when they repeated the task. Those who only watched someone without aprosthesis didn't improve at all."

More important, Wheaton noticed clear differences in brain activation of the two groups of individuals, with greater activation of the motor area for those who watched videos of people wearing the device, almost as if they were mentally rehearsing the movements.

Source: Cusack, Thach, Patterson, Acker, Kistenberg &Wheaton (2015); **www.sciencedaily.com/releases/2015/10/151008173519.htm**

Another question that begs an answer is whether we can consciously choose between open or closed loop control systems. Some insight has been obtained from studies that have analyzed the time components of a typical muscle movement. It is seen that a typical muscle movement takes about 200 ms comprising sensation (about 15 ms), perception (about 45 ms), response selection (about 75 ms), and response execution (about 15 ms). If each S-R adjustment takes about 200 ms and if we make ten such adjustment iterations, it would involve 2000 ms, or 2 s. This is often too long a time period for ongoing action that requires split second decisions. This is probably the reason why our motor control systems have evolved in a way in which we can perform each movement either with or without active, ongoing motor control. When demands for accuracy are high and there is sufficient time, we tend to use closed loop motor control mechanisms. You are driving and a pedestrian suddenly appears in front of the vehicle. There is absolutely no time to think and plan how to execute an action such that there is minimal loss. We simply perform the action automatically based on our stored motor programs using an open loop control mechanism. We rely then on what has been called "**muscle memory**." Thus, while the former is slower but produces greater accuracy, the latter is fast but may go off the target. This is the reason why open loop controlled movement is also called **ballistic movement**, that is, just as the speed and the direction of a ballistic missile cannot be altered once it is set off, in much the same way, the course of action for an open loop movement is predetermined, with no scope for any feedback, visual or proprioceptive. This is the distinction between the fast saccadic movement of the eye (the eye blink) and the slower pursuit or tracking movement of the eye (as when we are reading) (Goldberg, Eggers, & Gluras, 1991).

> ***Muscle memory***: *stored representations of movements made earlier*

> ***Ballistic movement***: *type of movement which cannot be altered once it has been initiated*

This possibility of alternating between open and closed loop movement, in fact, lies at the core of motor control. If we were to keep monitoring each of our muscle movements, it would become an extremely wearisome task. Normally, each new type of movement is based on closed loop motor control employing more and more open loop control as proficiency increases. Psychologists have been interested in learning about the ways through which closed loop motor control can be decreased, whereby the executive brain gets more free capacity to perform more crucial tasks. One factor that tends to deter open loop learning is the visibility of the limb or even visibility of the target that is to be moved (for example, a lever arm). Such visibility helps closed loop control so very necessary during the initial stages of movement learning, but if the limb or lever arm continues to be seen by the person, it impedes open loop learning (Sulzenbruck & Heuer, 2012). Cognitive neuroscience data, too, confirms that explicit verbal information disrupts implicit motor learning after a basal ganglia stroke but surprisingly not after a sensory-motor stroke (Boyd & Winstein, 2004)

This distinction is important for video game console designers. Gamers often play such games even under limited illumination conditions once the person has become an expert at manipulating the buttons on the console, that is, in using open loop control systems. This means that if the controls of the game were to be altered after each trial, the player would have to continue to rely on closed loop learning which, in turn, would slow down the entire process of gaming.

4.5.4 Intermittent Motor Control Systems

> ***Intermittent control****: use of a combination of closed and open loop motor control systems*

But are there really two distinct categories of motor control? Does the human brain choose between one and the other in a clear dichotomous manner? Vicario and Ghez (1984) are of the view that even extremely fast open loop movements may not be completely unmonitored and may manifest some type of overt correction known as **intermittent control** (first conceptualized by Craik as far back as in 1947 (Craik, 1947/1948). Real-life, pure open loop motor control may not even exist as seen from electromyographic signals of even extremely rapid movements (Gordon & Ghez, 1987). If one considers the task of writing, it is normally open loop as far as skilled writers are considered, changing to the closed loop type of motor control when accuracy demands increase. Some 60 years after Craik-Loram and his colleagues (2006) have indeed reported that continuous feedback is not the only feedback mechanism for motor control. Buchanan and his colleagues (2006), too, noted that as the task becomes more difficult, movement control shifts from preplanned cyclical control or open loop control to more ongoing control or closed loop motor control. However, has the science of engineering motion control given this serious thought? One example of this neglect of psychological principles would be in the case of laptops and computers where typing on the QWERTY keyboard becomes slower when illumination is low. This is simply because the task becomes much more difficult and after each press of the key you tend to look at the screen to see whether you hit the right keys. How much easier it would be if all the computers or laptops had backlit keyboards so that your reliance on external illumination decreases? But, it is not so and you still have to pay extra for having a backlit keyboard on your computer or your phone. Of course, in the case of the latter, the number of people affected by this problem has decreased due to the introduction of the touch screen. But even today there are countless people who continue to use the traditional cell phone with a keypad. Here is an example of a known psychological principle not paid much heed to by designers and manufacturers of technology.

Of greater concern is the introduction of gesture typing using the QWERTY keypad on touch phones. Engineers attempt to introduce innovative ideas into gadgets mainly to build competitive edge, but sometimes fail to realize that these ideas may create problems for the user. We are used to the QWERTY keyboard by tapping keys. However, when gestures are to be performed using this same keyboard, what had become open loop controlled once again needs to rely on closed loop control, slowing down the entire process. In a collaborative study, Brian Smith of Columbia University and two scientists from *Google Inc*, namely, Xiaojun Bi and Shumin

Zhai, have attempted to discuss the problems associated with gesture typing and how they can be resolved (Smith, Bi, & Zhai, 2015).

Even neurological data seem to support such an intermittent model of motor control. Medina, Jax, and Coslett (2009) have found evidence for a two-component model when subjects performed reaching tasks. According to their data, both feed-forward and feedback mechanisms are used to ensure accuracy of movement, with an initial ballistic type of movement guided by efferent copies maintained in human memory and a terminal stage in which sensory feedback is essential for accuracy. According to the authors,

> "These internal models store experience based information about the input and output characteristics of the motor apparatus to provide feed-forward signals to muscles and in turn use these feed-forward signals to estimate current and future location of the effector. Thus internal models are not static but are constantly being updated to ensure accurate functioning" (p. 222).

However, when we consider people with Multiple Sclerosis (MS), a very different picture emerges. These people are able to write in a very legible fashion if they are asked to scribble but when they are provided visual feedback, their writing becomes very slow, showing a shift from open loop writing to a closed loop one. Why is this so? It appears that the use of closed loop control is different for MS subjects as compared to their control group counterparts. Whereas, the latter use closed loop control when accuracy demands increase, the former use it to compensate for the newly acquired sensory-motor deficits (Schenk, Walther, & Mai, 2000). Just as insight regarding the visual system and visual perception is obtained by studying these processes in people with visual impairment, so too, we can understand more about normal writing processes from the writing processes used by those who have neuromotor deficits. Another problem that has been seen to follow the same pattern is that of **writers' cramps**, commonly noted in people who type on the computer over long periods. People who have been suffering from writer's cramp for as long as 7 years have been found to be able to return to normal fluent writing after training periods in which they were asked to scribble (Mai & Marquardt, 1996). So, here is a strategy to come to the rescue of those people who start getting finger and hand cramps due to excessive typing on the computer.

> ***Writer's cramp****: cramps in fingers and joints of fingers due to long hours of writing or computer typing*

4.5.5 *Lessons from Dancing*

If you believe that actions speak louder than words, dance is probably the best example that brings together physical, neural, and cognitive components. Being action-based, dance is a very effective medium, and goes a long way in providing highly enriching action-oriented expressions of our thoughts and emotions.

As mentioned in Chap. 1, psychology of technology will have to, by necessity, draw from the insights of various disciplines. An interesting example of such interdisciplinary research is one that had been undertaken by Bettina Blasing and her team of researchers in Sports Science at the Bielefeld University and the Palucca University of Dance in Dresden. Even more interesting is the fact that this research was part of the "*Dance Engaging Science*" initiative of the internationally known dance ensemble, *The Forsythe Company*, which tries to forge connections between dancers and researchers across disciplines in an attempt to foster interdisciplinary research.

In their research at the *Center of Excellence and Cognitive Interaction Technology* (CITEC), they have conducted several studies to examine the neural mechanism involved in movements (for example, Bläsing, Coogan, Biondi, Simmel, & Schack, 2014). We present details of one such study below.

Ever taken lessons in dancing? If you have, we are sure that you would agree that it is easier to master a dance sequence by watching a demonstration of the sequence than by listening to verbal instructions about it. Notwithstanding the validity of practical real-life experience, hard core researchers are of the view that even such well-known experiences need to be further validated by controlled laboratory experiments to unravel the role of extraneous variables which may be contaminating the phenomenon in real life.

The experiment was conducted in two parts. In the first part, the performers watched a video of a dance sequence five times and were given a chance to practice them before being asked to perform the sequence, which was then recorded. Thereafter, they listened to verbal instructions for the same sequence twice, and performed the sequence once again. In the second part of the experiment which used a different dance sequence, the order was reversed, with the verbal instructions (repeated five times) followed by the video recording of the sequence (two viewings). As in the first part, the learners had to perform the sequence after each stage of verbal and visual instructions. Box 4.3 gives a diagrammatic representation of the differences between the two conditions.

Ten days later, after the dancers had returned to Dresden, they were suddenly asked to perform both sequences. The results provided unequivocal evidence that the learners not only learnt better when they watched the sequence first than when they heard the instructions first, but that the sequences were also performed more

Box 4.3: Experimental Paradigm for the Study by Bläsing et al. (2014)

Condition I

Watch video (five times) → perform → listen to verbal instructions (two times) → perform

Condition II

Listen to verbal instructions (five times) → perform → watch video (two times) → perform

Ten days later: perform both sequences and provide written feedback

Box 4.4: Virtual K-Pop Dance Teacher Developed to Make Dance Learning Easier

> "Daijin Kim and his team at POSTECH, South Korea, have developed a virtual dance teacher that can make learning famous K-Pop dances easier at home by precisely tracking 3D body joints. They successfully developed the precise 3D tracking of 15 human body joints using big data of the human body and the relational information among human body joints and its tracking performance showed a 4.5 cm error on average among 15 human body joints."

Kim presented the paper at ICIP 2015 (*IEEE International Conference on Image Processing 2015*), regarding the use of big data of 15 human body joints from 100 dance routines performed by a professional dancer to develop the virtual dance teacher. They also demonstrated a score monitoring system based on the extent of similarity and deviation of each body part of a dance learner to the body part of the professional dancer.

Selected among the top 10 papers among the 1098 accepted papers is evidence of how outstanding the technology was considered to be.

Yet another example of how psychology of technology in general and movement psychology in particular can help in the development of technology.

Source: Pohang University of Science and Technology (POSTECH) (2015); ScienceDaily, 22 October 2015. **www.sciencedaily.com/releases/2015/10/151022103842.htm.**

"cleanly" in this condition. Written feedback from the learners also vouched for the fact that they enjoyed and learnt better in the first condition than in the second condition.

When Blasing with her coresearchers presented the work at the 2014 conference of the *German Society for Cognitive Science*, much to their joy and satisfaction, they received the first prize, their paper being judged the best paper of the conference. Surely, a prize for interdisciplinary research too! In general, while psychology seeks to gain greater insight about the dynamics of movement learning from such studies, dancers and dance instructors learn how to provide more efficacious training programs and technology uses both to develop innovative ideas. Box 4.4 on virtual dance instructors is one more piece of evidence of such synergy.

4.5.6 Lessons from Sport

> ***Choking under pressure****: not being able to perform in the face of external pressure*

One important lesson from sport is a common phenomenon termed "**choking under pressure**." There are many examples in the field of sport that closed loop motor control is not always the most effective strategy. World famous cricketer Sachin Tendulkar took, what seemed a never-ending time, to go from his ninety-ninth century to the hundredth century. Almost always, he would get out after scoring in the 80s and often even in the 90s. Many such examples exist in the real world of sport (basketball shooting, Hardy & Parfitt, 1991; golf putting, Hardy, Mullen, & Jones, 1996; piano playing, Wan & Huon, 2005). What causes this phenomenon? How can sportspersons be helped to overcome this pressure? This deterioration in performance is also observed in experienced video players if they realize that they are being observed (Baumeister, 1984). Normally performance pressure is associated with various symptoms such as cognitive and somatic anxiety (Jones & Hardy, 1989) and self-consciousness (Liao & Masters, 2002). But it is also apparent that anxiety does not always lead to performance deterioration. By allotting more attentional resources to the task at hand people can overcome such deterioration and it becomes visible only when these additional resources prove insufficient (Wilson, 2008). Under pressure to perform, what had become almost an automatically executed movement is once again being explicitly controlled by the person, taking up more of the once free working memory resources, or as Masters and Maxwell (2008) put it, explicit knowledge is being "reinvested" in the movement, disrupting automatic movement control. So, rather than allowing the automatic execution of the movement, the player consciously tries to control the movement step by step (Hardy et al., 1996). A strategy more effective than the normal stress management techniques is that of analogy building during the learning stage. An interesting example is that used by Liao and Masters (2001). They taught table tennis players to perform a topspin forehand shot using a right-angled triangle analogy. The players were asked to think that they had to move the bat along the hypotenuse of an imagined right-angled triangle. The results show that not only did this use of the analogy lead to reduced explicit knowledge regarding the movement but also produced less anxiety during times of pressure. The same analogy when used by Law, Masters, Bray, Eves, and Bardswell (2003) led to stress resistance while those who were learning under explicit knowledge conditions were adversely affected by stress. Similar results have been observed for other sports, such as golf (Gucciardi & Dimmock, 2008) and for basketball shooting (Lam, Maxwell, & Masters, 2009) using a very interesting analogy of finishing the shot thinking that "your hand is reaching for a cookie from the cookie jar." The mechanism underlying such choking under pressure becomes even more evident when one considers that decrease in the performance of subjects using closed or guided control was evident only for more difficult or challenging tasks (Ehrlenspiel, Wei, & Sternad, 2010). Altered perceptual processes also seem to produce choking and this can be overcome by getting the individuals to focus on some other aspect unrelated to the primary task, thereby reducing the choking effect (baseball, Gray, 2004; shooting, Vickers & Williams, 2007).

All these studies show the important role that executive functions play in both acquisition of motor skills and its execution. We are immediately reminded of the use of secondary tasks and its effect on working memory by Baddeley (1966), the conscious/automatic task differentiation proposed by Fitts and Posner (1967), Shiffrin and Schneider (1977), and the concept of limited attentional resources made famous by the noted Nobel Prize winner DanielKahneman (1973).

Performance under stress is also affected by changed kinematics (Pijpers, Oudejans, & Bakker, 2005; Higuchi, 2002; Collins et al., 2001) such that there is increased variability in joint coordination but reduced variability at the release point which may sometimes be offset by greater variability in joint angles (Cohen & Sternad, 2009).

The above have important bearings for our usage of technology especially when under stress. Think of the pilot trying to land his plane during a rough storm, the keyhole surgeon trying to operate on a patient who is showing signs of going into a coma at any time, or even you or me trying to remember one's password to open an urgent e-mail. The strategies that have been used for teaching players could well be used for the training of pilots or even surgeons in the use of high tech. Both the restriction of explicit knowledge regarding the movements to be made and the focus on a secondary percept could have ameliorative effects at times of stress, such that the pilot or the surgeon does not choke under pressure.

4.5.7 Lessons from Developmental Psychology

While dance and sport have provided us with invaluable insights, developmental psychology has also proved to be of great help in the understanding of the psychology of motor control. A key feature of the development of motor skills is learning how to climb up and down stairs. Most homes, as well as other buildings, require the child to navigate themselves through staircases and the majority of mothers find that carrying children up or down the stairs is the most difficult task and as such encourage the child to do so on his own. If one tries to visualize stepping up and down and compare the movement kinematics in both, stepping down is more difficult because of the greater fear of falling down. For stairs that we have got used to, most adults adopt an open loop type of motor control shifting to a combination of both open loop feed-forward and closed loop feedback strategies as the situation becomes more demanding in terms of unknown heights of stairs. Try going up a flight of stairs with your eyes closed and you will find that it is not too difficult if one is slow and careful. But try going down without visual cues and one sees how difficult and even frightening the experience can become. One aspect that has been studied in this respect is the age at which infants shift from closed loop to open loop control for stepping down. Research shows that when visual data are provided, children even as young as 3 and 4 years of age are able to plan their stepping down as well as adults are, as shown by the study of various kinematic indices such as knee-drop, toe-drop, etc. However, when visual feedback was removed, the three groups were differentially affected with the degree of effect decreasing with age. This clarifies that children, as young as 3 years of age, are able to efficiently use visual cues to plan their stepping down movements. The clumsy movements that are often witnessed in day-to-day life are not so much because of lack of motor control but more because the stairs that they are forced to navigate have not been constructed with their leg length taken into consideration. Once this aspect is considered, children seem to have no problems if visual feedback is present (Cowie et al., 2010). The implication of these findings is important when one considers play areas in schools, parks, and even in amusement parks. With so many new toys and gadgets being devised to enhance motor skill development in children attention to the size of the step and the provision of visual cues becomes mandatory and should be of prime concern to the designers of play areas and equipment.

The specific ways in which closed and open loop motor control help or hamper the effective use of technology is also clear. Efficient use of tools or technology implies that once one has had sufficient practice in the use of the tool, one should be performing these movements almost automatically. A skilled carpenter knows exactly how to turn his wrist in order to operate a screw driver. Normally what happens is that one takes into account the degree of movement of a body part such as the hand that leads to the desired movement of the tool. However, we normally do not watch our hands; we watch the action of the tool. We perform a mental inversion, that is, in order to move the tool through so many degrees or at this particular plane, this is the way my hand should move (Kleinsorge et al., 2003). When one considers modern tools vis-à-vis traditional tools such as a hammer or a screw driver, the position is completely different. Think of the computer mouse as a tool. Moving this tool effectively and efficiently is more difficult since the one-to-one correspondence between the body input and the cursor output is much less and less readily observable than in the case of the hammer. Another example is the set of tools found in the modern operating room of a hospital. Minimally invasive surgery has introduced a whole new set of tools such as the endoscope and the laparoscope, manipulation of which require a new set of motor skills that could be responsible for the life or death of the patient. Unlike traditional surgery wherein the surgeon was actually making the incision or suturing the skin, keyhole surgery challenges the surgeon's sensory-motor coordination by introducing a new computation whereby a movement of the hand results in the movement of the tool inside the person's body. The exact kinematic computations involved in the use of such tools in the operating room have been discussed at length by Heuer and Sulzenbruck (2009). Later studies from the same laboratory (Sulzenbruck & Heuer, 2012) show that in laboratory studies in which subjects had to perform movements similar to those performed in keyhole surgery, providing terminal visual feedback did lead to faster and straighter movements when concurrent visual feedback was also present. However, when such continuous feedback was not present, the subjects made larger errors. Since in surgery, especially laparoscope surgery, one never can be sure when visual feedback will be available and when it will not be available, these results provide important insights regarding the training of surgeons in the use of such tools and technology. While such surgery is becoming very popular because of the decrease in recovery periods and discomfort to the patient, the training of the doctors should preferably be such that they are able to form an internal representation of the movement required and should not always be dependent on the provision of continuous feedback.

4.6 Motor Imagery

> ***Motor imagery***: *mental simulation of movement without actually performing it*

While on the topic of learning how to dance, it would be appropriate to introduce another interesting aspect of motor behavior, namely, motor imagery, which has immense possibilities for rehabilitation work. **Motor imagery** refers to the mental simulation of a movement without actually performing the movement. So, even without actually moving your foot you can imagine kicking a ball and even generate the same amount of energy that would be needed to perform the movement.

> ***Overt changes***: *changes that are visible, such as changes in performance*

> ***Covert changes***: *changes which are not visible such as cognitive changes*

> ***Psycho-neuro-muscular theory***: *a theory which posits the conjoint action of these three systems and focusing on activation of muscles*

> ***Functional equivalence***: *equivalence at the cortical level between the actual movement and the associated motor imagery*

It is a well-known fact that the acquisition of motor skills induces both **overt changes** (improvements in performance) and **covert changes** (cognitive changes) over a period of time. The cognitive changes are generally seen in the form of mental representations of the movements involved in the motor skill and have received considerable attention over the past 50 years in cognitive sport psychology. It is important to remember that these cognitive representations are in the form of mental imagery which can be understood as the creation and recreation of real-world events in the absence of the actual sensory stimulus. It is this which differentiates imagery from perception (Morris, Spittle, & Watt, 2005; Annett, 1995). Early explanations of such imagery were based on what has been called **psycho-neuro-muscular theory**, implying the conjoint action of all these systems and focusing on the activation of the muscles during imagery. Neurophysiological findings starting with the classical work of Finke (1979) and Jeannerod (1995; 1994) to those more recently by Zhang & colleagues (2011, 2012), all point to a different explanation: that of **functional equivalence** between the actual movement and motor imagery at the cortical level. Studies have also revealed vast qualitative differences between the mental representations of novice and expert sportspersons, whether it be in volleyball (Velentzas, Heinen, Tenenbaum, & Schack, 2010), windsurfing (Schack & Hackfort, 2007), golf (Frank, Land, & Schack, 2013), or even dancing (Bläsing, Tenenbaum, & Schack, 2009). Similar differences have been obtained for stroke patients (Braun et al., 2007) and even for children in whom motor control is still under development (Stöckel, Hughes, & Schack, 2012). A description of a recent study will help to elucidate not only the role of motor imagery on both overt performance and covert representations but also the methodology for a controlled experiment in psychology (see Chap. 1 for more details). In an experiment by Frank and coworkers (2014), novice golf players were randomly distributed across four conditions, namely, physical practice, motor imagery, physical practice plus motor imagery, and no practice. When the performance of these four groups was compared, it was evident that those who used both physical practice and motor imagery surpassed their counterparts in the other three groups. Not only were there greater performance improvements but at the same time, there were functional adaptations in the cognitive representations, as measured by state-of-the-art statistical procedures. Based on studies such as those mentioned above, Moran and his colleagues (2012) provide very interesting insights in cognitive neuroscience for people in sport psychology in an article with an equally interesting title, *Re-imagining motor imagery: building bridges between cognitive neuroscience and sport psychology*, in the *British Journal of Psychology*.

The most amazing aspect about motor imagery is that there is a direct link between motor imagery and the actual movement at the biochemical, neurophysiological, and even the neuropsychological level (Jeannerod, 1997). PET and fMRI studies (for example, Lacourse, Orr, Cramer, & Cohen, 2005) show that the same areas of the brain are called into play for motor imagery as would be the case for actual movement. Motor imagery may even induce cortical plasticity comparable to those obtained under actual practice and this fact has been used in both sports and rehabilitation (Hall & Martin, 1997). We are sure that we would soon see technology being devised that would induce mental imagery of movements or at least stimulate motor imagery so that people who are temporarily unable to practice because of injury or other reasons can keep up their practice through the use of motor imagery.

One domain where actual practice of movements may not always be possible or maybe far too expensive to be feasible is space travel. However, mental practice is a feasible option and astronauts suffer from a variety of sensorimotor disturbances, spatial disorientation, and even difficulties in manual dexterity and postural instability after coming back (Clément & Ngo-Anh, 2013). To overcome this, Bock and his colleagues have devised a simple in-flight training program (detailed in Box 4.1 earlier in this chapter). The method requires little time and hardly any instrumentation and has been widely and successfully used in sport. Yes, it is none other than the use of motor imagery, which when performed repeatedly for the purpose of training is referred to as mental practice.

The effects of such mental practice are brought to the fore by a large number of studies. One example is a study by Clark and his colleagues (2014) in which they found that even 4 weeks of wrist immobilization reduced muscle force and impaired voluntary activation and was accompanied by corresponding changes at the cortical level. However, when another comparable group with the same immobilization was asked to perform mental practice, the muscle force was reduced by only about half as much, and the corticospinal inhibition was completely prevented. Similar effects were observed in a group of elderly patients with distal radius fracture (Schott, Frenkel, Korbus, & Francis, 2013). Mental practice has been used with success not only in sports, but also in a range of other disciplines such as surgery (Cocks, Moulton, Luu, & Cil, 2014) and music (Keller, 2012). It has also led to vast improvements in motor balance in the elderly and has proved to be a useful adjunct to orthopedic and neurological rehabilitation (Hamel & Lajoie, 2005).

Can motor imagery be used for rehabilitation of children who have suffered a stroke? Cerritelli and his colleagues (2000) found that motor imagery is seen even in children aged between 6 and 16 years of age. If this is so, motor imagery could be used as a rehabilitation technique for children who have suffered a stroke, much as it is used for adults. However when comparisons were made between children with acquired brain injury and their normal counterparts, it was found that while the former were able to imagine motor actions, they were not able to accurately predict the time it would take to perform these movements (Caeyenberghs, Wilson, van Roon, Swinnen, & Smits-Engelsman, 2009). This could act as a deterrent for the use of motor imagery for rehabilitation work.

The success of mental practice is dependent on a variety of factors. Probably the most well known is the PETTLEP model described in Box 4.1 according to which motor imagery should resemble the actual task in seven critical aspects (Holmes & Collins, 2001).

4.7 Application of Principles of Motor Behavior in Medicine

The degree of convergence between various hitherto unrelated fields such as medicine and motor skill learning is becoming increasingly greater with the onset of techniques such as those described above. In the journal, *Medical Education* (2011), Elliot and his co-scientists have presented an elegant account of how clinical procedures are learnt and taught in a paper entitled, *Action representations in perception, motor control and learning: implications for medical education.* They have drawn upon their own research and related literature to point out certain key principles such as the importance of minimizing distraction when learning motor skills and the need to make errors in order to learn. They have also shown the role of sensory inputs while learning perceptual-motor skills, the importance of variability, and the benefits of learning by watching and that by doing. Many key principles which have emerged from either literature on motor skill development and perceptual-motor coordination or from the field of sport and by observing players find an important place in the teaching and learning of medical procedures. Just as there is the need for variability in practice, say at ball throwing at different distances, so that he attains accuracy, so, too, the medical novice must get variability in practice, say, of inserting cannulae into veins of different sizes and shapes. This is important for the simple reason that there is wide variability in the types, size, and shape of veins that he will encounter in real life. But can such simulated practice be enough? Is it the same to practice on the wound on a simulated puppet as it is on the wound of a real person with real skin, flesh, and blood? While simulated techniques provide opportunities for practice, the medical novice must also realize that working on a real person is going to be different. To get over this effect of working on nonliving manikins and the degree of safety to make errors it presupposes, Tun and Kneebone (2011) suggest the use of what they call **hybrid simulation** in medical education. This uses simulated patients (professional actors) as proxies for real patients rather than manikins. The ways in which technology is developed to fit our needs is shown here in the use of "prosthetic technology to create high levels of perceived realism through 'seamlessly' attaching models to people" (p. 113). This compels the doctor to engage in human interaction while carrying out the medical procedure. There is a need to integrate real life with learnings from simulations in medical education. Thus, while no one doubts the value of simulated learning especially in training on medical and surgical procedures, maybe the more important aspect would be to focus on not whether simulations are needed or not but on how to make them more effective (Hatala, 2011).

> ***Hybrid simulation**: a technique which uses simulated patients (professional actors) rather than manikins*

4.8 Fitts' Law: Time Taken to Reach Targets

While the open and closed loop theories explain the **speed accuracy trade-off** seen in motor behavior, another commonly seen phenomenon is that speed and accuracy of getting to a target depends on both the size of the target and the distance between the source person and the target.

> ***Speed accuracy trade-off**: increases in speed lead to decrease in accuracy, so people try to reach a balance between the two.*

The movement of our hand towards the mouse in order to give some commands on the computer seems ever so effortless, but is it really so? Watch a person whose arm has been paralyzed for some time and is slowly recovering and you will realize that this task is not as effortless as it seems. Remember the oft quoted proverb, "there is many a slip between cup and lip." While this task does not seem difficult at all for adults or even older children, the mother of a toddler who is just starting to feed himself with a spoon soon finds that most of the food gets spilled before it reaches the mouth of the child. It takes considerable practice before this task becomes an automatic and effortless one. Psychologists have been interested in this phenomenon under the rubrics of **eye–hand coordination**.

> ***Eye–hand coordination**: being able to take the hand to the position that the eye is looking at*

Our brain almost unconsciously measures the distance between the present position of our arm and the mouse or any other target, for that matter, and adjusts the speed and force with which the arm must move accordingly. If you apply too much initial force, you may overshoot the target. If the force applied is too small, your arm may fail to reach the target. For example, suppose I am sitting and typing on my laptop. I, often, have to reach for papers, my cell phone and other odds and ends as I work. My cell phone has just started ringing but it is lying behind me. I reach out and can get my hand to it almost without even looking at it. How did I manage to do this? The sound of the cell phone ringing provided the distance and direction cue enabling me to get to it. If it had been a smaller object, maybe, my hand would have had to do some searching around (had I not been looking at it and, therefore, in the absence of visual cues). The same position could have occurred had the object been farther away. Thus, there seems to be some sort of a relationship between reaching out for a target, the size of the target, and the distance between the hand and the target. The exact relationship between these three aspects was worked out by Fitts and came to be known as the **Fitts' law** which mathematically states:

> ***Fitts' Law****: movement time is a logarithmic function of distance from the hand divided by the size of the object.*

Movement time = Log2 (distance/size)

Or, movement time is a logarithmic function that is directly proportional to the distance between the hand and the object and inversely proportional to the size of the object. In other words, the farther away the object, the longer will be the movement time. Also, the smaller the object, the longer will be the movement time (Box 4.5 and Box 4.6).

Box 4.5: The Classic Experiments by Fitts

To test his theory, Fitts performed a series of experiments using what is now known as the standard Fitts' task. Fitts conducted three different experiments, the reciprocal stylus tapping task, the disk transfer task, and the pin transfer task. He used two types of tasks, a continuous task and a discrete task. In the former, the subject had to move back and forth between two separate target regions without stopping and as quickly as possible. The total time taken was then divided by the total number of movements to determine the average movement time for a particular target size and distance. For the discrete tasks, the subjects were asked to stop after one movement and the time measured for that movement and target. Over the experiments, Fitts varied the size and the distance of the targets. When Fitts (1964) repeated the same tasks with discrete tasks, he found that the law holds for this type of task also.

Box 4.6: Create Your Own Example to Find Out Whether Fitts' Law Works

Create a 16 × 16 grid on your computer screen. Create a red box and place it in the second box diagonal to the extreme left hand corner.

Now, all you need to do is to move the mouse from the top left corner of the screen to the red square. From trial to trial, vary the square's distance from the top left corner and its size. For each trial, measure the time it takes to get from the left hand corner to the red square.

Now create a graph with movement times on the y-axis and size/distance on the x-axis. You get a plot of how RT (plotted on the y-axis), that is movement time, is a function of size and distance (plotted on the x-axis).

Study the plot and you will notice that as the distance becomes longer or the size of the square becomes smaller, movement times (known as reaction time in psychology) becomes larger, in line with Fitts' law.

Recent empirical work has shown that practice can help in shortening movement times without affecting accuracy. Boyle and his colleagues first conducted two experiments on young adults, both of which clarified that even on standard reciprocal Fitts' tasks massive improvements were obtained after the individuals were made to practice on targeting sine wave formations (Boyle, Kennedy, Wang, & Shea, 2014; Boyle, Kennedy, & Shea, 2012). Encouraged by their findings, they then performed the same experiment but with older adults (of 60+ years of age) and true to their hypotheses, these older adults also showed faster movement times and smaller dwell times, both without decrease in movement accuracy (Boyle, Kennedy, & Shea, 2015).

4.8.1 Validity of Fitts' Law

Fitts' law has been validated under a variety of conditions and appears to be one of the most robust relationships to be studied in psychology (Fitts & Peterson, 1964; Fitts, 1954). Variants of this law have also been postulated for a variety of situations (for example, Beamish, Bhatti, Mackenzie, & Wu, 2009). However, most of the studies have used situations in which the hand moves directly to the object, there being no obstructions in between. This is fine as far as laboratory experiments are concerned but in real life, we are constantly faced by situations in which there is some or the other obstruction. Try touching your left ear with your right hand; the nose obstructs a direct trajectory to the ear. Situations such as these were studied by Jax, Rosenbaum, and Vaughan (2007) using two-dimensional situations and Fitts' law seemed to be applicable with a few variations. Going a step further, David Rosenbaum and his colleagues attempted to validate the Fitts' law in three-dimensional space, that is, when the person has to move her hand or foot in a trajectory in which there are objects that will impede the straight line movement of the limb. In an interesting experiment, the researchers made subjects sit in front of a multilevel bookshelf that had different objects of different sizes placed on it. The task at hand was to touch the various objects in response to sound cues being provided. As predicted by Fitts' law, it took longer to reach for objects further away or smaller objects. The results verified that the posture-based motion planning model of Rosenbaum and his colleagues predicted the movement time more completely than Fitts' law when indirect movements were studied (Rosenbaum, Meulenbroek, Vaughan, & Jansen, 2001; Rosenbaum et al., 2009).

There are other conditions for which the Fitts' law has been proved to be very valid. Thus, Fitts' law seems to work equally well whether it is in actually reaching out to pick up an actual object or whether it is taking the mouse to the object on the laptop, or even pointing with the help of the mouse or joystick (Ifft, Lebedev, & Nicolelis, 2011). It is normally seen that movement accuracy increases with visual feedback from the moving limb, enabling both movement planning and online control of the movement (Mirabella, Pani, & Ferraina, 2008). It is also found that such visual feedback is more useful if it is regarding the target position than regarding arm position (Sarlegna et al., 2003). However, so robust is the prediction of Fitts' law, that even when visual feedback is restricted, Fitts' law holds for pointing movements (Wu, Yang, & Honda, 2010).

But does Fitts' law work as well in the laboratory as it does in real life is a question that has boggled many a scientist. As stated by the law, the wider the target, the more quickly we should be able to reach it. In games such as baseball or cricket, it is a known fact that after batting for some time, the ball appears to be perceptively larger to the batsman and hitting it with greater accuracy becomes easier. This is the reason why it is often said of batsmen in cricket that after being on the crease for a couple of hours or more, he is probably seeing the ball to be as big as a football and getting him out becomes more and more difficult as time elapses. Is this really a fact? Do batsmen really "see" the ball as being larger than it actually is? Well, Witt and Proffitt (2005) found that in American pitch softball, batters with a better batting average judged that ball as being larger, or that perception is related to action. In a follow-up study, even golfers were seen to judge the ball as being larger than it actually is (Witt et al., 2008). But laboratory results failed to verify this phenomenon (Zelaznik, Croxall, & Vanhooser, 2011). What seems to matter is the degree of salience of the stimulus as also the length of time over which the batting takes place. The performer must get a chance to learn the link between sensory-motor experience and the external world which builds up gradually over time (Noe, 2004). Since the laboratory had only 30 min, it could be too short a time for sensory-motor learning to take place.

4.8.2 Use of Fitts' Law in the Design of Technology

When Fitts' law was created, the idea behind its creation was to find a formula from which the speed of pointing movements could be predicted. Surprisingly, it accounted for both hand movements and mouse or joystick actions. While the author had the more bulky desktop computer in mind at that time, the advent of the laptop and going a step further, the tablets that we have today have to keep Fitts' law in mind if they want the user to be able to use not only a much smaller gadget but also a much more heavily loaded gadget in terms of what it can do effectively and at the same time efficiently. And now we have what is being termed the phablet, a combination of tablet and phone in terms of size and functions. All of the above have created a sea change in the field of what is known as Human–Computer Interaction (HCI).

4.9 Human–Computer Interface (HCI)

While we have always extended ourselves from the simple use of a hammer to enhance physical force or to the wearing of spectacles to enhance how far or how clearly we are able to see, these extensions have limited use, limited to their sphere of action. The invention of the computer has opened up a totally new realm of such augmentation of human capabilities. While the computer memory serves to extend our capacity to store easily accessible information, robotics enables even surgeons to perform operations with minimal invasive techniques.

***HCI**: point of intersection between human performance and computers*

Such uses of the computer suggest the following: that finally it depends on how well systems have been developed so that the human user can use them effectively. This is the area of **HCI** which is regarded as the point of intersection between human performance and computers. While the former can be understood through study of behavior or various behavioral sciences, the latter benefits from the study of computer sciences. The goal of HCI is to improve the interaction between the computer and the human being by making computers more usable and receptive to the needs of the user through an interdisciplinary approach involving diverse fields of knowledge ranging from computer graphics, operating systems, programming languages, etc. to communication theory, linguistics, social sciences, cognitive sciences, neurosciences, ergonomics, human factors in computer usage, etc. While an optimally designed system can provide many advantages, a faulty computer design can lead to calamities such as the Three Mile Island accident, a nuclear accident (described in an earlier chapter) that was at least partly caused by faulty design of the human–computer interface. Let us examine how Fitts' law helps in web designing.

> "The basic idea in Fitts's Law is that any time a person uses a mouse to move the mouse pointer, certain characteristics of objects on the screen make them easy or hard to click on. The farther the person has to move the mouse to get to an object, the more effort it will take to get to. The smaller the object is, the harder it will be to click on. Pretty simple, right?… The worst possible object is one that is very far away from the current position of the cursor, and very small in size" (Berkun, 2000, p. 1).

Have you ever thought why we place tool bars at the top or the bottom of the screen? Well, the top and bottom edges provide a boundary so that no matter how inaccurate you are in moving the mouse, you just cannot go past these edges. The same is with corners and Windows 8 uses the corners of the computer screen extremely effectively. Remember, Fitts' law states that it takes lesser time to move to bigger targets than to those that are smaller. As we work on our laptops, we can see that all the buttons on the tool bar can be clicked over a much larger area than the small area on which the words "home"' or "insert" is written. How much more difficult it would have been if we had to click just on the word "home"? Just think of how much time you save on Windows simply because when you right click on any target, the tool bar opens up in a window just adjacent to the target. Is this not a lot easier (and, of course, faster) than moving to the tool bar at the top of the screen each time you want to turn something into "bold" or "italics" or you want to change the font color? Fitts' law is immutable for web designing very much in the same way that construction engineers find many laws from physics being immutable.

You are no longer just pressing keys on a computer. This has given way to finger movements such as swiping, flicking, pinching, tapping, etc. Even the mouse pad on the laptop has changed from the single action circular pointing movements and the tap to multi-gestural movements. Now, we don't have to go to the font button to zoom in or zoom out the text. We simply pinch in or pinch out and lo and behold the size of the text changes. This, surely, is an advantage in terms of saving time because we no longer have to take the cursor to the menu bar at the top of the screen, point at the font button, click on it, and then tap it to get it to change the size of the text. Since the release of the iPhone in 2007, another action has become important. An on-screen object can be controlled by controlling the tilt of the screen. One can simply tilt one's phone to enter input rather than the normal finger action on touch phones. Can Fitts' law be applied in this case, too, so that the tilt is just appropriate Mackenzie and Teather (2012) used the new Samsung Galaxy tablet and found parameters based on Fitts' law which could be applied to design the tilt of the tablet so that tilt could be used as an input device much as a mouse can be used. Why has Apple been ever so successful? Why do people find it easier to use than other operating systems? A couple of examples would elucidate this. Compare the scroll bars on Windows with those on Mac (pre OSX Lion). The former has the "up" arrow at the top of the scroll bar while the "down" arrow is at the bottom of the scroll bar. Similarly the "left" and "right" arrows are at the left and right ends of the screen. You are scrolling down a page and now want to go back to something that you had been reading on an earlier page. What do you do? You use your mouse to take the cursor to the top of the page where the "up" arrow is and start clicking on that. In other words, Windows uses the mental model, that, it is natural to look up for going up and down for going down. While the mental model is fine, it does take time to navigate from the top of the screen to the bottom and vice versa. Compare this to the scroll arrows on the Mac: they are placed side by side, because according to Fitts' law navigating between them would be quicker. At first, this appears difficult because we may not be used to the Mac, but once the buttons have been committed to memory as we start operating the system, it is actually much faster.

4.9.1 Applications of Fitts' Law for Mobiles and Tablets

One common application for desktops and laptops is the principle relating size and distance. While this is easy for computers where you are working with a mouse, it becomes so much more difficult when designing for mobiles and tablets. One problem is that relating to gadget orientation. While mice are all held in the same way, you can hold a mobile in at least two different ways: horizontally and vertically. The way we hold the phone in the horizontal orientation would be different from our grasp when the phone is held vertically. Corresponding to this, the reach of our fingers and thumb would also vary. The problem is that most phones give the user the liberty of switching between vertical and horizontal orientations, leaving the designer in the queer position of having to design keeping both orientations in mind.

Also, unlike mice, it is our fingers and thumb that we use on the mobile screen and the reach of the fingers and the thumb must therefore be kept in mind. (Does this not remind you of the anthropometric measurements we introduced in Chap. 2?)

Box 4.7: Some Examples of Fitts' Law Applications for Cell Phones

Why is the iPhone menu kept at the bottom of the screen? Simple, it fits with Fitts' law: your thumb is already covering the bottom part of the screen due to the way you are holding the phone, reducing movement times to the minimum.

Why does the mouse cursor change as one gets to a clickable target? While this is a smart way of offering a reinforcement for the user, it should not force the user to wander all over the screen to finally come to points where the cursor changes from the arrow to a hand. So, the website developer creates a large area, all of which is clickable. Think of all the online shopping sites (for example, Amazon.com): as Christmas draws near, half of the screen is covered by a Christmas tree with "shop now" printed in the middle of it. Fitting into Fitts' law, the entire tree is clickable and the user does not get frustrated looking for small clickable spots.

Thus, the web designer must not only keep the Fitts' law in mind, but must also be knowledgeable about anthropometrics. This could be one reason for the popularity of iPhones. The size of the iPhone is such that the thumb is able to reach every part of the screen (Box 4.7).

Compare the 4.5 in. touch screen with the 7 in. or even the 10–12 in. screen of a tablet. What could be held and operated by one hand can no longer be done. We normally hold the tablet with one hand, or even place it on a table and then operate it. Just think of how design would have to change. While the tablet gives the designer a much larger area to place his icons, it also increases movement time. The reverse is true for the mobile touch screen: if the designer uses large icons, the space becomes crowded and the chances of accidental clicks is enhanced; if the size of the icons are reduced, crowding is also reduced but remember, it will take longer to reach a smaller target.

These are just a few examples to acquaint the reader with how principles drawn from psychology enhance user experience with technology. There are many more that an interested reader can glean from the vast amount of literature available on the topic of Fitts' law. Perhaps one of the most amazing pieces of writing on this topic is one that is entitled, *A Quiz designed to give you Fitts* (it can be downloaded from http://www.asktog.com/columns/022DesignedToGiveFitts.html).

4.9.2 Designing Devices for the Differently Abled

Another emerging area is the designing of touch screen based phones and tablets for those with visual impairments or even complete lack of sight. While Apple and some other cell phone companies use voice interfaces to help those with visual impairments, they continue to face problems to which the factor of stigma is added due to the noticeability factor when using a voice interface. Recently, there has been a spurt in studies on gesture usage by people who are visually impaired. The focus has been twofold: one, to find out the preference of this special population for certain gestures over others; and two, to evaluate the level of performance for gestures used to operate a touch screen based phone. A study by Kane, Wobbrock, and Ladner (2011) revealed that there is considerable variation in the preferences and performance for different gestures between those who are blind and their sighted counterparts and their study ends with several useful recommendations for designers of phone interfaces. Similarly, another study on people with visual impairments conducted by Buzzi and his colleagues (2015) in Italy found differences in the performance for different gestures. To and fro swipes and rotor gestures were found to be the most difficult, while simple swipes with one finger, one stroke, or in one direction were considered simpler to perform. Studies such as these suggest that new guidelines need to be established such that the phones can be used easily and comfortably by both people who can see and those who cannot.

Another group of people who find it difficult to operate computers, phones, and other such devices are those with motor impairments, especially those in wheelchairs. Imagine having to operate a phone while you are operating your wheelchair with one hand. Or, imagine even holding a device while sitting in a wheelchair. So, you are using a laptop, but where do you place it? Is it not ironic that your lap would be far too low for you to operate a laptop? Yes, the problem has been partially answered by having movable and adjustable trays in front of the person and attached to the wheelchair. But then where do the wires, the battery, other small devices go? There is just so much space on a tray. Make the tray larger and the maneuverability of the wheelchair is affected. Because of these problems wheel-chaired people tend to own fewer devices (Kaye, 2000) and are less likely to use the internet (Fox, 2011). Carrington and his associates (2014) have come up with a novel idea: that of **chairables** (taking inspiration from **wearables**!) as against wearables which are attached to the body of the person, some examples of which are the *Shoesense* which captures gestures using a camera on the shoes of the user (Bailly, Müller, Rohs, Wigdor, & Kratz, 2012), or wearables on the belt (Brewster, Lumsden, Bell, Hall, & Tasker, 2003), the wrist or the hip or the pocket (Ashbrook, Clawson, Lyons, Starner, & Patel, 2008). Carrington and his group of researchers (ibid) are of the view that there is considerable space on the wheelchair which can be used to hold these devices and their accessories. It is when technology takes advantage of currently underutilized spaces on the wheelchairs that they term it "chairable" technology to contrast it "wearables" which are worn on the body. Not only will the person have to carry less weight, but it would also be possible to conceal the wires, batteries, etc., which, in turn, would go a long way in reducing the stigma caused by assistive technology that is visible to all. Since stigma is a major barrier to the use of assistive technology (Shinohara & Wobbrock, 2011), this could be one way of reducing such stigma. In other words, technology aims at "designing out stigma" (Bispo & Branco, 2008).

> ***Chairables***: *devices attached to the wheelchair*

> ***Wearables***: *devices attached to things that can be worn*

4.10 Brain–Computer Interface

Almost all standard HCI require the use of pointing devices operated by the hand, the fingers, or the thumb. However, what if the person has motor impairments due to paralysis or Parkinson's disease, or is severely handicapped as far as movements are concerned simply because of age? It is for people such as these that new input devices are being worked on. These include **pointing with the eye**, or even more recently, **pointing with one's thoughts**. While face recognition and eye movements have been used since the 1990s (for example, Jacob, 1991) and did you know that the eyes provide the fastest muscle set in the body, a more recent entry is the use of thoughts to replace the mouse as an input device.

> ***Pointing with the eye:*** *using the eye as a pointer, by gazing at what one wants to point to*

> ***Pointing with one's thoughts:*** *pointing by simply thinking about it*

All such technology presupposes that there is some kind of via media through which the brain can be connected to the computer. Normally, a sensor records the brain waves and is able to translate them to give commands to a computer. Amazing isn't it! But this state-of-the-art technology has already seen the light of day though often called by different names: **BCI** (brain–computer interface), **BMI** (brain–machine interface), or **MMI** (man–machine interface). No matter what the name, the above refers to a set of technology that can help connect man's brain to a machine and by so doing, augment the functionality of the individual. Ways in which one eminent scientist, Stephen Hawking, is using it has been explained in Box 4.8.

> ***BCI:*** *use of a sensor to record brain waves and translate them into commands for a computer*

Box 4.8: Stephen Hawking: Noted Physicist and Cosmologist, Moving, Talking, and Writing Through Machines

Shortly after his 21st birthday, a young Englishman Stephen Hawking suffered from a rare fever which left him with a lifetime suffering causing paralysis of a slow progressing form called Amyotrophic Lateral Sclerosis (ALS), also known as Lou Gehrig's disease or motor neuron disease. He gradually lost control over all his muscles and was finally confined to a wheelchair. This was in 1963 and doctors gave him 2 more years to live. However, he is not only still alive but also very active. He was Lucasian Professor from 1979 to 2009 at the prestigious University of Cambridge, and has written a number of well-known books on physics and cosmology. He continues to give talks all around the world on various aspects of cosmology, and is associated with a large number of movements and organizations that attempt to spread awareness regarding ALS.

(continued)

Box 4.8 (continued)

What is amazing is how he manages to do all of this. The only muscle that he is able to move is a single cheek muscle. But that does not deter him. With his immense drive and help from technology, he communicates through the twitches of this cheek muscle which is connected to a speech generating device. In a report on the Cosmic Log, Ted Warren (2012) describes how Hawking communicates.

> "To produce the words for text or speech, British physicist Stephen Hawking currently uses an infrared sensor mounted on his eyeglasses, visible here during an appearance this month in Seattle. The sensor picks up twitches from his cheek, which are translated into the desired letters or words."

It may take Stephen half an hour to twitch out a couple of sentences in answer to a question, but he is satisfied. Scientists have recently been able to develop a brain–computer interface which may allow him to directly translate his thoughts onto the speaking computer, but as of now he says that he feels that his cheek twitches work better for him.

> "The system, developed by San Diego-based NeuroVigil and known as *iBrain*, uses a head-mounted receiver the size of a matchbox to pick up different types of brain waves. *iBrain* employs a computer algorithm called SPEARS to analyze the brain emanations and encode them for a text-based speech reader."

The amazing product translates imagined movements into waves which are caught by the interface. Hawking was asked to try moving his legs or flexing his wrist. While he could not perform the movement, even the thought of these movements generated readable brain patterns. What a hope for the countless number of people suffering from ALS and similar conditions! Low and Hawking say their work "opens the possibility to link intended movements to a library of words and convert them into speech, thus providing ALS sufferers with communication tools more dependent on the brain than on the body"…It would "give ALS sufferers mobility—sort of like a real-life version of the Stephen Hawking robotic exoskeleton proposed in an Onion parody 15 years ago"

Source: Warren (2012); **http://cosmiclog.nbcnews.com.**

4.10.1 Types of BCI

- **Invasive BCI**: In which BCIs are implanted directly into the grey matter of the brain through microsurgery and can be used to provide vision to those with visual impairment, hearing for those who cannot hear, or motor neuroprostheses for those with paralysis.

Invasive BCI: BCI planted directly into grey matter of brain

- **Noninvasive BCI**: These are easy to wear noninvasive implants, for example, a glove, which is connected to a machine. Thus, people have been trained to self-regulate their EEGs by wearing electrodes on their heads.

Noninvasive BCI: easy to wear implants which are not inserted through surgery

- **Partially invasive BCI**: In this case the implant is inside the skull but not inside the brain. It, in fact, rests on the brain. Current technology uses electrodes that sit on the surface of the brain below the skull and the dura mater and as such there is no need to pierce the cortex.

Partially invasive BCI: implant is inside the skull but not inside the brain

While BCIs are useful adjuncts in the life of people who suffer from motor impairments, one problem has always plagued the technology provider: how to achieve the optimal trade-off between the invasiveness of the signals and its robustness. While EEGs are noninvasive, with the electrodes being placed directly on the scalp, the obtained signals lack high resolution. In contrast, fully invasive BCIs such as intracortical electrodes do provide a much higher degree of accuracy but they obviously require complicated microsurgery. Moreover, the quality of the signals is found to decay over time as the electrodes tend to get encapsulated in brain tissue. The best option, as of now, appears to be partially invasive BCIs: not requiring surgery as complicated as that for intracortical electrodes and at the same time providing a degree of accuracy much higher than that obtained through EEGs.

4.10.2 Electrocorticography (ECoG)

Electrocorticography: *a partially invasive technique in which electrodes implanted in or outside the brain are able to sense brain impulses*

One technology that has been used for such partially invasive BCIs is electrocorticography or ECoG. The ECoG recording is performed from electrodes placed on the exposed cortex. In order to access the cortex, a surgeon must first perform a craniotomy, removing a part of the skull to expose the brain surface. This procedure may be performed either under general anesthesia or under local anesthesia. Electrodes are then surgically implanted on the surface of the cortex, with placement guided by the results of preoperative EEG and magnetic resonance imaging (MRI). Electrodes may either be placed outside the dura mater (epidural) or under the dura mater (subdural). ECoG electrode arrays typically consist of 16 sterile, disposable stainless steel, carbon tip, platinum or gold ball electrodes, each mounted on a ball and socket joint for ease in positioning. These electrodes are attached to an overlying frame in a "crown" or "halo" configuration.

The technique was pioneered by Wilder Penfield and Herbert Jasper, neurosurgeons at the Montreal Neurological Institute more than half a century back, in the 1950s. They used it to treat patients with severe epilepsy and to map the cortex to isolate the origin of the seizures and thereafter surgically remove those parts. Now that ECoG has emerged as a promising technique for brain–computer interfaces (Shenoy, Miller, Ojemann, & Rao, 2007; Leuthardt, Schalk, Wolpaw, Ojemann, & Moran, 2004), scientists from the Department of Biomedical Engineering and the Schalk Laboratory have demonstrated that ECoG can not only be used as a training paradigm for humans (Leuthardt et al., 2004), but also that the patients learn how to control a computer cursor in minutes whereas it requires months or even years when EEG is used for similar training (Wolpaw & Mcfarland, 2004). The last decade or so has seen a flush of such studies which substantiate that ECoG can be used for BCI (summarized in a Report of the Schalk Lab, 2015).

Just think about it: a technology first used for monitoring epileptic patients has thrown up a host of possibilities. Before the dawn of ECoG-based BCIs, when the patients were not having seizures, the electrodes would just lie there, implanted, waiting for the person to have the next seizure. What was better than using these patients as subjects for the creation of a new type of BCI intervention? It is truly another exaptation in the making (see Chap. 2 for further details on exaptations) and forced Leuthardt and colleagues (2004) to posit that this was surely the future of neuroprostheses.

In a volume published very recently (October 2015), entitled *The Brain Electric: the dramatic high tech race to merge minds with machines*, Malcolm Gay traces the literal race to develop BCIs, from the early start by Leuthardt to the contributions of Philip Kennedy, Niels Birbaumen, John Donoghue, Miguel Nicolelis, and many others. The veritable progress that has been made can be gathered from the fact that when they started 20 years back, they were using equipment that required a full room and at best they could tap just 20 neurons. Today, they are able to tap close to 20,000 neurons.

4.10.3 Modus Operandi of a BCI

The basic method by which BCIs operate is to ask a person to think of performing some type of movement. The presence of mirror neurons in the cerebral cortex makes it possible for the brain cells to get activated much the same way as to when the person is actually performing the movement. These neuronal activations produced by the thoughts are then picked up by a sensor and transmitted to a computer software for interpretation and the movement would be actually performed. The only difference: rather than the movement being performed by the use of muscles, it is being performed by thinking or mentally simulating the movement (as in the case of mental imagery discussed earlier in this chapter).

Any BCI (Brain–Computer Interface) or BNCI (Brain Neural Computer Interface) requires at least four components. At least one sensor must detect brain activity. (In a BNCI, the sensor could detect other signals from the body, which might reflect activity from the eyes, heart, muscles, etc.) Next, a signal processing system must translate the resulting signals into messages or commands. This has then to be sent to an application on a device, such as a web browser on a monitor or a movement system on a wheelchair. Finally, there must be an application interface or operating environment that determines how these components interact with each other and with the user.

4.10.4 BCI and Fitts' Law

As scientists delve into the neural basis of movement control (for example, Ifft et al., *Cortical correlates of Fitts' Law*, 2011), there could be considerable advancement in the much needed field of neuroprosthetics. Fitts' law has been modified for use in the design of brain–computer interfaces and studied using able and motor disabled subjects (Felton, Radwin, Wilson, & Williams, 2009). They have found the Fitts' law to be robust even when the computer cursor is controlled by neural signals.

The widespread usability of Fitts' law has propelled people into action even at the governmental level. The *International Organization of Standards* requires compliance of Fitts' law optimized performance on all non-keyboard input devices, quantified by both bits/seconds and error rate (ISO 9241-400, 2007).

4.10.5 Commercially Manufactured BCI

It all started in 2011, when a woman, Cathy Hutchinson, used thoughts to operate a robotic arm to bring a flask of coffee to her lips. She was able to do it through *BrainGate*, the brainchild of researchers from Brown University collaborating with the Massachusetts General Hospital, the US Department of Veteran Affairs and researchers from Stanford University and the Case Western Reserve University. *BrainGate* consists of a total of 96 electrodes that pick up signals from nearby neurons and send them to a computer that translates them into action. It is a BCI developed for severely motor-impaired individuals and is based on technology that can "sense, transmit, analyze, and apply the language of neurons." Basically, there is a sensor that is implanted in the motor cortex of the brain and a device that analyzes brain signals based on neuroscientific findings that with intact brain functions, brain signals are generated even though they may not be or cannot be sent to the limbs. However, these signals can be collected and used to move a cursor on a computer. The beauty of the technology lies in the fact that you can move the cursor on the computer by simply thinking about these actions without actually performing them.

One of the biggest problems being faced by BCI developers is the brain itself, consisting as it does of soft, "squishy" material (Underwood, 2015). This causes the organ to have a tendency to shift inside the skull which displaces the electrodes implanted in the brain. As a result, the patient has a hard time keeping the cursor on course and the researcher has to recalibrate the system every so often, maybe every ten minutes. In a new study on the BCI, *BrainGate*, scientists have been able to develop a system which is capable of auto-correction for neuronal noise when the person stops typing. This enables the patients to type not only faster but also for longer periods of time.

In two separate tests, two patients with Lou Gehrig's disease or ALS (the syndrome suffered by Stephen Hawking, too) used the technique to move a mouse cursor by simply thinking about it. And, more importantly, they could do it twice as fast as Cathy Hutchinson could do it (Woolaston, 2015).

Another commercially developed BCI is *i-Brain*. According to the neuroscientist Low who is responsible for developing *i-Brain*, it was already into version 2 by the year 2012 and version 3 had been planned for the next year, as reported by Warren (2012). The *i-Brain* is about the size of an American quarter and can be used by people to measure brainwaves much as you would measure blood pressure. From where did the idea originate? Surprisingly, Low says that it is based on an algorithm he developed for analyzing brain patterns of bird movement. As pointed out in the earlier chapters too, animal behavior has helped not only develop greater insights into human behavior but has also helped in the development of technology to augment human capacities. Another interesting example is given in the Box 4.9.

Box 4.9: Drawing Inspiration from Animals

a. Learning from earthworms

Earthworms creep along the ground by alternately squeezing and stretching muscles along the length of their bodies, inching forward with each wave of contractions. Snails and sea cucumbers also use this mechanism, called peristalsis, to get around, and our own gastrointestinal tracts operate by a similar action, squeezing muscles along the esophagus to push food to the stomach.

Now researchers at MIT, Harvard University and Seoul National University have engineered a soft autonomous robot that moves via peristalsis, crawling across surfaces by contracting segments of its body, much like an earthworm. The robot, made almost entirely of soft materials, is remarkably resilient: Even when stepped upon or bludgeoned with a hammer, the robot is able to inch away, unscathed.

Sangbae Kim, the Esther and Harold E. Edgerton Assistant Professor of Mechanical Engineering at MIT, says such a soft robot may be useful for navigating rough terrain or squeezing through tight spaces.

The robot is named "Meshworm" for the flexible, mesh-like tube that makes up its body. Researchers created "artificial muscle" from wire made of nickel and titanium—a shape-memory alloy that stretches and contracts with heat. They wound the wire around the tube, creating segments along its length, much like the segments of an earthworm. They then applied a

(continued)

Box 4.9 (continued)

small current to the segments of wire, squeezing the mesh tube and propelling the robot forward. The team recently published details of the design in the journal *IEEE/ASME Transactions on Mechatronics.*

b. **Tiny dancers: Can ballet bugs help us build better robots?**

Is it not difficult to believe that the inspiration for designing robots can come from the creatures of Mother Nature? Well, this is true and even spider crickets can help us. This has been borne out by researchers at the Johns Hopkins engineering professors and students who have spent more than 8 months studying the movements of spider crickets and presented their work in Boston at the 68th annual meeting of the *American Physical Society's Division of Fluid Dynamics*.

So, what were the learnings? The researchers were able to get clues regarding how these wingless creatures can leap distances equaling even 60 times their body length and even land safely on their feet. They found that the way they accomplish this is by using their limbs to stabilize themselves. Moreover, the movements were not only extremely artistic but also provided lessons in aerodynamics.

Knowledge such as the above would help scientists design tiny robots that would be able to traverse all kinds of terrain, such as helping dig out people after an earthquake. In addition, it will expand our understanding of motor behavior.

Source: Will Kirk, Johns Hopkins University (2015)

4.11 New Directions in BCI

4.11.1 Neuro-gaming

Technology has moved ever so far, that it is sometimes beyond belief. Can you imagine playing a computer game without keying in options or without having to operate a joystick? Yes, it is now possible, through what is known as **neuro-gaming**, whereby one operates a console not by one's fingers but through noninvasive BCI such as gloves connected to brain EEGs. While some use brain waves, there are others that depend on heart rate, expressions, pupil dilation, and even emotions to complete tasks or affect the mood of the game.

Neuro-gaming*: one operates a gaming console through noninvasive BCI connected to brain EEG*

4.11.2 *The Double Bind: OPTT (Optimizing Participation Through Technology) Comes to the Rescue*

There is one fact of life that is irrevocable: aging. And, to add to the problems faced by all those who are aging is to have mobility becoming limited either due to Parkinson's Disease, paralysis, limb amputation, or muscular dystrophy. Such people definitely face a double bind: growing old and in a wheelchair.

For these people, it has been seen that not only does the normal aging process gets accelerated but that they also face upper extremity pain (because of having to operate everything, plus the wheelchair with one's arms), rapid decline in bone mineral density, coupled with issues such as those of urinary control and breathing. Being unable to use their lower limbs, they also suffer from various pathologies of the shoulder which undergoes all the strain of vigorous and repetitive pushing. This is where the psychology of motor behavior has proved to be very useful.

The problems associated with rehabilitation become even more exacerbated if the person with upper extremity problems also suffers from neurological impairments. It is but natural that the entire process becomes slower due to fuzziness in thinking, slower learning and executing of movements. Recent research has provided considerable insight into the ways that can be adopted to help in the rehabilitation process. Some of the aspects considered are skill acquisition (Gentile, 1972, 2000), the role of trial and error in the learning of new skills (Muratori, Lamberg, Quinn, & Duff, 2013), and the ways of improving motor flexibility (Schmidt & Lee, 2005). While teaching the motor skills is important, it is equally necessary to devise ways of calibrating changes that may have taken place (because teaching per se does not always lead to learning and retention). For this Magill (2011) has suggested ways by which we can measure the amount of motor learning at each of the three phases of acquisition, retention, and transfer. Earlier in this chapter we have focused on two types of movement: discrete movements and continuous movements. Research clarifies that there are clear differences between the learning of these two types of movements and this proves to be useful since these patients have to be taught both types of movements (Huys, Jirsa, Studenka, Rheaume, & Zelaznik, 2008). The role of open and closed loop motor control is also helpful for rehabilitation work (Gentile, 2000). It had been clarified earlier that motor imagery and mental practice of motor acts help in the acquisition of skills. The role of mental practice (Magill, 2011; Malouin & Richards, 2010; Dickstein & Deutsch, 2007) and that of modeling and manual guidance (Sidaway et al., 2008) in the rehabilitation of patients in wheelchairs has also been brought to the fore.

Apart from these aspects of motor behavior which prove to be salient, there are principles drawn from the psychology of learning that are useful. These include the role of practice and especially the role of different schedules of practice such as whole versus part learning and practice, massed versus distributed practice, and constant versus variable practice (Kruisselbrink & Van Gyn, 2011). Similarly, the role of feedback including types and timing of the feedback in movement acquisition has been studied (Wulf, Chiviacowsky, Schiller, & Ávila, 2010; Sidaway et al., 2008).

Probably the most important finding is that movement therapy produces brain plasticity in stroke patients (Laible, Grieshammer, Seidel, Rijntjes, Weiller, & Hamzei, 2012; Treger, Aidinof, Lehrer, & Kalichman, 2012) leading to regeneration of brain functions lost because of the stroke.

Drawing from the research evidence of the like presented in the sections on motor learning and motor control in this chapter, considerable advances have been made in gaining insight into how best to rehabilitate people with upper extremity problems.

> ***OPTT**: developing devices to help wheelchair users perform daily living tasks without discomfort*

Now as we move from the perspective of psychology to that of technology, the question is: has such research been only of academic value or has it led to the development of tools and aids to come to the rescue of people with impaired mobility? While considerable still needs to be done, there has been some progress. For example, the *Rehabilitation Engineering Research Center (RERC)* is a boon, attempting to address problems of people with mobility issues. The objective of the *RERC* is ***OPTT** (Optimizing Participation Through Technology*) by developing adaptations to devices to help wheelchair users perform tasks of daily living with minimal discomfort. A major concern of this Center is the redesign of the wheelchair so as to make it more comfortable and reduce the load on the shoulders while pushing different types of propulsion devices. Other important objectives are the development of clinical guidelines to teach users how to get in and out of the wheelchair more easily and to even customize car configurations for wheelchair users. According to Requejo, Director of the *Rehabilitation Engineering Program* at the *Rancho Los Amigos National Rehabilitation Center* in Downey, California, USA, as far as rehabilitation of people with upper extremity problems is concerned, there are **two Es** that are of utmost importance. The first is **exercise** and the second is **education**. Exercise of the rotator cuffs of the shoulders reduces shoulder problems while education not only of the wheelchair user but also of the family and the general public goes a long way in ameliorating many of the problems faced by the patients. By studying movement patterns, the ways in which the person gets in and out of wheelchairs, the pushing techniques he is using, and even how the person maneuvers herself through different kinds of terrain helps the trainer to educate people in the correct way to perform these tasks, thereby preventing both fatigue and injury (Brackens, 2011). Several new devices are now available commercially, such as the *Tennis Ball Walker Glide* and the *Go-Go Folding Scooter.*

> *Two Es of OPTT: Exercise and education*

4.11.3 Shared Control Techniques

Jose del R Millan from the Swiss Federal Institute of Technology, Lausanne, Switzerland presented another technological wonder at the recent meeting of the *Cognitive Neuroscience Society* at San Francisco. The technology is described in a report. Taking inspiration from natural motor control, he and his colleagues have designed new prosthetic devices that can replace limb functions in an amazingly new way. They have designed wheelchairs to artificial limbs to even robots, all using what has been called "**shared control" technology**, in which the robot's sensory capabilities are used to interpret the user's commands. The technology decodes brain signals to find out what the user wants to do and then uses advanced robotics to do what would be normally performed by the spinal cord to orchestrate those movements.

Shared control technology*: robot's sensory capabilities are used to interpret user's commands*

Millan has recently tested the technology on a large number of people with motor impairments, some of them with severe impairments on a variety of tasks ranging from writing to navigation. Amazingly, the results show that these people could perform the tasks as well as the healthy control subjects. At the same time, the technology is very simple and easy to learn, the patients being able to use them after as few as nine sessions.

According to the report from the Cognitive Neuroscience Society (Millan, 2015):

> "Future neuroprostheses—robots and exoskeletons controlled via a BCI—will be tightly coupled with the user in such a way that the resulting system can replace and restore impaired limb functions because it will be controlled by the same neural signals as their natural counterparts," Millán says. "This is no longer science fiction; the questions now are which are the key components to guarantee reliability and long-term operation of neuroprostheses, and when they will be part of the clinical portfolio available to motor-disabled people" (Talk entitled, "The rise of neuroprosthetics: the perception-action closed loop" at an invited Symposium at the Cognitive Neuroscience Society Annual Meeting, San Francisco, March, 2015, by Millan).

4.12 The Eye as a Pointing Device

Tracking the movements of the eye is not new to psychology or even for technology as a matter of fact. Ever since the 1980s, eye tracking has been used for cockpit pilots. What is, however, a more recent development is the use of eye movement as an input device for operating a computer. This is especially important when one considers the large number of disabled people who are unable to use the internet or the computer only because they are either amputees or are suffering from upper limb paralysis.

We must, however, keep in mind that the success of any technology is determined to a very great extent by the limits of the human body. Let us begin by trying to understand why the eye has been considered as a pointing device and whether it can or cannot be so used. There are a variety of reasons why the eye can be used as a pointing device. For one thing, the eye provides a natural device because the individual would already be looking at the target even if the eye is not being used for pointing. Thus, using the eye leads to parsimony of effort, since only one set of muscles are used for both input and output rather than two. At the same time, eyes have not only the fastest muscles in the body but also fatigue less easily, and thus movements using the eye as a pointing device would lower the RT. Thirdly, the eye provides an alternative via media when the hands are busy (Smith, Vertegaal, Sohn, & Cheng, 2005).

Yet, despite all these apparent advantages, one wonders why the eye has not been exploited in computer interfaces at large. One reason is that while people are used to the eye as an input device for learning about the environment, they are not used to it as an output device. Secondly, saccadic movements of the eye are extremely fast and so it is not very useful for any graphical user interface which requires a longer gaze. Thirdly, eye movements are sensitive to head movements. This, of course, has been offset recently: one example is the LC Technologies' *EyeGaze System* which now features head movement tolerance to correct for head movements.

> ***Midas Touch Effect:*** *each eye glance is converted into a command*

But perhaps the most difficult problem faced by designers of eye trackers is what has come to be known as the ***Midas Touch Effect***. Remember the King who had once wished that whatever he touched should turn into gold and had his wish granted? The net result: his food became gold, the water with which he wanted to take a bath turned into gold, along with so many other things, that life became impossible. This legend has been used by eye tracking researchers to describe a fundamental problem in eye tracking. Each glance of the person is converted into a command because the eye has evolved in the way it has so as to be able to view the environment, not to manipulate it. Once the eye is used for pointing, it must differentiate between (a) a gaze intended to gather information and (b) a gaze intended to activate a specific command. Designers of computer interfaces have attempted to get over this problem by using dwell time and blinks as clicking devices, but do we not blink otherwise too? Do we not stare at a page even when we don't want it to activate anything?

Otherwise too, our visual system is being put to great lengths because of the overuse of computers and cell phones not to talk about social networking sites. Do we want Pupil Tunnel Syndrome akin to the Carpal Tunnel Syndrome that people are suffering from today? Thus it seems that as of now, we shall be able to use the eye as a pointing device only to assist the user, that is, by creating a hybrid design which combines the hand with the eye. Maybe the next million years will see a new biological exaptation of the eye which in turn will enable us to use the eye as a true pointing device.

4.13 Augmenting Motor Functions in Humans

In the previous chapter we had focused on how cognitive functions of humans can be augmented, and the ways in which psychology, neurosciences, and technology have come together to make such augmentation possible.

Much in the same way, our understanding of the psychology, anatomy, and physiology of motor behavior has enabled considerable advances, making motor augmentation a reality. Though many aspects are still at the clinical trial stage, it will not be long before this becomes a reality and ready for general use.

4.13.1 Exoskeletons

The first exoskeleton was built in the late 1960s by the US Office of Naval Research Development and was named *Hardiman*. It was massive, weighing 680 kg and was built by General Electric Global Research. Its very weight caused its death and it had to be abandoned. The idea however did not die. In the year 2000, the **HULC** (*Human Universal Load Carrier*) was designed and built with the help of engineers from the University of California at Berkeley and aided by the US defense agency, DARPA. In 2011, Lockheed Martin was ready to test an updated and lighter version of the *HULC*. It enabled soldiers to carry weights upto 90 kg over a distance of 20 km on a single battery charge. However, this **exoskeleton** too proved to be a failure mainly because of the inability of the scientists and engineers to understand the normal biomechanics of walking. There was a lack of coordination between the man and the machine, causing the user to walk in an unnatural manner. As a result, the *HULC* was seen to cause the soldiers to become even more exhausted in carrying out the tasks than when they performed the same tasks without the *HULC*. In fact, it has been easier to design prosthetic aids than these exoskeletons. As one engineer from MIT, Hugh Kerr puts it, "it's very difficult from a design perspective to augment human walking and running, because we're so good at it." In most cases the problems occurred because neither the physiology nor the natural rhythm of the body had been addressed, so that at best what was developed was a mere exercise machine.

> ***HULC**: Human Universal Load Carrier: a device for augmenting motor functions*

> ***Exoskeletons**: a system of cables, motors, and fabric which the user wears to augment motor functions*

> "A lone soldier stands in a dark alley, eyeing a door. Even though he's covered in bulky armor, he charges forward and bursts through, and is engulfed in a barrage of gunfire. Rather than retreat, the soldier stands tall as bullets ping off him harmlessly. This isn't a trailer for the latest superhero movie. It's an animation produced by the U.S. military, designed to show off its vision for a brawny robotic exoskeleton that it hopes to deploy with elite commandos" (Cornwall, 2015, p. 270).

The ***TALOS (Tactical Assault Light Operating Suit)*** is funded by a multimillion dollar research project which became the need of the hour when a US commando was killed in a hostage attack in Afghanistan. The name *TALOS* is inspired by Greek mythology from the name of the metal giant which guarded the island of Crete and could encircle the island three times a day almost effortlessly. Rather than being heavy, *TALOS* is a lightweight suit which uses fabric, flexible cables and

> ***TALOS**: Tactical Assault Light Operating Suit*

Box 4.10: Augmenting Human Motor Functions Using *TALOS*

Warren Cornwall, the reporter mentioned earlier, was given an opportunity to try the *TALOS*. It took a full 45 min to get "dressed" for the event which meant getting strapped and latched to a variety of cables, boxes, and a backpack. Then Cornwall was ready to mount the treadmill and the treadmill was started. What occurred is best described if it comes from the horse's mouth:

"Ignacio Galiana, one of the engineers, starts the treadmill. I'm walking at a pace of about 5 km an hour. My first step is met with a surprisingly abrupt yank on my heel. It lets go and almost immediately my other leg is tugged up and back. I keep my balance and settle into a brisk walk, the tiny electric motors and gears keeping time with a frenetic whirring. They retrieve and release the wires with every step, synced to my pace by microprocessors and the motion sensors. Even after a few minutes, each pull is slightly jarring, a bit like being a marionette with four wires controlling my legs. Am I walking in the suit, or is it walking me?

'We're doing a significant percentage of what your body needs,' Galiana explains. 'It takes a little bit to get used to these additional forces and be fully relaxed.'

After 12 min on the treadmill, he turns off the exoskeleton as I keep walking. Something unexpected happens. My legs suddenly feel slower, the boots heavier. There is less pep in my stride"

Source: Cornwall (2015), **http://www.sciencemag.org**

small motors and weighs a mere 9 kg and uses only 140 W of electricity. It seems to be working in the trials that have been undertaken (see Box 4.10). As shown by a recent test on seven people, they were able to carry loads equal to 30 % of their body weight and were 7 % more efficient than without the suit. According to present plans, it should be ready for general use by the year 2018, though it is still beset by many problems, mainly because engineers seem to lack a thorough grasp of biomechanics. Once again, as pointed out earlier too, technology can be effective only if the anatomy, physiology, and psychology of the user of that technology is understood fully. And, can this be possible without psychology of technology?

Since working on a treadmill does not provide full validation regarding its usefulness in real life, or ecological validity as psychologists would put it, the *TALOS* has been tried by the army under the supervision of scientists and engineers in the backwoods of USA. The trials have been successful though not without a few hitches. The scientists still warn against overexpectations. There are many challenges, the most important being that the exoskeleton is designed for walking not for running, especially in rough terrain. Secondly, there are individual differences in the adaptation time. And thirdly, the suit will have to be integrated with all the other equipment that forms a part of the normal gear of a soldier. The army has other concerns, such as, whether using the exoskeleton could cause its own set of injuries.

4.13.2 Noninvasive Brain Stimulation (NIBS) Techniques

> ***TMS**: Transcranial magnetic stimulation which uses a focused magnetic stimulation*

> ***tDCS**: transcranial Direct Current Stimulation: light direct current is delivered at specified part of cortex.*

Another technology that is being developed to augment human motor function, but this time for those who are suffering from motor impairments, is NIBS (Noninvasive Brain Stimulation). There are two methods that are being used for NIBS: one is **TMS** or Transcranial Magnetic Stimulation while the other is **tDCS** (transcranial Direct Current Stimulation). As the name suggests, both are forms of noninvasive technology through which either a focused magnetic stimulation (in the case of TMS) or a light direct current (in the case of tDCS) can be delivered at a specific part of the cortex. A magnetic/electric field is created which excites the brain cells and activates them. In fact, TMS is used for a variety of purposes. While it had been developed to map the brain, to measure changes in brain function after an injury or lesion, and even to show how the brain reorganizes itself after the injury, it is proving to be extremely effective in the rehabilitation of stroke patients (Freitas, Mondragon-Llorca, & Pascual-Leone, 2011; Benninger et al., 2010; Webster, Celnik, & Cohen, 2006). It appeals to clinicians especially because of its ability to excite brain cells and to create brain plasticity in a manner that is noninvasive and safe, and when combined with motor tasks is seen to facilitate motor learning (Tretriluxana et al., 2013; Takeuchi et al., 2008). It has been found to help improve a variety of motor functions including improvements in gait (Danzl, Chelette, Lee, Lykins, & Sawaki, 2013; Kakuda et al., 2013; Wang, Tseng, Liao, & Wang, 2012), finger movements (Kobayashi, Hutchinson, Théoret, Schlaug, & Pascual-Leone, 2004), hand muscle force (Vaziri et al., 2014), and hand dysfunction (Theilig et al., 2011). TMS has also been found to help children with cerebral palsy for improving their motor control (Allford, 2015).

The usefulness of NIBS is undermined by certain caveats. For one thing, the dosage is important as also safety guidelines. As such the training of clinicians and support staff becomes very important. There is also the need to integrate many of the therapies that are today being used independently (Chollet et al., 2014; Nilsen & Dirusso, 2014). A good review of the techniques used in stroke rehabilitation is provided in an excellent paper by Caflin and his colleagues who are of the view that used with care and understanding, these techniques would go a long way in augmenting "spontaneous neurologic recovery and modulate neuroplastic changes after stroke" (Caflin, Krishnan, & Khot, 2015).

As we complete this chapter, it becomes clear that there is so much more to be written. But every volume has its constraints in terms of the balance between chapters. However, it should be apparent that the psychology of motor behavior is of utmost importance as far as psychology of technology is concerned. It has, and still is, helping in the design of various types of technology ranging from that used for cell phone and computer interface design to brain computer interfaces and technology that will help augment the motor functions. A thorough understanding of this aspect of human functioning will go a long way in not only designing prosthetic aids for those with motor impairments but also in designing devices for those with visual impairments. As such it cannot be neglected or remain the Cinderella of modern psychology.

Summary

While Chap. 3 focused on the role of the executive brain functions of perception and cognition, the present chapter deals with yet another important area, that is, the motor system and motor control. This oft neglected Cinderella of modern psychology is of utmost importance, not only for life in general but also for the use of technology. As technology changes, so must our motor functions, clearly exemplified by the much higher degree of convergence seen between the hitherto unrelated field of medicine and motor skill learning, necessitated by the onset of techniques such as endoscopy, laparoscopy, and cyber-surgery. Modern state-of-the-art technology, whether used by the manufacturing industry, medicine or even players of computer games and other forms of human–computer-interaction (HCI) require not only greater precision of movements but also a strong motor imagery. This can be achieved by a clear understanding of our motor system, coupled with knowledge of the executive functions gained through the last chapter and current research on visuomotor coordination, the intricacies of motor development, motor imagery, open and closed loop motor learning, and Fitts' law.

Suggested Readings

Cocks, M., Moulton, C., Luu, S., & Cil, T. (2014). What surgeons can learn from athletes: Mental practice in sports and surgery. *Journal of Surgical Education*, 71, 262–269.

Bertucco, M., & Cesari, P. (2010). Does movement planning follow Fitts' law? *Neuroscience*, 171, 205–213.

Parasuraman, R., & Rizzo, M. (Eds.) (2007). *Neuroergonomics: The brain at work*. New York: Oxford University Press.

Wulf, G., & Mornell, A. (2008). Insights about practice from the perspective of motor learning: A review. *Music Performance Research*, 2, 1–25.

Chapter 5
Behavior in the Virtual Environment

"We are challenged to ask what such things augur. Some people are looking for robots to clean rugs and help with the laundry. Others hope for a mechanical bride. As sociable robots propose themselves as substitutes for people, new networked devices offer us machine-mediated relationships with each other, another kind of substitution. We romance the robot and become inseparable from our smart phones. As this happens, we remake ourselves and our relationships with each other through our new intimacy with machines. People talk about Web access on their BlackBerries as 'the place for hope' in life, the place where loneliness can be defeated. A woman in her late 60s describes her new I-phone: 'It's like having a little Times Square in my pocketbook. All lights. All the people I could meet.' People are lonely. The network is seductive. But if we are always on, we may deny ourselves the rewards of solitude" (From Introduction to *Alone Together*, Turkle, 2011b, p. 14).

Yes, while Turkle warns us of the dangers of always being 'on,' Box 5.1 reveals what a day without data would be like. Before we can think of the all-pervading effects that data transmission is having upon us, it would only be salient to pinpoint who is transmitting the data and what type of technology is being adopted for this end. Seeking an answer to the second question first, the most important of all the technologies being adopted today, as far as the common man is concerned, is ICT, that is, Information and Communication Technology. Probably no technology has changed, and is still changing, at the pace at which ICT is changing. Moreover, the growth in the number of users is increasing exponentially even in the remotest of remote areas. Riding piggyback on satellite technology, the industry is marked by an ever-increasing variety of uses. No longer is it just a means of communication, entered as it has, all fields, ranging from education to medicine, entertainment to social relationships, travel and tourism to political campaigning, banking and insurance to meteorology and many more. You name it and there it is, omnipresent and sometimes even omnipotent. Probably, no one is left untouched by its effects, either as a direct user or as a beneficiary of systems that use it. And, it is these technological wonders of the last 50 years or so that have caused the data explosion we are in the midst of.

V.K. Kool, R. Agrawal, *Psychology of Technology*,
DOI 10.1007/978-3-319-45333-0_5

Box 5.1: A Day Without Data

Data are so pervasive in our daily lives that we are heavily dependent on it to power ourselves and our systems. But have you ever stopped to think what will happen if data came to a standstill for just 1 day, just 24 h?

Here is what an *Economic Times* survey found (The Economic Times, 2015):

There would be at least six fields affected, namely, communication, social media, E-commerce, entertainment, shipping, and transportation. The degree, to which each of these would be affected, as of June 2015, is given as follows. Of course, with each passing day, these figures would go on increasing, often at a rate beyond imagination.

Communication

- 1.75 billion smartphone users will be affected
- 182 billion email messages will not be received/sent
- 4.9 million Skype users will not be able to spend two billion minutes

Social media

- 500 million tweets would not be shared
- 4.75 billion FB posts would not be shared
- 70 million photos would not be shared on Instagram

E-commerce

- Retailers will lose $2 billion by way of sales
- 244 million members on Amazon would be affected
- 149 million members on eBay would be affected

Entertainment

- 903.3 million Pay TV subscribers will lose access
- 55.8 million music hours will be lost by Pandora users

Shipping

- 58 million UPS tracking information requests will not be submitted
- 50 million FedEx tracking information will not be available

Transportation

- 16,320,000 cars will not be able to use GPS systems
- 87,000 flights will not be able to use air traffic control

The previous list gives us an idea about how the receiving and transmitting of data has become such an important adjunct to our lives.

Just think of yourself. You probably wake up not by a normal alarm clock but by the alarm set on your mobile phone. Rubbing your bleary eyes, you immediately try to see whether you missed any messages or calls while you slept. As you go about your morning chores, you probably have your iPhone or iPod plugged into your ears, listening to music while you get ready. Oh, you forgot something: you didn't check your email! What a blunder! You check it without losing any time, even at the cost of your morning cuppa getting cold. While you wait for your pickup to work, you pass the time playing a short game on your cell phone, or making out your to-do list for the day, again on your cell phone. And of course, once you get to work, and get logged into your computer, you probably have half a dozen windows open, switching from one task to another, often seamlessly, or so it appears. While you take a break at work, you go on to perform other chores: booking tickets and hotel for the next vacation, paying utility bills, shopping for your friend's birthday, chatting with long lost school friends, reading email, listening to music, even making a call through Google: so much is possible with just one gadget—either the computer: desktop or laptop, tablet or phablet or even your smartphone.

Coming to the first question now: who is adopting ICT? What is the age group, their gender, their socio-economic status, their nationality or religion? ICT adoption seems to be cutting across all boundaries, swiftly, but surely. While we did have, and still do have, a **digital divide** that separates those who have access to computers and those who do not, this divide is fast vanishing. Technology is changing this world, albeit certain groups have been showing resistance. The Amish have long been known for the ways in which they have tried to limit the use of technology in their homes and at work. But such restrictions cannot last very long. The Amish culture is also changing and it has become difficult for them to maintain their simplicity or to keep their youth in control. Be it in remote Africa, or the corners of Siberia, from Korea to Bangladesh, from India to the United Arab Emirates, from the remotest rural areas to the crowded metropolitan suburbs, the computer and its half-sister, the mobile phone is being adopted at a rate beyond imagination. Women are no longer behind their male counterparts, while children often outdo their parents in the use of these gizmos. Grandparents, too, have joined the fray, using technology for a variety of purposes, otherwise difficult to manage. Not only are these older adults adopting technology at a phenomenal rate but they now constitute one of the fastest growing segments of internet users, using these devices for communication and social support, including contact with grandchildren, dealing with mobility and grief, seeking information, health information and medicine monitoring, leisure, and hobbies (see review by Wagner, Hassanein, & Head, 2010).

Digital divide: *the line which separates those who have access to computers and those who do not*

While the previous chapters attempted to give the technology designer and developer clues regarding the human apparatus, both mental and physical, this chapter will try to enable greater insight into what happens to us when we adopt certain technologies, with special emphasis on the adoption of Information and Communication Technology. As far as the designer of any technology is concerned, its use will be, without any doubt, advantageous for the user. Then, why is it that

Table 5.1 Internet users around the world by region (as of June 2016)

Region	Percentage of users
Asia	49.6
Europe	17
Latin America/Caribbeans	10.7
Africa	9.4
N. America	8.9
Middle East	3.7
Oceania/Australia	0.8

Source: Internet World Stats; www.internetworldstats.com

certain technologies fail in the market? Why is it that many features of gadgets remain unused? Even more important, are anonymity and the ease of change of identity through the use of twenty-first century communication gadgets the reason for rising crime rates, especially rates for rape and cyber crime? Does the use of ICT technology lead to problems in daily life? Are we becoming different people as we interact with such technology? Is it affecting our self-concept? Are our relationships being damaged? At the same time, what are the advantages to the use of ICT technology? How is it helping us cope with the ravages of twenty-first century life? We will try to examine these and many more such issues and hope to provide answers to at least some of them in the rest of this chapter.

If we go by the statistics provided by the *Internet World Stats* (as of June 2016), there has been a growth of 900 % in total internet usage around the world in comparison to what it was in the year 2000. Looking at the regional breakup of internet usage as of June 2016, Asia accounts for the largest number of users (49.6 %). Table 5.1 gives a continent wise breakup while Table 5.2 provides data on the demographic characteristics of internet usage by different groups in the US.

The previous data (Table 5.2) clearly reveals that as far as internet usage is concerned, demographic variables play a very important role. But are people also engaged in a host of other activities? What are these? As far as children and youth are concerned, we do have some insights. Witt, Massman, and Jackson (2011) reported in their NSF funded study that over their 3-year longitudinal research on young children, text messaging had increased and overall, its volume had gone up several hundred times in recent years. The most interesting conclusion of their study is the fact that the modern youth is likely to spend almost half of his day in using his communication devices: 90 min watching television, 46 min on internet, 73 min on video games, 150 min listening to music, 60 min on the phone and text messaging for 30 min, 25 min watching movies and emailing for 20 min: grand total = 554 min (Witt et al., 2011). Beyond the time spent in these activities, the teenager also needs time to sleep, eat, bathe, and commute to and from school. How is all this managed during the limited period of 24 h? Observing the ways of life of this generation and how it manages life surrounded by all sorts of newfangled gadgets and gizmos, Larry Rosen (2010) wrote:

> "Just peek into the bedroom of any preteen or teen and you will see at least six forms of media engaging their attention at the same time. Our research shows that they are likely to

Table 5.2 Internet users in the US as of 2014 (Pew Research Center Report, p. 12)

All adults	81 %
Sex	
(a) Men	80
(b) Women	81
Race/ethnicity	
(a) White	83
(b) African-American*	77
(c) Hispanic	71
Age group	
(a) 18–29	89[d]
(b) 30–49	86[d]
(c) 50–64	84[d]
(d) 65+	56
Education level	
(a) High school grad or less	66
(b) Some college	89[a]
(c) College+	94[ab]
Household income	
(a) Less than $30,000/year	65
(b) $30,000–$49,999	84[a]
(c) $50,000–$74,999	92[ab]
(d) $75,000+	96[abc]
Community type	
(a) Urban	81
(b) Suburban	81
(c) Rural	79

Among adults, the % who use computers at workplace, school, home, elsewhere use computers

Source: Pew Research Center Internet Project Survey, January 9–12, 2014. $N = 1006$ adults. Note: Percentages marked with a superscript letter (e.g., a) indicate a statistically significant difference between that row and the row designated by that superscript letter, among categories of each demographic characteristic (e.g., age)

*$n = 94$ for African-Americans. PEW RESEARCH CENTER

have the TV on; have music coming from an iPod, CD player, or computer; have the Internet running with multiple windows showing one or two social networks; be IMing at least three or more friends; and either be talking on the phone or, more likely, having a rapid string of back-and-forth text messages. Add to that a dash of You Tube, Twitter, and a plate of food and you have the typical teenager consuming a hefty daily diet of media" (p. 12–13).

There is no doubt that the i-Generation kids born around 1990 are extremely familiar with the latest technology. They may have well over 200–300 "friends" on Facebook or MySpace with whom they not only talk, but also seek advice from and share their feelings with. Despite knowing that these cyber friends are different from those defined in the traditional way, they feel that "their cyber world is a place to explore their identity" (Rosen, 2010, p. 15) and thus serve a very important purpose in their lives.

> "If you are in a shipwreck and all the boats are gone, a piano top buoyant enough to keep you afloat may come along and make a fortuitous life preserver. This is not to say, though, that the best way to design a life preserver is in the form of a piano top. I think that we are clinging to a great many piano tops in accepting yesterday's fortuitous contrivings as constituting the only means for solving a given problem"—Buckminster Fuller (1895–1983; www.qotd.org).

If this was the case in the mid-1990s, it certainly is not so today—technology is changing faster than the human is able to adjust to those changes. Moreover, rather than clinging to old piano tops, technology designers often create newfangled gadgets that can be at best described as a piano-top shaped life preserver—how else do you explain the curved LED TV introduced only recently but already coming in for a lot of flak.

Faster channels of communication show corresponding changes in the activities of the generation that receives such communication. Teasing out information from Toffler's work (The *Third Wave*, 1980) on generational change, it is clear that each new wave of change is coming more rapidly than the previous one:

- The agricultural society: 3000 years
- The industrial society: 300 years
- The computer society: within just a few years, for example
  - Baby boomers: Face-to-face communication, telephone
  - Generation X: email, mobile phones
  - Net-generation: Text messages, Facebook
  - I-generation: Text messages, WhatsApp, Twitter, Skype, Facebook, i-chat, iPhone, and more

While it took so much time to experience change from the agricultural to the industrial era, as far as living with computers is concerned, from baby boomers—who are retiring or coming close to retirement—the process of change has been amazing. Computers were developed and introduced for computing and teaching purposes, but today, they are being used for almost everything we do.

With each coming generation, from the key pressing generation who worked on typewriters and calculators and is now past 50 years of age to the screen-swiping youth of the twenty-first century, the exposure to technology has changed and so has the dynamics of its adoption. Each generation has, in fact, adopted or has been forced to adopt technology, but adaptation to technology may be a different matter. The ever-increasing pace at which technology is changing only adds to the problems.

Way back in 1954, Piaget attempted to understand cognitive development among children and concluded that it is marked by a series of adoptions and adaptations,

assimilation and **accommodation**, as he preferred to call it (Piaget, 1954). Technology demands a new set of assimilations and accommodations, with each successive generation finding it easier to do so even as older generations struggle. The rest of this chapter will attempt to explain the dynamics of these assimilations and accommodations to technology and the differences therein due to age, gender, and type of technology along with the interaction between them.

> ***Assimilation***: *gathering new information from the environment*

> ***Accommodation***: *making changes in one's cognitive system in order to adapt to the information assimilated*

5.1 Instant Messaging

> ***Texting***: *using mobile cell phone to send text messages*

One of the latest fads among teenagers and youth is instant messaging, **texting**, IMing, or SMSing. Little did mobile phone creators realize that this feature, that they had originally devised to facilitate communication with customers, would take the world by storm. Ask any teenager and they would say that they text much more than they actually call, with most text messages being sent to friends. One of us recently asked a group of 20-year-old students the number of text messages they sent each day—the reply was that it could very easily cross the figure of 200! And these are figures when, in most countries across the world, the market penetration of mobile telephony for the younger age group is lower than even 50 %. What we are witnessing is, thus, apparently only the tip of the proverbial iceberg, because as more and more people start using mobile phones, the number of text messages will also be on the rise. As a survey in the UK points out: 46 % of people within the age range 25–43 find it difficult to even imagine living without their phones, yet they hardly used it for its initially designed purpose: telephoning while on the move. Instead, it was used more for texting (Ananova, 2003). Almost 10 years later, a study on how people react to restrictions on texting shows that it leads to high levels of both measured and felt anxiety (Skierskowski & Wood, 2012). What is even more surprising is, unlike our general view, that men tend to use technology more than women, it is the latter who engage in texting more than the former, refusing to switch off their cell phones even at night and having the tendency to feel lost if they ever forget it at home and venture out (Ananova, 2003). Maybe, it provides them a sense of security in a world beset by various types of exigencies, not the least of which is sexual harassment.

For some people texting becomes almost a compulsion, so much so, that they engage in sending and receiving SMSs even in situations where such behavior would definitely be considered out of place. Harrison, Bealing, and Salley (2015), from Penn State University, report that people are unable to resist texting even while attending a funeral or a church service, or while using the washroom. On May 23, 2016, CNN showed a female West Point cadet texting while marching into graduation. Can an explanation be offered for such inappropriate behavior? Is it simply the thrill of being connected? Harrison, who is trained as an evolutionary psychologist, is of the opinion that the 'buzzes' and the flashing lights on the texting device act much in the same way as danger signals do. Probably, a remnant from the days when survival depended on the ability to detect threatening signs at the earliest, we are unable to resist attending to the signals accompanying the message. Thus, even if it is a telemarketer's message, which in all probability, would be a useless message, we cannot refrain from looking at it.

So great can the obsession for texting become that it seems to be one of the causes of rifts in romantic relationships (McDaniel & Coyne, *Brigham Young University Release*, 2014). The survey brought to the fore that 62 % of the respondents felt that technology interferes with the free time meant to be spent solely on a one-to-one basis; 35 % of the respondents mention that partners will pull out a phone mid-conversation (often without even an 'excuse me') to either read an SMS or send one, while, 25 % responded that active texting occurs even during face-to-face conversations. What is the effect of such texting? The authors sum up by saying,

> "Texting is like a circular process that people become trapped in where allowing technology to interfere, even in small ways, in one's relationship at least sometimes causes conflicts, which can begin to slowly erode the quality of their relationship."

Though caught unawares, the earlier findings show that psychologists and sociologists alike have already become aware of this new phenomenon, named "**digital sociality**" (Thompson & Cupple, 2008), referring to the melding of technology to social interaction to such an extent that one cannot be thought off without the other.

Digital sociality: *melding of technology and social interaction*

5.1.1 Why Do People Text? Texters Versus Talkers

An important question is 'why do people text?' The advent of the facility to send text messages through mobile telephony was probably to enable speedy transmission of short messages during emergencies. Creative as man is, he soon found new uses for this feature and, today, it has become a tool for managing social interaction, a function about which the designers of the feature had not even thought of. Truly, it is another example of technological exaptation. Almost a decade back, Donna and Frazer Reid (Reid & Reid, 2004) made a distinction between people whom they called "**texters**" and "**talkers**." As the name suggests, the former are those who prefer to use their mobile phone for texting while the latter prefer to use it for talking. Their study, along with other studies (e.g., Rettie, 2007), clarified that it is not the preference per se rather the reasons for the preference that are important. First, texting allows them to create a new social environment, very different from the real-world social environment. Texters maintain interconnections with a small well-knit

Texters: *those who prefer to use mobile phone for sending text messages*

Talkers: *those who prefer to use mobile phone for talking*

circle of people who text each other sometimes even hundreds of times a day. The reason for this is that the text circle provides them a haven wherein they may even maintain a self-image incongruent from their real self. The anonymity and asynchronicity of texting creates new affordances, one in which they have more time to frame their messages, while phone calls are like face-to-face conversations where the auditory cues, such as tone of voice, give away more than what is actually been said while at the same time, the lack of visual cues provide very little information about the caller. This double bind is avoided by texting, being more commonly used by people who are shy, hesitant, or for some reason suffer from phone aversion. Similarly, adolescents who suffer from social anxiety (Pierce, 2009) or even language impairments (Conti-Ramsden, Durkin, & Simtin, 2010) feel that texting provides them more time to respond and at the same time relieves them of social pressures. Not surprisingly, texters avoid talking to people over the phone in the presence of others and may even switch off their phones in public places. Talkers, on the other hand, can engage themselves in a phone call which may last for hours together.

In their article, *To text or not to text*, Skierskowski and Wood (2012) have very aptly discussed another facet of texting. A major finding is that the younger generation tends to use different media for communicating with different people, based on the strength of the relationship. Texting is reserved for peers with whom they have strong social ties, while with people with whom ties are weak, face-to-face communication or social networking sites are thought to be better, and email is reserved for parents and people of authority (Van Cleemput, 2010). Texting is used more for maintaining links with existing social ties than for developing new ones (Bryant, Sanders-Jackson, & Smallwood, 2006), unlike social networking, probably because texting requires knowledge of the phone number of the person to whom the message is being sent. Email ids are easier to obtain and, in general, people are more wary of revealing phone numbers than email ids.

Another recent addition to texting is what has been called **sexting**, that is, the adding of sexual pictures and messages to the text. Not only is it being used by people who experience attachment anxiety (Weisskirch & Delevi, 2011) but also by cyber stalkers and for online sex solicitation. It is the latter that is frightening. Attempting to delve into the earlier findings, Drouin and Landgraff (2012) report the results of a study relating texting and sexting to attachment styles. They noted that people, who manifest a secure attachment style, showing neither anxious attachment nor avoidant attachment, find texting good enough for maintaining romantic relationships. However, people characterized by **insecure attachment** styles, that is, those who are either overly anxious about the relationship or are not sure of themselves use sexting to a greater extent. Also, men were found to resort to sexting to a much greater extent than women. Of course, this could simply be a function of what is considered socially acceptable behavior for women.

Sexting*: adding sexual pictures to the text message on mobile phones*

Insecure attachment*: attachment marked by anxiety and lack of self-confidence*

With so many communication channels open to them, another question is why the choice of instant messaging? Does IM offer advantages not found in mobile telephoning? It certainly seems so. For one thing, it is much cheaper than placing a call. Second, call dropping, a phenomenon so common in mobile telephony is avoided because IM is based on only one person's network. You send a text message whenever you have the time and the inclination; it will get delivered as and when the other party's network responds. You do not breach anyone's private time, the receiver can look at the message when she wants to and has the time for it. As Faulkner and Culwin (2005) put it, it is more like email. Third, and probably more importantly, it is private and discreet (Davie, Panting, & Charlton, 2004). While calling a person often tends to run the risk of revealing what is being talked about to onlookers, texting reveals nothing, neither the person nor the message. This becomes an important consideration when one realizes that only one-third of the messages serve any practical or functional purpose, the rest are for a variety of other functions such as friendship maintenance, social and even romantic relations (Thurlow, 2003).

Last but not the least, doesn't, every message alert provide a thrill? You may talk to a person only once a day, while you may interchange hundreds of messages during the same time giving you the feeling of being connected at all times and enhancing your sense of belongingness. Every "ping" of the cell phone gives a new high to the receiver, you are being remembered. At an age when establishment of self-identity is ever so important and is determined to a great extent by the degree of social acceptance in the peer group, texting seems to be playing an important psychological purpose (Walsh, White, & Young, 2009). Even marketers have taken note, with every new model of mobile phone being advertised for its sleekness and as a fashion statement, an extension of the self (Ling, 2004), to make young people feel ashamed about their "old-fashioned" phone, even though it may be just a few months old. Apple, too, seems to have caught on, as we can see from their colored iPhones. Phones now even come with interchangeable shells so that one can change them according to one's mood or to make them appear new.

5.1.2 Effects of Texting

> ***Social anxiety:*** *anxiety caused by the thought of meeting people*

In view of the heavy usage of IM facilities on the cell phone, especially by teenagers and youth, it is but natural to start thinking of the possible effects on the senders and receivers of such messages. One salient question would be whether interacting with friends through text messages helps to overcome **social anxiety**? Do these teens gradually improve on their social skills? While the empirical work cited earlier does clarify that it is people who are high on social anxiety who tend to be texters, we do not know whether anxiety relieved through text messages helps to make the individuals generally less anxious. But it does seem probable, because of the very fact that texting does help many of them to resolve the identity crisis, which has been seen to be an important precursor to the lowering of social anxiety. There are, however, other aspects on which empirical results are unequivocal such that conclusions can safely be drawn. Some of them have been discussed later.

5.1.2.1 Texting and Linguistic Skills

It is interesting to note how communication among teenagers has undergone change due to texting or using what has come to be known as **textese**. For example, "you" has been replaced by a short form "u"; I is replaced by i; 2nite for tonight, and so on. Does it affect formal English writing? At least this writer has just made an error while typing this manuscript. I did type "tonite," that is, my texting habits got my formal writing habits confused. The importance of texting for the development or deterioration of linguistic skills is clear from the fact that an entire issue of a journal, the *Journal of Computer Assisted Learning* has focused on it (2011, 27-1). Surprisingly, and much against the widely held myth that using abbreviations while sending SMSs causes deterioration in the linguistic skills, the bulk of the authors note that generally there is a positive relationship between textese use and literacy skills across a variety of samples drawn from different countries and across ages (Kemp, 2011). They contend that textese helps students understand sound patterns or phonetics thereby helping them in their spelling. At the same time, Kemp notes that "as technology develops, the communication behavior of technology users of all ages will change as well, and so may the links between this behavior and traditional literacy skills" (p. 3).

5.1.2.2 Texting and Mobility

> ***Textese**: abbreviated words and short forms of words used while texting*

> ***Cognition first strategy**: giving greater priority to cognitive tasks than to physical balance while walking*

> ***Posture first strategy**: giving greater priority to act of physical balancing than to cognitive tasks*

It is not only communication skills that are changing. The hazards of texting while walking have only recently been understood, though considerable effort has been devoted to the study of mobile cell phone usage and especially texting while driving (e.g., Drews, Yazdani, Godfrey, Cooper, & Strayer, 2009; Hosking, Young, & Regan, 2009). In a recent article (January 22, 2014), Susan Schabrun and her colleagues, from the University of Queensland have been able to demonstrate the dangers of texting while walking. Based on their own work and that of others (e.g., Schwebel et al., 2012; Demura & Uchiyamal, 2009), they conclude that people who perform cognitive tasks while walking are, in fact, multitasking and are at great risk of falling. Cell phone usage changes the way we walk (Lamberg & Muratori, 2012) since the arm which normally helps us regain our balance in case we trip (Pijnappels, Bobbart, & van Dieen, 2005) is now being used to hold the phone. Schabrun contends that under normal

> ***Posture first strategy**: act of physical balancing getting priority over cognitive tasks while walking*

> ***Cognitive restructuring**: changing the ways in which one thinks and analyses situations*

conditions, a "**posture first" strategy** is used, or, that the act of physical balancing gains priority over cognitive tasks. However, when physical hazards appear low, a "**cognition first" strategy** could be used. This hypothesis is further strengthened by the finding that in the aged, though physical balancing tasks normally get priority over cognitive tasks, but based on postural reserve, hazard estimation and expertise, cognitive tasks may get priority over the act of physical balance. While under normal circumstances, the hazard of falling is not too great, once the person becomes habitual of giving priority to cognitive tasks over physical tasks, unexpected hazards could pose a problem. This would be especially true for the younger generation who are still in the prime of health and do not feel that walking has any associated hazards. Unfortunately, this is also the group that indulges the most in texting while walking. As Beth Ebel of the University of Washington puts it, "they are as if on autopilot. Your eyes are off the road and your brain is also off the road." While this hazard has been found to be associated with all types of social distraction, it is most pronounced while text messaging (Thompson, Rivera, Ayyagari, & Ebel, 2012). The interplay between gait, falls, and cognition has become a major cause for worry not only among the aged who are most prone due to normal aging processes, but also among healthy children, youth, and young adults who are in a habit of texting while walking. The degree of concern is so great that psychologists are of the view that interventions need to be planned and success rates of these interventions worked out (Thompson et al., ibid). In older adults a combination of motor and cognitive therapy is seen to lead to fewer falls and injury (Segev-Jacubovski et al., 2011) but interventions for youth would probably need to be based more on **cognitive restructuring** than anything else. They need to realize the dangers of texting while walking and that despite the fact that they generally do not trip or fall, one can never predict what will be lying on the road, or where there could be just a root bulging out of the ground or a small pebble, on an otherwise smooth pathway, but enough to make one trip when one is busy typing a message on one's cell phone. Overconfidence can lead to more accidents than underconfidence. Just as there are banners proclaiming the risks of driving while drinking, or, there are statutory warnings on packs of cigarettes, so too greater awareness needs to be created regarding the dangers of messaging while moving.

5.2 Mobile Apps and Their Use

What do you use your phone for? If it is a smartphone, and in all probability it is, you would use it for a myriad other purposes apart from using it to telephone people. You may be playing games while you commute; you may be checking the weather report before you leave for work in the morning, or even while you are at work (one doesn't have to be hooked on to a TV set any longer for finding out about the weather); you may be paying your utility bills, booking tickets for your vacation, navigating your way through a busy town, communicating over social media sites, watching a baseball match or getting live updates about a football match you could not attend, or watching a movie, etc. The list would actually be much longer and goes on getting larger every day. According to a report in the *New York Times* some 5 years back, Corasaniti (2010) gives us some idea as to how people have been using their smartphones. As many as 59 % use smartphones to download apps, 61 % to play games, and 55 % to check weather. Have you ever thought of how your mobile phone enables you to do all this and so much more? The answer lies in that three letter word 'app' denoted as the '*Word of the Year*' by the *American Dialect Society* in the year 2008.

> ***App**: a computer program which enables smartphones to perform different activities*

> ***Cognizant computing**: understanding the user by collection and collation of historical data*

'**App**' is a short form for application software. It is basically a computer program which enables your smartphone to perform a whole host of activities, often much more than your computer can, and, even more so, it is extremely user friendly. From downloading the app to using it, it is child's play and most children of today are able to do it, often faster than their parents and, of course, much faster than their grandparents. These mobile apps started appearing in 2008, with the first app store being the Apple Apps Store for iOS users, to be soon followed by almost all other mobile companies introducing their own apps. Today we have the Google Play, the Samsung Play Store, the Windows Phone Store, and the Blackberry App World, to name just a few common ones. As we all know, some of these apps are free while others have to be bought. By 2012, there were 650,000 apps available for iOS users alone, being downloaded by 200 million users (Indvik, 2012). Today, Apple boasts of over one million apps. A report in a well-known daily newspaper, the *Guardian* pegged app revenue at $26 billion by the year 2013 (*The Guardian*, 2013), while according to the leading IT analysis company, Gartner Inc, the total number of app users by 2017 will be 268 billion and will cross revenue figures of $77 billion (Gartner Inc., 2014), making it the most used computing tool across the globe. It also predicts that apps will not be restricted to mobile phones but will impact home appliance, cars, and even wearable devices. Using what is known as **cognizant computing**, apps will be developed to collect historical data of the user and based on that be able to perform tasks as simple as the turning on of a water heater to calling a doctor or a rescue team without the user

Box 5.2: Cognizant Technology: Can a Computer Judge Your Personality Better Than Friends and Acquaintances?

John Bohannon (2015a) provides evidence to show that Facebook data can be used by a computer to understand and predict a person's personality often to a much higher degree of accuracy than some of our closest acquaintances. The origin of the idea for the study is as interesting as the findings themselves.

Two researchers from the University of Cambridge, UK, one a psychologist, Youyou Wu, and the other a computer scientist, Michael Kosinski, watched *Her*, a sci-fi movie about a man who fell in love with his computer's OS. According to Wu, "by analysing his digital records, his computer can understand and respond to his thoughts and needs much better than other humans."

They decided to see whether this is possible in real life. Using a Facebook app, *myPersonality* they started collecting data. In a paper published in the *Proceedings of the National Academy of Science* (*PNAS*) in 2013, the researchers provided evidence which proved that the pattern of Facebook likes is enough to predict many of the respondents' personality traits. So much so, that the results showed that Facebook likes were almost 15 % more accurate than assessments by friends. Only spouses were better than the computer. By January 2015, the app went viral and had over four million users.

Psychologists who were not part of the study note that this very clever use of machine data shows very clearly how personality affects life choices. It is not difficult to envisage that soon marketers will catch on to the idea and use Facebook and other social network site data to promote their products.

Source: Bohannon (2015a)

initiating the action. An idea of how cognizant technology can help is elucidated in Box 5.2.

Is it not surprising that even with such widespread usage, there is not much that we know about the ways in which apps affect us, or how softly, yet surely, it preys upon us; impacting us in ways we are not often aware of? As you use the apps on your phone, you may have noticed that apart from what the app is actually helping you to do, it also contains advertisements. This is one reason for why app companies want to make the app popular: it provides revenue for the app store through getting people and companies to advertise on it. Since most of these advertisements stay on the screen for as long as the app is on, you become a captive audience to it, and psychological research shows that even if you do not directly focus on it, your cognitive system does become aware of it and even processes it and saves it. Much of this is because of the small size of the screen on your mobile device, due to which the entire screen can be looked at through a single glance without any head or eye movement. So, as your system takes in other information that is relevant to you, you are forced to take in what the app provider wants you to look at. Another effect is that the mobile operating system and mobile company is able to collect personal data about the user which it can then pass on to other companies and agencies that use it for their own advertising purposes. With touch phones so much in vogue, one may often find that one has touched a link inadvertently and that starts downloading an app. Before you realize the app has been saved on your phone and you start receiving messages not only telling you of the usefulness of the app but also beguiling you through rewards and incentives for using it. With phishing and other fraudulent uses of personal data being on the rise, should we not have awareness drives for app usage too?

5.2.1 Developing a Mobile App

In order to be useful, an app must cater to the specific user for whom it is meant, it must meet the needs of the user, her preferences, and the context in which it is to be used. A wide variety of factors have been seen to affect user experience as far as mobile apps are concerned. Figure 5.1 presents one model which can help the designer of an app.

Today's app user is not looking for features in the app but rather for apps that are functional and easy to use. From Fig. 5.1, one can see that user experience is a complex phenomenon depending on the interaction between not only user characteristics and product characteristics but also sociocultural factors and the context of use. Since the number of antecedents is so large, specific methods have to be devised which will be able to capture all the information. The authors of the model are of the view that two methods that would prove both viable and useful are the interview method and the observation method. Once all the information has been gathered, one can start designing the app. But how does one go about it? A model that has been successfully used in Bahrain for the development of an app for mobile banking is described in Box 5.3. Another model developed by Paul Lin (2015) uses a five step user-centric design process formulated by the Stanford University Design School:

- **Empathize**: we must create a persona of the user keeping in mind the exact characteristics obtained from the University of Finland model (Fig. 5.1).
- **Define**: the purpose of the app and the operating system which will support the app have to be clear. As has been pointed out in the foregoing paragraphs, apps can be used for a variety of purposes and for a host of operating systems.
- **Ideate**: once you have the user clearly in mind, along with the purpose and the operating system, the next step is to bring these ideas to the table. In other words, it calls for the collation of all the information gathered in the previous two steps.
- **Prototype**: a prototype fulfilling the earlier can now be designed.
- **Test**: the crucial phase is this phase, where it will become clear as to whether the app fulfills all the criteria decided upon.

Box 5.3: A Model for M-Banking App

One of the more recent uses of mobile apps especially in developing countries is in various government services and in banking. The main reason why banks, the world over, are encouraging net banking and mobile banking is that as footfalls to the bank decrease, the total staff requirement diminishes, cutting down on the total cost of services and thereby increasing the profitability of the bank. However, one hurdle is getting customers to use mobile banking services. This becomes even more difficult when a significant proportion of the population is only barely literate or semiliterate as in a large number of developing countries. In such cases, as against in mature economies, understanding the essentials of factors that determine the adoption of mobile banking services can provide the management useful insights.

As far as Bahrain is concerned, the total penetration of mobile phones is 17 %. Despite this, in comparison to the total penetration of banking services, that for mobile banking (M-banking) is extremely small. There is thus considerable scope for increasing M-banking usage in Bahrain. In a recent issue of the *International Journal of Managing Information Technology*, Alsoufi and Ali (2014), present a model that can be used to understand the antecedents of M-banking user adoption. According to them,

> "Mobile application in Bahrain has been increasing rapidly, and expected to increase in the coming years in the banking sector. Most of the banks have started to launch mobile banking services. With intensive bank competition and the popularity of mobile device use, there is an urgent need to understand the factors that would entice customers to adopt mobile banking" (p. 1).

Based on classical work in the area of attitude development, and especially the Theory of Reasoned Action (Fishbein & Ajzen, 1975), the authors aimed at extending and testing the Technology Adoption Model (TAM) to incorporate the role of factors influencing customer's perception toward M-banking adoption. While the model was extended to include perceived cost and perceived risk, the test of the model revealed that the intention to adopt mobile banking is mainly affected by two factors, namely, Perceived Usefulness and Ease of Use.

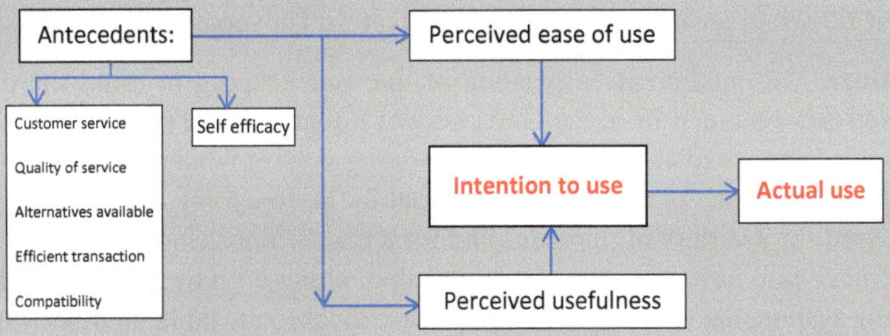

Source: adapted from Alsoufi& Ali, *International Journal of Managing Information Technology*, (2014)

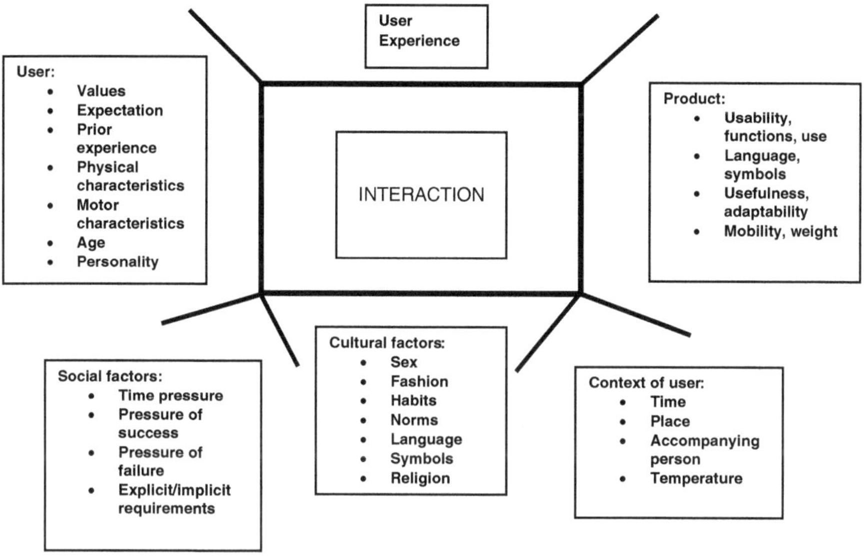

Fig. 5.1 Diagrammatic representation of a model for user experience (source: adapted from Arhippainen & Tähti, 2003, University of Finland)

5.3 Online Communication and Use of Social Network Sites

Another form of technology that is in common use today is online communication. Starting with the advent of the lowly email, it now includes a variety of related forms, such as chat rooms, Facebook, Twitter and LinkedIn, WhatsApp, and blogging. Not only does the youth use it, its reach and range has caught the fancy of even advertisers. Political leaders are not to be left behind: President Obama of the US used it for his election campaigning and thereby added a new dimension, helping to widen the Net even further. And then, we have Donald Trump, who seems to have an insatiable need for tweeting! Yet, it is its usage by the younger generation that has propelled psychologists, educationists, and policy makers to make attempts to understand the phenomenon better.

Research by the Pew Internet and American Life Project had found that by the year 2005, 87 % of teenagers in the US were hooked to the net and as much as 51 % were using it daily for a variety of purposes ranging from gaming to gathering news, online shopping, and generally researching on topics of interest (Lenhart, Madden, & Hitlin, 2005). By the year 2015, the same organizations, namely, the Pew Internet and the American Life Project, reveal the following in their *Teens, Social media and Technology Overview,* 2015 (Lenhart, 2015):

- Despite the availability of a large number of social media sites, Facebook continues to the most popular site.
- At the same time, 71 % of the respondents said that they used multiple social media sites.
- 92 % of the teens report going online daily, while 24 % are online almost constantly.
- Girls seem to dominate the social media while boys prefer to play video games.
- The typical teen sends or receives 30 messages per day through social networking sites.

The author of the report is also of the view that much of this 'frenzy' has been caused by the easy access that teenagers now have to mobile phones and especially to smartphones and also the fact that many internet providers offer social media site messaging for free.

Another major factor that has caused the rapid growth of the use of online communication channels is the need to connect with others and to explore their identity (Boyd, 2006). Online communication makes it possible for an individual to communicate with others without disclosing one's identity. In fact, the communication might continue for a long time without even knowing the real identity of the person. There is another vital difference between communication in the virtual setting and that in face-to-face communication. It is possible to keep a record of all the communication that had taken place. There are many examples of how this advantage can become a source of anxiety when the authenticity of the communication transpiring over the electronic medium becomes suspect or if the record is misused as in the case of **cyber stalking**. However, the technology also affords an opportunity, without any backlash, to participate in online forums without requiring any evidence of physical identity such as race, gender, age, or disability. You may call it social inclusion at the highest.

> ***Cyber stalking:*** *using information gathered through social media sites to threaten and harass a person in virtual space*

It seems as if the teen of today has rewritten the rules of socialization. While in the past, the teenager would be focusing on his personal looks and attractiveness in an effort to buoy his self-image and self-esteem, the teen of the twenty-first century is more preoccupied about how to present oneself to his ever-increasing band of online friends (his '**digital profile**') and what Boyd calls the "**digital public**" (Boyd, 2006). With adults becoming over restrictive, the online media often provides a safe haven, away from the eyes of peering adults and yet within the physical space allowed by these very same adults.

> ***Digital public:*** *people who are using the social media sites and the internet*

> ***Digital profile:*** *personal profile that a person posts on the net and on social media sites*

Using Erikson's theory of psychosocial development, Valkenburg and Peter (2011) have developed an interesting model regarding the features of online communication and its relation to tasks of adolescent development. According to Erikson (1959), the most important task for the adolescent is identity formation accompanied by a deepening sense of intimacy and the development of sexuality. The immature teenager starts experimenting and teasing out the behavior of others in response to his own self-disclosure and gears his self-presentation accordingly. This leads to the development of self-esteem in the child, with resolution of the identity crisis as the end result. Conversely, self-disclosure and consequent self-presentation can go awry, vitiating social behavior and devastating the still forming identity of the child. According to Valkenburg and Peter (2011) one crucial reason for using and abusing the internet is as an aid to this very important developmental task, namely, psychosocial development, by making self-presentation and self-

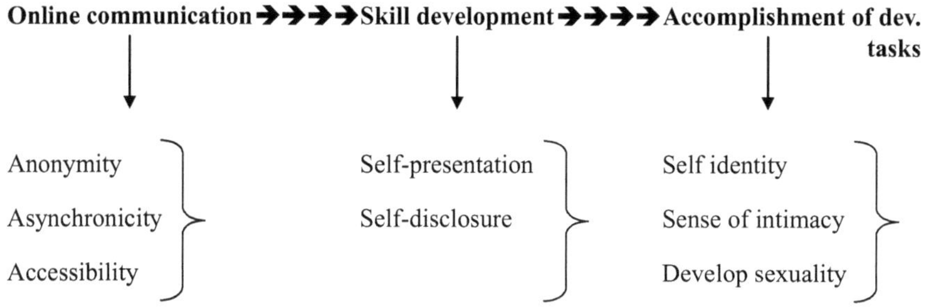

Fig. 5.2 A diagrammatic representation of effects of online communication on development of the self (adapted from Valkenburg & Peter, 2011)

disclosure less stressful. The **three A's of online communication**, namely, anonymity, asynchronicity, and accessibility provide the child with that degree of comfort that face-to-face communication often fails to do. One can choose to disclose what one wants to and present oneself the way in which one wants to: you can change not only your name, but also your nationality, your location, or even your gender—who will know? Even more importantly, body language and gestures that often serve as a giveaway of the feelings of the person can be edited, or even washed out. One can just imagine what this does for people who are otherwise shy or people who want to talk about and discuss topics normally considered social taboo.

> ***3 A's of online communication**:: anonymity, asynchronicity, and accessibility*

To put it more succinctly, a diagrammatic representation has been given in Fig. 5.2 based on the model by Valkenburg and Peter (2011).

The model is in line with a large bulk of the research (e.g., Jackson, von Eye, Fitzgerald, Zhao, & Witt, 2010; Pierce, 2009) which points out that high internet use and especially use of the social media is associated with low self-esteem, hesitancy in making friends, and the fear of social failure. The personality correlates of Facebook use have also been looked into (Orr, Sisic, Ross, Arsenaeault, & Orr, 2009) and findings unequivocally reveal that low self-esteem and the hesitancy to make friends is the major cause of the high use of social media sites and even cyber crime (such as **cyber bullying**, cyber stalking, and online sexual solicitation (Tokunaga, 2010).

> ***Cyber bullying**: using social media sites to bully people*

5.3.1 Effects of the Internet and Computer Technology on Children

One of the disadvantages of high internet usage, especially among children, is that their information processing habits are changing. As Rowlands et al. (2008) put it, the new generation is "hungry for highly digested content" and their "information seeking behavior can be characterized as being horizontal, bouncing, checking and viewing" (p. 294). By horizontal, the reference is to the skimming activity so common today. Gone are the days when children would pore over not only books but also encyclopedias in the hope of finding some information to help them. With search engines offering options for advance searches, one can go on refining one's search terms to the narrowest of narrow slices of information, and no longer need to sit for long hours in a library. This has resulted in what can be called "**power browsing**." Children are becoming increasingly tech-savvy, but their ability to organize, synthesize, and apply information to create new knowledge is decreasing. What Google could be doing to us has been analyzed by Nicholas Carr (2008) in his thought provoking article, *Is Google making us stupid?* Not only is deep reading becoming difficult but as he puts it,

> ***Power browsing**: limiting internet search activity to only that which is most relevant*

> "My mind now expects to take in information the way the Net distributes it: in a swiftly moving stream of particles. Once I was a scuba diver in the sea of words. Now I zip along the surface like a guy on a jet ski" (www.thealantic.com).

However, there are others who feel that researchers such as Carr are simply making too much ado about nothing. For example, Vezina (2011) is of the view that the internet is just one more extension of Wegner's "**transactive memory**," a collective social memory which has existed from time immemorial, from where we can draw information as and when we need it. We do not try to remember everything but instead try to remember the sources for various types of information. Thus, rather than remembering the *Iliad*, one remembers that it is a Greek epic and remembers the name of the friend who specializes in Greek poetry. Is that not easier, and is it not what we constantly seem to be doing? Also, not having to bother about minutiae, we can use the saved mental resources to look for the larger picture (Vezina, 2011).

> ***Transactive memory**: a memory that is not that of a single person but of a social collectivity*

What will be the net effect of the above is yet to be seen. Whether new models would be created, new theories be posited, will be clear only when these netizens reach adulthood.

How about children who are much smaller? Is the electronic media affecting infants, toddlers, and very young children? It certainly seems possible when one thinks of how devices such as the TV and the iPad and iPhone are constantly being used by busy and overworked parents to soothe children. The effect of electronic media on young children under 6 years of age is, however, unclear. Examining this effect strictly from a neurodevelopmental context, it is clear that the embryonic form of the brain at birth undergoes radical changes with sensory bombardment from the external world. During the first 24 months, it is believed that active manipulation of objects, not passive watching of television programs, is important for the growth of the brain architecture. The American Academy of Pediatrics (AAP) recorded in their study, *Zero to Six*, that 59 % children watched TV on an average of about 2 h per day and that their parents believed that it was helping them. On the other hand, AAP assessed the negative effects of such sedentary activity and concluded that the perceived benefits of technology are fewer than the losses, including the probability of increased aggression and fear, sleep disturbances, and other developmental issues. In short, they argued that TV exposure should be limited to a maximum of 2 h per day (Ghose, 2013). Again, with children as young as 2 and 3 years of age playing with the iPad, another possible ill effect could be the permanent impairment of vision. While interacting with Facebook friends helped in developing the identity of the teenagers, to the extent that real-life empathy was positively correlated with online virtual empathy, high FB usage was also related to more depression, anxiety, and narcissism. Of course, it could well be that students who were high on social anxiety depended more on FB than those others who felt comfortable with offline friends (Rosen, 2011).

5.3.2 Computer Use by Older Adults

> "It may be said that the twenty-first century is characterized by two "megatrends"—the emergence and widespread diffusion of new media, and the rapid aging of the population as a result of better health services and improved quality of life in most developed nations. The rapid aging of the population is pervasive, enduring and has no parallel in human history. Moreover, forecasts suggest that the twenty-first century will witness even more rapid aging than did the previous century" (The United Nations' Department of Economic and Social Affairs, 2009) (cf. Nimrod, 2013, p. 46).

Recent surveys show that in both developed as well as developing countries, life expectancy has increased, though with wide differences, ranging from a life expectancy of 45 years in Afghanistan to 83.5 years in Japan (UN, 2010). Naturally, this increase in the life span has led to a corresponding increase in the percentage of the population that can be categorized as elderly. To get some idea, one can focus on the figures provided by a UN report, which predicted more than a decade back, that the elderly population will increase enormously by the year 2050, reaching a staggering figure of 9.1 billion (UN, 2004).

Not enough attention has been paid to either the ways in which older adults use the computer or the problems they face in the use of computers. This is saddening as it is clear that one of the fastest growing segments of internet users are people beyond the age of 50 years (Pew Internet & American Life Project, 2012; Hart, Chaparro, & Halcomb, 2008). Developers of ICT would have to keep this ever-increasing population in mind and design devices in keeping with the changes in the mental and physical capabilities with age, not the least of which is attitudinal change. We do, however, realize that this is no easy task, compounded as it is with the fact that people of the same age group (i.e., cohorts) may show differences in most motor, sensory, and cognitive abilities, which may or may not be related to their chronological age. So while they may be of the same functional or chronological age, they may differ on perceived age (their mental perception of their abilities), social age (societies differ in the age they consider people as elderly), and even cognitive age (how they rank on mental faculties as compared to other people of the same age). One method by which this problem may be managed is by using what has come to be known as **gerontographics**, a method based on the segmentation of the aged population on the basis of similarities in attitudes, needs, and preferences. Galit Nimrod of the Ben-Gurion University, Israel shows us how gerontographics can be used to understand internet use by the elderly (Nimrod, 2013).

***Gerontographics**: a method based on the segmentation of the aged population on the basis of attitudes, needs, and preferences*

The elderly are commonly depicted as **technophobic**, but is this really so, or is it only a stereotype? Current evidence shows that not only is this a stereotype but also that such stereotypes regarding the elderly and their use of ICT are not a characteristic of the Western society alone, but can be seen across cultures, both of the East and of the West. In an article in the *Journal of Community Informatics*, Barbara Neves and Amaro (2012) describe the results of a survey carried out in Lisbon, Portugal on the use of ICT by the graying population. The authors are of the view that,

***Technophobic**:a person who fears, dislikes or avoids new technology*

> "The elderly have traditionally been an excluded group in the deployment of Information and Communication Technologies (ICT). Even though their use of ICT is increasing, there is still a significant age-based digital divide. To empower elderly people's usage of ICT we need to look at their patterns of usage and perceptions" (https://www.researchgate.net/publication/265208632).

Empirical research (e.g., Malta, 2008) shows that not only are the elderly willing to use ICT but that a large number of them are also quite proficient at its use. They have been found to be keen users, using it for communication and for social support (Thayer & Ray, 2006), and also for entertainment and health monitoring information seeking. However, their motivations for using the internet and the problems they face are vastly different from those of the younger age groups. As the Lisbon study cited earlier (Neves & Amaro, ibid) found, that, despite their difficulties, most senior adults do not consider themselves to be technophobic.

Information technology is being seen as a useful adjunct by health service providers, especially those that focus on the elderly. While many aged people would like to use technology such as *Interactive Health IT* for managing health-related issues, the area is still fraught with constraints. An issue of the *Journal of Community Informatics* focuses on the use of technology by the older generation and includes studies from around the world to show how if designed properly, ICT has a very important role to play not only for communication but also for improving the quality of life of the elderly (Gene Loeb, 2012). Similarly, a meta-analysis undertaken by Jimison et al. (2008) reveals many of the barriers and drivers of *Interactive Health IT* use by the elderly (see Table 5.3). Being aware of these factors would help the designer of the systems to improve their services in line with the needs of the user population.

Probably the most consistent finding of the meta-analysis cited earlier is that the effectiveness of the system was a function of the degree to which it provided a complete feedback loop, including assessment of current health status, interpretation of

Table 5.3 Barriers and drivers of interactive health IT use by the elderly (adapted from Jimison et al., 2008)

Barriers	Drivers
Lack of perceived benefit	Perception of benefit to health
Lack of convenience	Convenient access to technology
Overly cumbersome data entry procedures	Amount of time required for interaction
Low clinician participation	Familiar devices
High cost of using ICT	Clinician involvement
	Gradual introduction to technology

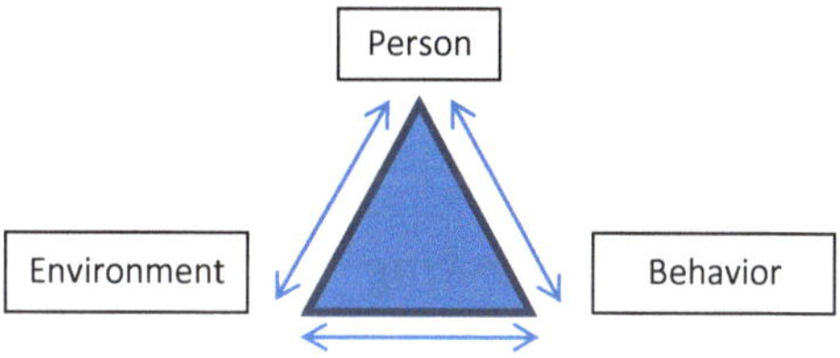

Fig. 5.3 Triadic reciprocity between person, environment, and behavior

this status by the clinician, treatment plans, and communication with patients regarding recommendations and advice which was repeated over time. At the same time, the elderly need to be involved in the design of the ICT product and be given adequate training, since lack of functional literacy and general education was found to be a key factor for not using the internet, computer, or even the mobile phone. But the encouraging aspect is that most of the respondents of the Lisbon study had a positive attitude (62 %) and did not feel that they were too old for technology. The majority also agrees that ICT is essential for the development of the country (77 %) (Neves & Amaro, 2012).

Another issue faced by the older generation is regarding attitudes toward computer use. While many feel that computers serve a very useful purpose, helping them to manage their lives better, there are still as many who see no use of learning how to use computers and in several countries, may even feel that they are too old to learn how to use it. Most elderly people perform a loss–benefit analysis regarding perceived usefulness vis-à-vis perceived risk of using the ICT. This is borne out by various studies (e.g., Liebana-Cabanillas, Sanchez-Fernandez, & Munoz-Leiva, 2014). The complexity of the problems and its multidimensionality suggest that we would probably need some over-arching model in the framework of which, the needs of the older group can be addressed. One such framework (Fig. 5.3) and could be the social cognitive theory posited by Bandura (1986) which sees behavior in the context of a triadic reciprocity between the person, the environment, and the behavior could be useful to understand the needs and problems of computer use by older people (Wagner et al., 2010).

5.4 Computers and Multitasking

Are computers helpful to children? On the surface it certainly seems so because it challenges their brains and they are forced to put various mental capacities to use. Unfortunately, we also have research evidence to the contrary. When we are learning, the input needs time to settle down, and, if we overload our mental operation system, it forfeits downtime that helps to store information and create new ideas (Nettersheim, Hallschmid, Born, & Diekelmann, 2015). The irony is that modern game makers have been creating short duration games to lure customers into using them during brief breaks such as while commuting in a train, waiting for a bus, etc. Little do they realize that these short spells of idleness without multitasking are very much needed and help the brain to recuperate and avoid further fatigue.

5.4.1 Multitasking

> ***Multi-tasking****: the ability to switch between different tasks so as to be seemingly performing all of them simultaneously*

Carrier, Cheever, Rosen, Benitez, and Chang (2009) assessed the amount of multitasking among three generations, namely, the Baby Boomers (those born between 1946 and 1964), the Generation X (those born between 1965 and 1979), and the Net generation (those born between 1980 and the present). The authors of the study were not surprised to note that the younger generations reported not only more multitasking but task switching between a greater number of alternative tasks than the older generations (Carrier et al. 2009). But do these youngsters realize the psychological costs of multitasking? In reality, we hardly ever perform tasks simultaneously, most of the time we switch between tasks, and, each switch is associated with a switching cost (Yantis, 2010). It is because of these costs that every person is not able to be a supertasker, and starts showing sloppy work or begins to show a rise in the number of errors.

Rosen, who has been researching and writing on the psychological effects of technology for over 25 years, laid out some of the negatives of multitasking at the 119th annual convention of the *American Psychological Association* (Rosen, 2011). When observed over even short 15 min periods of study, teenagers were barely able to concentrate at the task at hand. Rather, they were found to be off-task almost every 3 min. Moreover, they observed that the longer the lapse of time, the more the number of open windows appeared on their computers, peaking at 8–10 min and on-task performance declining correspondingly (Rosen, 2012). The researchers also noticed that when the students stayed on the task, their performance was better than when they toggled between windows.

> ***Tech breaks****: period of time when one has to switch off all gadgets or refrain from using them*

In the same lecture, Rosen provided tips on how to overcome such obsessive task switching in the classroom. One idea was to provide what he called '**tech breaks**.' The teacher can start with a 15 min slot of time, when children would have to put away their phones with the idea that they will get a chance to check their FB account after that. Amazingly, when children were given this option, they refrained from checking their phones during the on-task time, probably because they knew that they will be given time for that later. Rosen also suggested that the teenagers be enlightened about the workings of their brain and how and why multitasking reduces performance on the task at hand. Some of the discussion later could well be shared with students helping them gain a greater insight into what they do to themselves when they indulge in this seemingly harmless activity of checking emails, WhatsApp, SMSs, or FB messages.

What are we actually doing and how do we cope with such task switching? A typical multitasking situation involves the ability to switch between two or more tasks while maintaining one's focus on the key task, despite the distractions offered by the other tasks. What also has to be kept in mind is that at any one point of time, a certain activity may be the key activity. But as we switch to another task, task 2 becomes the key activity while task 1 now acts as a distractor. Thus, as we multitask, we are constantly shifting our focus from one key task to another, but distractors would always be present. A good multitasker would, therefore, be one who can continue to work on several tasks seemingly simultaneously. However, we should also remember that our cognitive system has limited attentional resources and as such when we focus on one task, we would often be left with insufficient resources to concentrate on another (see Chap. 3 for greater details). It is because of this limited capacity cognitive system that it is often said that multitasking is actually a misnomer.

One way to overcome switching costs is to attain a level of expertise such that the task can be performed without the voluntary use of attentional resources. Kahneman (2011), in his book, *Thinking, fast and slow* has, through his studies, clarified how we can deal with such a situation. According to him, tasks may be performed either automatically (**automatic processing**) or consciously (**conscious processing**). When we start on a new task, it gobbles attentional resources leaving hardly any for other tasks. Thus, when we were first learning how to drive, even a word uttered by the person sitting next to us would distract us and mistakes would be the result. On the other hand, an experienced driver can be driving her car almost automatically and at the same time, be consciously discussing an important problem with her colleague. A housewife may be cooking, and at the same time, listening to music, chatting to a friend, and, keeping an eye on the children. Thus, our attentional system is such that we can divide the available resources in a manner that is most parsimonious for the tasks at hand, with more and more tasks being performed automatically and the limited attentional resources being used only where necessary.

> ***Automatic processing:*** *performing tasks without using attentional resources*

> ***Conscious processing:*** *performing tasks with the help of attentional resources*

Research also clarifies that there could be such a phenomenon as too much multitasking, but the problem is apparently finding out how much is too much. In fact, constant immersion in multiple tasks can lead to fractured thinking due to which one can have problems in shutting out irrelevant information; deep thinking about a problem becomes near impossible and learning becomes less flexible. In their book, "*The Invisible Gorilla*" Chris Chabris and Dan Simons (2011) have discussed how multitasking is a good example to show that humans hardly understand how their brains work. Everyone thinks that they are part of that 3 % of supertaskers, little realizing the costs associated with it.

> ***Executive attentional network****: that part of the cognitive system which controls all other cognitive subprocesses*

Our understanding of the "why" and "how" of multitasking has come a long way with a recent article by Mary Rothbart and Michael Posner (2015) entitled, *The developing brain in a multitasking world*. Rothbart and Posner attempt to explain the brain systems and processes behind this seemingly amazing ability. The prime system appears to be the **executive attentional network** that enables effortful control despite distracting stimuli. It is this system that directs our orienting system and our alerting system (both subparts of the attentional system) toward the goal at hand. At the same time, it controls our behavior and our emotions through a series of simultaneously operating inhibitory and excitatory processes, inhibiting competing stimuli from drawing our attention and at the same time arousing our attention toward the task at hand. Multitasking skills also requires constant updating of the executive system, especially that of the working memory, and spatial skills such as mental rotation (Mantyla, 2013), all of which are normally found to be better among men than in women. This could be the reasons why the former are often better at multitasking than the latter.

It would be expected that since both general multimedia usage and everyday multitasking involve switching between tasks, there should be sufficient transfer of learning from one task to another, and a positive relationship between multimedia usage and multitasking is warranted. However, this is not found to be true. The paradoxical finding is that people who are heavy users of multimedia do not do very well on attentional switching tasks. A study of Stanford undergraduate students divided into groups in terms of their reported multimedia use shows that those who reported high multimedia use did poorly on task switching in comparison to those who reported low usage (Ophir, Nass, & Wagner, 2009). According to Rothbart and Posner (2015), there could be a number of reasons for such a finding. One could be that people who are attracted toward multitasking are lower on attentional control and as such have problems focusing on the task at hand and more easily distracted by exogenous stimulation (Lin, 2009). It has also been seen that multitaskers are high on the trait of sensation seeking (Jeong & Fishbein, 2007), making them more prone to pay greater attention to novel stimulation. In other words, their high multimedia use is not because of their ability to switch between tasks effectively but because they get bored and switch from one multimedia source to another, much like Attention Deficit Hyperactive Disease (ADHD) children who have trouble sitting still and show a constant shift from one task to another. A totally different hypothesis is that rather than multitasking increasing working memory capacity, it operates the other way around, with people having higher working memory capacity finding multitasking not only easier but also more enjoyable (Garcia, Nussbaum, & Preiss, 2011).

5.4.2 *Training at Multitasking*

While high multimedia users may not necessarily be good at multitasking, the working of their executive attentional system can be enhanced through training. These very same action video games that children indulge in all the time can lead to improvements in both sustained attention and divided attention tasks, especially when the task processing load is high, or has to be consciously processed (Green & Bavelier, 2012), and may produce long lasting changes in the cognitive system (Maynard, Subrahmanyam, & Greenfield, 2005). Not surprisingly, a number of recent studies show that there is a transfer of training from game playing to task switching (Chiappe, Conger, Liao, Caldwell, & Vu, 2013; Green & Bavelier, 2012; Cain, Landau, & Shimamura, 2012), but this positive effect may depend on the type of video game being played with all genres of games not leading to the proposed effects. More interesting is the finding that this type of training produces changes in the brain systems (Bruya, 2010), such that with practice, there is increased connectivity between the neurons excited by that activity and a consequent decrease in switching time. This could also be the reason for habitual multitasking, because the brain now takes lesser arousal for divided attention tasks than that needed for focused attention tasks. These ideas on ways of using multitasking and video games for increasing cognitive capabilities have become the foundation for what have been called **serious games** (discussed in a later section of this chapter).

> ***Serious games****: video games that have been developed to aid some kind of learning*

5.5 Pathological Media Use and PIU

In his book, *iDisorder: Understanding our Obsession With Technology And Overcoming Its Hold On Us,* Larry Rosen (2012) has argued that many forms of behavior patterns associated with the use of technology run the risk of being labeled as pathological behaviors. Citing an example of our obsession regarding checking emails, personal messages, or other information on our devices, Rosen seeks to compare it with other habits that are, or have the potential to be, called obsessive–compulsive disorders (OCD). The pervasive nature of our anxiety with technological gadgets can be assessed by asking simple questions such as follows:

- "I have experienced phantom vibrations from my phone.
- I get irritable when I am not near my technological devices.
- I cannot go on vacation without checking my cell phone or email.
- I become highly anxious when I can't check my text messages, cell phone calls, or social networking account
- I feel tense and nervous when I am online or when I am using my cell phone" (p. 59).

Rosen concluded that if the earlier forms of behavior go out of control, one should seek the help of a professional. The good news in this direction is that clinical psychologists are beginning to realize the degree to which computer usage can become addictive. Even in the developing countries such as India, the National Institute of Mental Health And Neurosciences, under the guidance of Sharma has started a bimonthly *SHUT (Service for Healthy use of Technology*) clinic for the healthy management of technology addiction (IndianPsychologists Portal, 2014).

Additionally, internet provides an instant source for the expression of narcissism. The **Narcissistic Personality Disorder** (DSM) describes several features of a narcissist: self-importance, unlimited power, feeling special, needing admiration, unreasonable expectations/entitlements, exploitative nature, lack of empathy, envy, and arrogance. Now if you take this stuff seriously, that is, if you frequently email regarding what you have been doing or achieving, would it not make you believe that you are narcissistic? Not if those emails were not checked excessively or just displayed a few examples from the several behaviors listed earlier. Rosen contended that it would be simply human to display such behaviors, but if it obstructs routine life, especially performance at work, such behaviors would bring a geek closer to the therapist's office.

> ***Narcissistic personality disorder****: a mental disorder in which a person shows heightened self-importance and power, lacks empathy, and becomes arrogant*

We had mentioned earlier that technology has changed our social life and it certainly has pushed us in the direction of isolation. Citing an example of a computer programmer, named Alan, Rosen described his tendency to manifest withdrawal behavior very similar to that seen in people with a schizoid personality. Typically, Alan showed no interest in his colleagues and relatives, avoided going to office, barely left his home, hardly cared about his appearance and comments by others, ordered his meals to be delivered at his home, and remained engrossed in his computer. Rosen posed a question regarding the classifying of this behavior as a precursor to schizotypal personality disorder that is operationally described when a person shows:

- Nonnormative behavior
- Odd thinking
- Paranoid behavior
- Lack friends
- Indifference
- Heightened anxiety

Concluding his research on the role of technology in making this form of behavior vulnerable to pathology, Rosen wrote,

> "We discovered that the total daily use of the media and technology, as well as, more specifically, hours spend online and playing video games, were all associated with schizoid disorders in both the I-Generation (those born in the 1990s) and the Net Generation (those born in the 1980s)" (Rosen, 2012, p. 173).

That the personality of the person is also important in deciding how the internet will be used has been pointed out by many. Papacharissi and Rubin (2000) found that one personality trait that tends to create a difference is extraversion–introversion. While those high on the former tended to use the internet to gather information, introverts used it for social purposes, once again pointing to the key role of social shyness. Similarly, Landers and Lounsbury (2006) found that high internet users were lower on two of the **Big Five** personality dimensions, namely, agreeableness and conscientiousness. As such internet deaddiction centers would do well to focus on the personality of the individuals who come to their clinics.

Big-Five; a standardized test of personality describing personality in terms of five factors

At the same time, we should not forget that as students, we were fascinated by the writings of the founding father of modern psychology, William James (1890), who wrote in his classic book, *Principles of Psychology*, that life without habits would be strenuous and boring. With the learning of, and, performing on computers, we become habitual of using it. It creates vistas of knowledge and opens doors for new relationships on Facebook and Twitter. And later with mastery in multitasking, we begin to enjoy the flow of the technological world around us. Does it really matter if we become addictive of technology and begin to live a life based on our own choice?

5.5.1 Problematic Internet Use

With the total amount of time being spent by people on the internet rapidly increasing, there is a growing concern about the effects of excessive internet use on psychosocial well-being (e.g., Magsamen-Conrad, Billotte-Verhoff, & Greene, 2014; Rosen et al., 2014). Iacovelli and Valenti (2008) are of the view that despite large-scale increase in communication because of the computer, the "**internet paradox**" is very much there, in that, internet is hindering the social development of people. While virtual circles and groups may be on the rise, people are actually spending lesser time with other people in the real world, bringing in its wake, more loneliness, stress, and depression (Nie, Hillygus, & Erbring, 2002). So great could be the deleterious effects of excessive internet use, that there could be offline repercussions also (Suler, 2004), which in turn could force people to gravitate even more toward the internet (Hardie & Tee, 2007), the internet thus becoming a cause as well as an effect of loneliness and depression.

Problematic internet use*: when using the internet starts hampering our psychological well-being and happiness*

Internet paradox*: the fact that rather than being useful to man, it is creating problems*

As Nuccitelli (2014) puts it:

> "The information age technocentric concept of being connected is a paradox of disconnections causing us to lose control of our instinctual drives for social cohesion, allegiance and selflessness. As our dependency upon ICT grows, spreading throughout our collective human consciousness, the less we care for our neighbors and the more we delude ourselves into thinking that online communications are far more valuable than reality based relationships" (Nuccitelli, 2014, www.ipredator.com).

Specific internet use: *using the internet for a certain purpose*

Generalized internet use: *over use of the internet with no specific purpose in mind*

Cognitive behavior therapy: *a psychological therapy method which attempts to change the person's ways of thinking*

But the question that arises is whether all types of internet use has deleterious effects on well-being and whether control of internet use would solve the problem? Davis' (2001) cognitive behavior theory of problematic internet use (PIU) can be used to understand this issue. According to him, internet use can be categorized into two broad streams: **specific internet use** and **generalized internet use**. While the former refers to the use of the internet for specific purposes such as stock trading or even gaming, the latter refers to a multidimensional overuse of the internet coupled with purposeless surfing on the internet. Specific internet use is not a big problem because if the internet were not there, the person would find some other path to perform the same activity. Treatment of this type of PIU merely leads to the indulging in the same activity through some other path. More problematic is generalized internet use, wherein, internet represents the vital lifeline to the outer world and resources which would be used for personal, social, or professional purposes are wasted on purposeless internet surfing, etc. Davis (ibid) has developed a scale, the GPIU Scale, to assess the degree and type of internet use. Thus, no matter what the line of treatment, the proper assessment of the type of PIU would go a long way in mitigating the problem and even in the counseling the students. Second, it becomes imperative to help enhance the self-esteem of the person and to help him get over social fears. This is clarified by the results of the Iacovelli and Valenti study (2008), mentioned earlier, showing that excessive internet users rate themselves as being more depressed and socially inhibited than average users. One treatment method commonly used for overcoming low self-esteem is **Cognitive Behavior Therapy** (CBT) and would therefore go a long way in helping youth manifesting PIU.

5.5.2 *Cyber bullying*

Cyber bullying: *bullying people using the internet or social media sites*

Cyber stalking: *persistent intimidation, vilification, and taunting through the social media network*

Cyber sexual predation: *motivating or threatening and harassing people through the social media for the ultimate aim of meeting and engaging in sexual activity*

Another aspect of problematic internet use is **cyber bullying** and its ramifications taking the form of **cyber stalking**, **cyber sexual predation**, and even online sex solicitation. While a relatively new phenomenon, its implications are so important for developmental aspects of children, especially teenagers, that the need for its study and understanding is imperative. Before we can do anything about it, the dynamics of cyber bullying has to be clear. Two recent articles, one by Law, Shapka, Domene, and Gagné (2012) and the other by Slonje, Smith, and Frisen (2013) help us to understand the differential dynamics of traditional bullying and online bullying either by using the internet or through the mobile phone. As our understanding of cyber bullying increases, it is also becoming clear that the two are not the same, either in terms of its dynamics or its manifestations (Werner, Bumpus, & Rock, 2010).

While Smith (2012a, 2012b) differentiates between the two on the basis of seven characteristics, there are certain aspects that are clear. First and foremost, the power base is different. While in the case of traditional bullying, physical power and even social status differentials determine who will be the perpetrator and who the victim, cyber bullying is not so determined. Rather, if any, the power differential could be internet skills, with people with more advanced knowledge of internet more likely to engage in deviant internet and mobile activities (Vandebosch & Van Cleemput, 2008). In fact, Dooley, Pyzalski, and Cross (2009) are of the view that it is not the perpetrator's possession of power but the victim's lack of power that leads to being bullied online. Since the material exists in cyberspace, it is more difficult to either remove or avoid it, making the victim feel even more powerless. The perpetrator has to simply post the material; it, then, seems to take its own course, either by being sent to others by bystanders or the victim looking at it again and again and each look increasing the perceived victimization exponentially.

Another differentiating factor is time and place, with no restrictions being placed on either. While traditional bullying can be done only at school, at home or on way to and from school, or at work and to and from work, cyber bullying is done in cyberspace, for which there are no boundaries, either physical or temporal giving the bully a much wider reach that may extend to even the bedroom of the victim (Tokunaga, 2010). Traditional bullying also differentiates between the perpetrator, the victim, and the bystander. In cyber bullying, these distinctions often fade into oblivion, with the victim finding it much easier to retaliate online and the bystander becoming the next perpetrator, making cyber bullying much more reciprocal in nature than anyone can dare to think of as far as traditional bullying is concerned (Law et al., 2012).

Who is more likely to engage in cyber bullying? Empirical evidence points out that there is a curvilinear relationship between age and victimizing, peaking at around seventh to eighth grade (around 13–15 years of age) (Tokunaga, 2010). Not surprisingly, gender fails to emerge as a differentiating factor, with girls being as prone to engage in it as their male counterparts (Smith, 2012b). It could well be that people who are not able to bully in real life or who are victims of real-life bullying find cyber bullying an easy way to take revenge. While the personality of the person seems to be important in predicting cyber bullying, physiological factors could also be responsible, in that, areas of the brain responsible for empathetic behavior may have failed to develop, making the person less empathetic, even in offline communication (Iacovelli & Valenti, 2008), such that bullying in any form becomes easier to engage in. Some of the common forms used have been detailed in Box 5.4.

While children and youth have developed their own coping strategies, ranging from not divulging their email ids, changing one's online identity, to not looking at anonymous mails or even tracing IP addresses (Smith et al., 2008), most feel that

Box 5.4: The Perils of Cyber Bullying: Helping Children at Risk

A New York licensed psychologist and forensic expert, Michael Nuccitelli has founded, *i-Predator Inc*, an internet safety company, to help people and organizations deal with the perils of cyber bullying.

First and foremost, both parents and educators and the general public need to understand that cyber bullying can take a number of forms. Want to see creativity and intelligence being misused? Just see the ways through which cyber bullying now takes place! Some of the common methods listed by Nuccitelli have been given as follows:

- **Bash boards**: it is almost as if the bully wanted to bash up the target child. Chat rooms are used to post derogatory and defamatory information about the child being bullied so that the information reaches other people who also use the chat room.
- **Blogobullying**: the bully starts a blog, the central subject of which is the person being targeted. The idea of course is that as others read the blog, they will not only read the derogatory information but maybe also add their own.
- **Cyber bullying by proxy**: as the name suggests, here we have a bully who doesn't do any harm directly, but persuades and encourages others to bully the target through posts and messages on various social media sites.
- **Cyber stalking**: when a bully constantly intimidates, vilifies, or taunts the target through the ICT, it is then called cyber stalking.
- **Cyber harassment**: not as strong as cyber stalking, this method is used to harass the child by sending hurtful messages.
- **Instant messaging attack**: rather than using Facebook or any other social media site, the bully uses the general instant messaging service provided by the telecom provider. The obvious advantage is that the reach of the bully is undermined only if the target shuts off his mobile device.
- **Interactive game harassment**: it is commonly seen today that children form communities in which they compete with each other while playing games. Many of these are interactive games that allow a dialog between two or more players. While they play, the bully may use this medium to verbally abuse others, use profane language, or harass the target child in other ways.
- **Flaming**: this method tries to provoke the child by using vulgar or profane language so as to arouse the anger of the child.
- **Phishing**: a common method today, used with both adults and children. Efforts are made to persuade the person to reveal personal or financial information about oneself or one's loved ones. This information is then used to harass the child or even to use credit/debit card information to make purchases or withdraw money from bank accounts.

(continued)

Box 5.4 (continued)

At the same time there are a whole host of stronger methods that make matters even worse, for example, thieving a person's password, creating videos about the person (MMS), and web page assassination, to name just a few.

It is time that parents and educators make children aware of the perils of using social media sites and especially of divulging personal media to people not known otherwise.

Source: Nuccitelli (2014)

little can be done to stop it completely. Schools, too, have attempted to take the initiative and certain programs such as the *KiVa* in Finland or the online support through a website called *CyberMentors* do give some relief. As suggested earlier, the proper assessment of the type of PIU and the personality characteristics of the users would go a long way in mitigating the problem.

5.5.3 Technostress

Technostress*: stress brought about by technology or by using or trying to use technology*

In their book, *Technostress: coping with technology @ work @ home @ play*, Weil and Rosen (1998) pointed out that the twentieth-century human is not alone in thinking about the perils of technology. Way back in history, Socrates started the tech-scare in his book, *Phaedrus* wherein he lamented the invention of books. Instead of thinking for themselves, Socrates felt that the new reader was blindly trusting in "external written characters" and that the "library was ruining the mind." What do we have to say of the cell phone memory and the computer memory because of which we have stopped trying to remember phone numbers, email ids, and even very important bits of information? Now with cloud computing and cloud storage, the ways in which our memory systems are becoming almost defunct and obsolete is worth noting. Yet, technology causes stress and this can be seen in the reactions of people when they lose their mobile phone which had all the telephone numbers stored on it or when the laptop with tons of important material crashes. Technostress is also seen in people who are unable to cope with the onslaught of technology, causing them to change habitual work patterns.

5.5.4 Technology Overload

Technology overload: *overload in our cognitive systems due to technology*

The old typewriter gave some relief as the operator shifted from one line to another, but computers are not so kind. You may go on typing without any pause,

Box 5.5: Sample Items from a Short Questionnaire to Measure Technology Overload

1. Do you always check your email before doing other things?
2. Do you frequently find yourself anticipating the next time you'll be online?
3. When you're online and someone needs you, do you usually say "just a few more minutes" before stopping?

Source: Parker-Pope (2010)

unless, of course, you run out of battery power. Prolonged work on technological devices has resulted in a variety of modern era ailments. A commonly known physical damage is **carpal tunnel syndrome**, an operational problem leading to excessive pain in the hand and neck areas. Disorientation, eye strain, and illusory sensation are other common effects.

Carpal tunnel syndrome*: a medical problem affecting the nerves and muscles of the arm, shoulders, and neck*

According to Parker-Pope (2010), there are seven signs of technology overload. She suggests the use of the short questionnaire (sample items from which have been given in Box 5.5) to understand the extent to which one is under technology overload.

If you do feel that you are undergoing technological overload, two methods have been detailed in the Rothbart and Posner (2015) paper mentioned earlier in this chapter (Sect. 5.4.1). The first is, taking a holiday from using the internet which would undo brain wiring caused by high and perpetual internet use and that it would then be easier to get deaddicted. The other method is ***Integrated Mind Body Training***, a form of meditation that has been successfully used to help people quit smoking. On a study of smokers with no intention of quitting to smoke, this type of training led to a 60 % reduction in both actual smoking and the craving for tobacco. Interestingly, this took place at times with the subjects not even aware of the fact that there has been a reduction in either actual smoking or the craving for a cigarette (Tang, Tang, & Posner, 2013). Since internet addiction is similar in many ways to substance addiction, it is contended that the same technique could work with them too.

Integrated Mind Body Training*: a meditation technique to help people with various types of addiction*

5.6 Video Gaming

One of the ways in which children and youth spend considerable time on the internet is by playing games. With these games also being available on all types of mobile phones, the ease of access to them has increased tremendously. Boys, regardless of race, play more video games than girls. Also, gaming was found to be associated with lower behavioral self-concept and self-esteem (Jackson et al., 2010). We must of course remember that this was a correlational study and the association simply shows that the two are associated. It is probable that rather than low self-esteem being an effect of gaming, it is actually a cause of gaming, with such children being more prone to solitary activities in comparison to social playing which might hurt their self-esteem. Moreover, since success at gaming is not dependent on social factors, it often provides an avenue for the fulfillment of the need for achievement which could remain unsatisfied due to negative remarks from the peer group in day-to-day life.

What is it in video games that not only sucks people into it but often even causes addiction much akin to other forms of substance addiction? So great has this problem become that deaddiction centers are being established to help gaming addicts get over this seemingly wasteful habit.

In a seminal paper entitled, *Why Games Work*, Curtiss Murphy (2011) proposes that the principles underlying game development are similar, if not the same, as the principles of learning. Little did Thorndike realize that the laws of learning proposed by him more than a century ago (Thorndike, 1898) for the benefit of education givers would find their way into the laboratory of game developers. As Murphy (ibid) puts it, "with games, learning becomes a drug," while Jesse Schell (2008), says that "game design is more art than science, more like cooking than chemistry" (p. xxvi) and the idea is to create conditions in which the laws of learning operate in a context that the player finds difficult to leave. A "good" game, the type that people get addicted to, shows seven characteristics (Murphy, ibid). These have been briefly described as follows:

1. **Simplicity**: most games are simple. The goals are clearly highlighted, for example, the goal maybe to clear the screen in as short a time, or to get rid of as many monsters, or locate as many mines as possible within a stipulated time. By keeping it simple, the basic principle involved is the avoidance of cognitive load. So the player seems to be relaxing, even though the cognitive system is fully active.
2. **Feedback**: most games provide short-term feedback for both success and failure in the sense of increasing points, or, time-up. At the same time, the encouragement to repeat the game provides holistic feedback leading to the development of strategies based on metacognitive factors.
3. **Practice**: we all know that practice leads to mastery and this principle has been exploited to the fullest in video games by encouraging the player to go on and on. Games, in fact, encourage replay, sometimes even providing hints on how to improve. They even use failure to teach the player by pointing out the errors made in the previous try. A psychologist can very easily see the link between encouragement of practice and the ways in which feedback is provided, but the beauty of the game lies in the way in which the two are so deeply intertwined that the player is simply drawn on.
4. **Choice/involvement**: games are nothing but a series of interesting but meaningful choices. Again, the game developer takes into account the psychological reality that both too much and too little choice is deleterious and tends to decrease

the level of interest. As such, choices in games are kept at an optimal level so that just the right amount of arousal is maintained (does this not remind one of the omnipresent Yerkes–Dodson law relating performance to levels of arousal?)

5. **Immersion and engagement**: while both the terms may appear synonymous and refer to deep involvement, the differences between them far outweigh any similarity. The former is passive involvement while the latter is active. While playing most computer games, one tends to get carried away by the sheer force of compulsion but at the same time, the game calls for active involvement of the executive attentional system and the working memory mentioned earlier. One has to use the right strategy, change strategies, and attention has to be at its highest. Even a few milliseconds of attentional lapse leads to losses since the total play time is often very short.
6. **Fun**: this factor is probably at the root of all games, computer, or otherwise. They give you pleasure in that you get the feeling that no serious effort is required; there is the pleasure of doing, and there is pleasure in the sense of triumph, all leading to gaming being associated with strong positive feelings.
7. **Flow**: all of the above lead to the creation of what has been called **flow**, a state of complete absorption in a task such that nothing else seems to matter (Csikszentmihalyi, 1998, 2014). The clear, simple task and regular feedback, the balanced, challenging, yet attainable goals, the degree of concentration demanded—all lead to create an intrinsically motivating condition. It certainly seems that game developers are far ahead of psychologists, in general and teachers in particular, in understanding the psychology of young players and ways by which sustained interest can be had.

***Flow**: a state of the mind when one is so engrossed in an activity that nothing else matters*

The above can be summarized in the form of what Curtiss Murphy (ibid) has called the **elemental tetrad of gaming**, an activity based on four elements, namely, story, esthetics, mechanics, and technology. Each affects the other and all are important. The important idea is that if we simply add one more element, namely, content, one can use a game for pedagogy, and the learner will probably learn without the pain of traditional learning. There is currently considerable interest in designing games that have pedagogic value. Figure 5.4 gives one simple way of accomplishing this.

***Elemental Tetrad of gaming**: the principle in which games are created on the basis of four interacting elements, esthetics, story, mechanics, and technology*

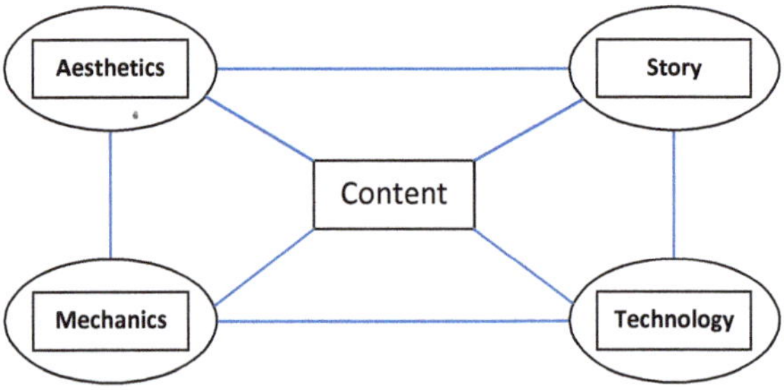

Fig. 5.4 Elemental triad of gaming (adapted from Murphy, 2011)

5.6.1 *Effects of Gaming*

Impetus to research on the effects of playing games on the computer was provided by the startling revelation that the unfortunate shootout at Sandy Hook Elementary School in December 2012 was carried out by a student who played shooter video games, leading President Obama of the US to request the American Congress to allocate as much as ten million dollars for research on the effects of violent media, especially violent video games (Obama & Biden, 2013). Similarly, the shooters at the Columbine High School in Colorado, USA, were found to be regular gamers, playing the shooter game, *Doom* (Trent, Bai, Glick, Annin, & Keene-Osborn, 1999). However, violence has not always been a part of gaming. The first game, *Pong*, was nonviolent. Aggression started in the second generation with the game *Breakout*, but there, too, there was no human aggression. It was in the next generation of games such as *The Empire Strikes Back* that human aggression became a pervasive aspect of gaming. By the late 1990s, almost 80 % of the games had aggression as the chief objective. It is but natural that practitioners, parents, and educators alike be concerned about the possible negative effects of violent games.

While not discounting the negative effects of violent gaming noted by researchers such as Bushman and Anderson (2002) and Anderson and Dill (2000), Granic, Lobel, and Engels (2013) are of the view that looking at current controlled study findings on gamers, we need to take a more balanced attitude. The main reason for this is that in recent years the nature of games has changed considerably, becoming not only more complex but also more diverse, realistic, and socially oriented (Ferguson & Olson, 2013) and vis-à-vis TV, books, or even movies, they have become more interactive in nature, forcing active engagement of the brain. Also, since these games can be played along with other online friends, they have important repercussions for the socioemotional development of the child, providing children and youth with a new kind of experience hitherto not possible in the real world. However, they are very clear in that the effects are often noted for one genre of games and not others, so that one must be very careful in drawing conclusions. Granic et al. (2013) point to four specific areas in which positive effects of gaming are seen, namely, cognition, motivation, emotion, and social skill development.

5.6.2 *Cognitive Gains*

First and foremost we must not forget that games, whether video or other traditional games, have the common denominator of being part of the play life of children, the developmental implications of which have been documented by many psychologists, ranging from Piaget to Vygotsky to Erikson. **Evolutionary psychology**, too, points to the adaptive functions of play

> ***Evolutionary psychology****: a subdivision of psychology which attempts to study the adaptive roots of behavior in the course of evolution and natural selection*

whether among animals or humans (for an excellent review, the reader is asked to see Bjorklund & Pellegrini, 2010). Even play fighting among rats has been found to lead to the release of certain chemicals important for the development of those brain areas that are responsible for the social cognition of the animal (Pellis & Pellis, 2007). Controlled experiments with gamers and nongamers on games such as *Halo 4* or *Grand Theft Auto IV* show that in comparison to nongamers, gamers were better at the allocation of attentional resources, visual processing, and even mental rotation abilities (reviewed by Green & Bavelier, 2012), comparable to improvements noted for formal high school and university courses specifically geared at the enhancement of such skills (see meta-analysis by Uttal et al., 2013). Needless to add that these skills have long-term implications for career development as has also been clarified by a 25-year longitudinal study (Wai, Lubinski, Benbow, & Steiger, 2010). Other cognitive effects noted were enhanced problem-solving skills (Prensky, 2012) and higher creativity (Jackson, Witt, Games, Fitzgerald, von Eye, & Zhao, 2012). How important some types of games may prove to be is pointed out in an interesting article, entitled *Carrot sticks or joy sticks: video games improve vision* (Caplovitz & Kastner, 2009). The empirical findings showing that nudging a joy stick could cause the same type of improvements that eating carrots could cause. Specifically, playing action video games has been seen to induce long lasting improvements in contrast sensitivity, a basic visual function commonly seen to deteriorate with age. But we also have to remember that the improvements were seen only for those who played action video games. For those who played other types of video games there was no improvement. Once more, it is psychologists who have provided insights regarding the types of games children should be allowed to play, and clearly points to the relevance of psychology of technology. Studies such as these and many of the others mentioned earlier help not only caregivers of children but also provide game designers with inklings regarding the types of games that should be developed.

5.6.3 Motivational Benefits

What is important is not what motivates people to indulge in gaming but whether there are any long-term effects on the motivational style of the individual. The subtle ways in which video games balance the levels of success and frustration coupled with the optimal types of intermittent reinforcement suggested by Skinner produce not only learning but also create conditions that seem to promote the development of an effective motivational style, having at its base, persistence and effortful involvement (see review by Dweck & Molden, 2005). Using the concepts of **entity** and **incremental theories of intelligence**, these researchers point to the differences in the ways in which children start perceiving their own intelligence. While praising a child by saying "how intelligent you are" seems to develop, what they

> ***Incremental theories of intelligence****: a personal feeling that intelligence is malleable and depends on efforts of the child*

call, an **entity theory of intelligence**, praising the child for her effort develops an incremental theory of intelligence such that the child starts thinking that intelligence is malleable and depends on the efforts of the child. This feeling of mastery over the environment protects the child from both **learned helplessness** (Seligman, 1972) and gives him mental strength to cope with failure in the real world. In this context, video games provide just the right mix of challenge and frustration with enough experiences of success (Sweetser & Wyeth, 2005) to help the child develop persistence which is seen to generalize to situations outside the gaming situation too (Ventura, Shute, & Zhao, 2013).

Learned helplessness*: a tendency to avoid situations where the person has felt in the past that he cannot control*

Entity theory of intelligence*: a personal feeling that his/her intelligence is unchangeable*

5.6.4 *Emotional Benefits*

One reason why people indulge in any type of activity is that they are having fun, or, in other words, they are enjoying themselves. It is this enjoyment that turns even work into play and studies by Curtiss Murphy (discussed earlier) show that gaming is definitely associated with this spirit of fun. In fact, several studies point to mood enhancement with the playing of games such as *Angry Birds* or *Bejewelled II* (Russoniello, O'Brien, & Parks, 2009). The mood enhancement is so pronounced that McGonigal (2011) even contends that gaming could produce the type of positive experiences that help to broaden one's vista of possible experiences. The theory, known as **Broaden and Build theory** (Fredrickson & Cohn, 2008; Fredrickson, 2001), has identified upward and downward spirals leading to happiness and well-being and technology seems to reinforce such spirals. For example, one factor, namely, joy, relates to the intricacy and affordability of technology around us.

Broaden & Build theory*: a theory which posits that positive experiences that broaden one's vista help to build and enhance social relationships*

5.6.5 *Social Benefits*

Over the last decade or so, the most dramatic change that has occurred in the nature of games is that they have become social. Rather than games being played by solitude loving people, games are more often played in the multiplayer context sometimes running into virtual communities the size of which could run into millions. Thus, the Entertainment Software Association (2012) has found that the *World of Witchcraft* boasts 12 million regular players. Whether played in the spirit of cooperation or competition, these virtual communities help develop a variety of social skills such as whom to trust and whom not to trust which may even generalize to nongaming situations (Gentile et al., 2009). Another social skill that tends to develop, especially when games are played cooperatively (and surprisingly this even includes violent games), is **prosocial behavior** (Ewoldsen et al., 2012; Ferguson & Garza, 2011), while playing a game that involves civic duties such as Guild War II has been found to enhance **civic engagement** in children (Lenhart et al., 2008).

Prosocial behavior: *a tendency to help other people*

Civic engagement: *the tendency to display civic sense and engage in community activities*

5.7 Virtual Reality

If you have seen movies like *Brainstorm, Matrix, Avatar*, or *Virtual Reality*, you would have had the experience of being placed in an illusory environment that has three-dimensional effects that seem almost surreal. For a person not familiar with technology, such experiences appear to be an extremely clever manipulation of reality. If passive watching of a movie has such effects, imagine what the experience of a person would be when she can actively manipulate tools in computer games. Looking at the excitement of young kids playing games in a shopping mall, one can feel that they are engrossed in a world which is different from the physical reality of the room in which they are playing. It seems to create the feeling of being in some other world.

It was in the 1980s, that Jaron Lanier, an American writer, computer scientist, and composer of classical music, first used the term **virtual reality** and in 2010, was nominated to the *Time* magazine list of 100 most influential people. But what exactly does the term mean? It generally means that one is in a technologically mediated environment rather than in the immediate physical environment. Virtual reality creates such a strong sense of existence in this new surrounding that

Virtual reality: *extent to which one feels present in the computer-mediated environment than in the physical environment*

(a) We lose sense of our own surroundings,
(b) We begin to feel our existence inside the virtual environment, and
(c) We tend to focus on this new environment as we seek opportunities in the virtual reality setting.

While we are connected to only a few devices on the internet at present, the number of our choices is increasing rapidly. A rough estimate shows that we will have 50 billion devices available for connection to the internet by 2020. Be ready to use biosensors that would help you adjust the room temperature and receive instructions upon opening the refrigerator regarding the management of your calories by selecting the appropriate available food. Once we live in this technology-immersed environment, it is but natural that we will move closer to accepting the significance of the virtual environment.

How is virtual reality created? The feeling of being part of the virtual world is caused by presenting pictures frames at a rate of 20–30 per second. When this manipulation of motion shifts our experience from the current physical environment to another environment, it is referred to by several names, all implying living in a virtual world:

- Cyberspace, a word that has its origin in science fiction
- Artificial reality
- Augmented reality
- Telepresence

Scholars of today, however, prefer to use the term virtual environment rather than any of the terms listed earlier. With prolonged exposure and interaction, an individual begins to feel that she exists in the virtual world and is affected by the conditions and demands of this environment.

5.7.1 The Turing Test

> ***Turing test****: a test for intelligent machines based on whether an individual is able to distinguish between human and machine intelligence*

Let us begin by trying to understand the basics of virtual reality. With its ability to be highly interactive, the computer has proved to be an important tool for studying social behavior. In 1950, Turing worked out the proposition that if a person starts believing that a computer is another person, then the computer has achieved human intelligence. In other words, if a person is not able to distinguish between the responses of a human and that of a machine, the machine is said to have passed the **Turing Test**. Turing (1950) devised several experiments to test machines and in a seminal paper, entitled, *Computing Machinery and Intelligence*, was able to wash away all arguments against his proposition that it is possible to design machines having human-like capabilities. Later, it was this line of argument that led to the growth of the very important field of artificial intelligence.

The irony is that even when the person knows that he is interacting with a computer, s/he is unable to stop social responses that would have been given while interacting with a real person (Reeves & Nass, 1998). This weakness of the human, or one could call it this strength of computer software, is what has fueled the growth of computer games wherein the player is able to change the focal character in the game according to his likes/dislikes and then see how the computer reacts in return. Examples include games such as *Second Life*, *The Sims*, and the *World of Warcraft*, to name just a few.

Further, when we start identifying with the virtual environment to such an extent, that we become oblivious of our immediate surroundings, we can be said to be "immersed" in it. The last few years have seen a spate of studies on not only the extent of such **immersive behavior** but also on the 'why' and the 'what' of this phenomenon. The *Academic Search Premier* database shows that during the past four decades, there has been a phenomenal growth in the amount of work on virtual reality. From just one peer reviewed article on virtual reality published between the period 1975 and 1990, there were 305 articles during the next decade, that is, during the period 1991–2000, and between 2001 and 2009, the total number of articles was as high as 1610 and has been multiplying exponentially ever since that last count.

Immersive behavior*: the extent to which one identifies with the virtual environment*

5.7.2 Avatars and Agents

In 1976, a game *Maze War* was developed in which players would shoot at one another while running around in a maze. The first popular adventure game, *Adventure* was also created around the same time but it operated only with some predefined goals. In 1978, Roy Trubshaw and Richard Bartle invented the first ***MUD*** at the Essex University in England. Short form for Multi User Dungeon, *MUD*s became immensely popular for several reasons. First and foremost, it was an adventure game that was available for free. Second, it was a multiplayer game. Third, it was a role playing game in which players could take on roles that were mainly characters from the world of fantasy, including characters such as wizards, warriors, prince, and shaman. These characters could acquire and lose magical powers by way of slaying monsters during the course of the game. Fourth, it contained yet another interesting aspect: there was a dungeon master who basically set up and ran the game. While this role was played by the computer itself, it added to the suspense and the thrill. Soon variants of the game started being developed and we had games such as *AberMUD, Tiny MUD*, and *LPMUD*, to name just a few. The growth of the internet during the 1990s spurred the development of such games, including *MUD*s and the *Kingdom of the Winds*, to new heights and the focus began to shift to a genre of games involving user–computer interaction. This was the beginning of another life for the player: not well defined at that time, but later known as Avatar, a character based on the Hindu concept of reincarnation. As players got an opportunity to ascribe physical and mental characteristics to the virtual character with advances in technology, these avatars or second life characters began to establish authenticity and they started appearing life like. The versatility of *MUDs* was so great that not only were they being developed for entertainment but also for educational purposes.

MUD*: short form of the popular video game, Multi User Dungeon*

> ***Avatars**: video game characters that can be controlled by the player*

> ***Agents**: video game characters that are controlled by the computer*

> ***Hybrids**: video game characters who are partly modifiable by the player and partly by the computer*

> ***Player-avatar identification**: the degree to which the player feels and identifies with the avatar*

Most virtual reality games use what have been called **avatars** and **agents** and even combinations of the two, **hybrids** as they are called. The difference between the two is that whereas the avatar is controlled by the player, the agent is controlled by the computer. Hybrid avatars and agents have characteristics of both, that is, they are partly modifiable by the player and partly computer controlled.

Coulson, Barnett, Ferguson, and Gould (2012) are of the view that people start having real feelings for virtual people to the extent that they even form emotional attachments to them. However, it is not all types of virtual characters to which such attachments are developed. While the authors do not specifically refer to Bandura's theory of social learning, the findings are clearly reminiscent of Bandura's theory of identification, in that the level of identification seen and the extent of emotional attachment arise "from a complex blend of the player's personalities and motivations and the virtual character's level of physical attraction, friendliness and general usefulness within the game situation" (p. 176). In a study, Li, Liau, and Khoo (2013), clarify that the degree, of what has been called, **Player-Avatar identification** (PAI) can be conceptualized and even measured in terms of four factors:

- Feelings during play
- Absorption during play
- Positive attitude toward the avatar
- Importance to identity.

Can avatars influence our experience? This is one question that has been debated for a long time. Sherry Turkle of MIT argues that the avatar is not just about manipulating the virtual environment but it is also an embodiment of one's personal traits. It becomes isomorphic with one's own persona as the intended outcomes are projected onto the avatar. Writing about the effects of games such as *MUDs*, wherein one can chose the character one wants to play, Turkle (1995) remarks that it is not that these people are simply playing a game; more importantly, the game is changing them in ways not envisaged by the developers of the games (Box 5.6). Just think of it: multiple players, each at their own machine, vying with each other, creating virtual guilds, collecting rewards and awards, and joining communities that exist not in physical space but only in virtual space. She often found college students playing 3–4 different characters, not just role playing or enacting but 'living' different lives, while at the same time working on their assignments. She asks a very important question that the developers of the game had probably not foreseen: will this not lead to an identity crisis, much in the Ericsonian sense, involving as it is, youth who are still trying to establish who they are, their identity (Turkle, 1997)?

Box 5.6: The Second Life

The MUD software gave a major boost to the creation of the avatar. In 2003, *Second Life* was launched and its technology helped users to socialize with other people and create objects. From enacting virtual marriage to making buildings, the participants could perform any number of activities. According to Turkle (1997), *Second Life* also invited your personal user identity to create an avatar.

The first experience of *Second Life* is the amazing real-life setting and the navigator feels almost instantly being drawn into this virtual world. Initial control of the avatar appears difficult, but with continuous navigation in the educational island one begins to feel connected. Through repeated teleportation, one gets a sense of digital embodiment with the avatar as if it is an extension of oneself. The interesting point here is that unlike other video games, it has no end, no winning or losing streaks. After a while, this scenario is experienced as a user-defined sociocultural environment where we begin to crave for another visit. It is this which has been nomenclatured *Second Life*.

The question becomes even more important when one considers the finding that the level of intimacy and emotional involvement with the avatar is so high that there are manifestations of it at the brain level much to the same extent as would be seen when interacting with close 'others' (Ganesh, van Schie, de Lange, Thompson, & Wigboldus, 2011).

While human beings have always enjoyed interaction with inanimate objects such as puppets, the history of virtual reality technology is not very old. At the same time, it has led to the rapid development of a culture with its own symbols and other artifacts. Within the virtual world one finds communities, cultures, and societies much as one finds in the face-to-face world. Keeping this in mind, it is of interest to not only psychologists but also to sociologists and anthropologists to see how these virtual cultures grow; what are their symbols, their rites, and rituals; and the nature of the social interaction therein. To gain answers to such issues, methods much akin to those used by anthropologists generally (namely, ethnography) have been devised with the prefix virtual added to refer to the nature of interaction being focused upon. **Virtual ethnography** is a highly interactive process that involves observations of computer/device-mediated cultures through a variety of non-face-to-face methods (Harrelson, 2011). Recent studies in virtual ethnography reveal that the online culture cannot be considered to be disparate from the offline culture. Both are seen to interact with each other to produce a totally new environment. This is one of the reasons why the present elderly generation find it so hard to understand this new generation that spends more time on Facebook and WhatsApp than in interacting face to face with people.

Virtual ethnography*: a study of communities and cultures and their artifacts that are computer/device mediated*

5.8 The Technoself

> ***Technoself***: *changes in the personal identity as a result of technology adoption*

> ***Self-ing***: *creating one's personal identity in line with what gains popularity in a digitalized society*

The term **technoself** was used by Luppicini in 2013 in the *Handbook of Research on Technoself: identity in a technological environment*, focusing on the changes in personal identity as a result of the adoption of technology. The basic question being probed by Luppicini was how technologies change the way people define and even present themselves in society. In this era of digitalization, when names, last names, and even names of parents or date of birth are no longer considered adequate for proving one's identity, when the stability of the person is undermined by a fast changing landscape of technology, when biometrics, DNA analysis, and other genetic assays are considered a necessity, then, how do we, the inhabitants of this internet society, define ourselves? Our friends are often those whom we have never met face to face, but 'know' each other through Facebook, LinkedIn, or WhatsApp, where every post is not only read but often disseminated to a much larger online public, then, how do we present ourselves to our fellow netizens? Ever since the invention of the World Wide Web, virtual reality has often become more real than reality itself. It would be interesting to see whether the traditional theories of self-development (e.g., Erikson's theory of psychosocial stages, 1959) still holds true. Or, whether, Kohlberg's theory of moral development (Kohlberg, 1958) follows the same pattern isolated by him way back in the 1950s? In fact, Robinson (2007) has raised the key issue: how does '***self-ing***' take place in this digitalized society?

As discussed earlier, it was the interactive adventure games of the 1970s (e.g., *Adventure*, *MUD*) which first offered players a chance to project themselves online. More significantly, it was probably the only forum for such a projection of the self. However, from 1978, when the first *MUD* was introduced to 2015, a sea change has taken place in the demography of not only video game players but also of internet users in general. In the 1970s and even till the 1990s, the average gamer was an American male teenager, who was able to live a kind of life online that he would never be able to do offline. Keeping in view cultural gender norms, male avatars were presented as muscular honchos, capable of showing great strength, while female avatars were designed in keeping with the standard social preference of a lithe, delicate physique, manifesting feminine behavior. When one looks at the gaming population at the turn of the century, the demographics point to a more heterogeneous group. Even if one considers the American population, we see that the percentage of female users of the internet parallels that of males; the racial diversity is also greater; and, most of all, gaming was no longer the primary reason for using the internet (The UCLA Internet Report, 2000). By the time the 2002 report was published, the percentage of users who expressed an interest in playing games had dropped to 26.5 % (The UCLA Internet Report, 2002). As a result, the *MUD*ders of the 1990s now formed only a small proportion of the people using the internet.

A whole host of other cyber activities such as blogging, Facebook and Twitter, chat rooms, and LinkedIn have become more popular. Yet, self-ing continues though in different ways. The *MUD*der was more interested in creating an online self-image that helped him/her to enact scenarios and personae not possible offline. These *MUD*ders often created multiple online selves to escape from offline realities with new online identities (Turkle, 1995) and often comprised males who were unsure of themselves and their offline social skills and therefore relied on the fantasy world of the *MUD* to create an idealized online self (Chen, 1998). Most of them were also technically proficient college going students (Robinson, 2007).

The technoself of today departs from fantasy but certainly engages in impression management. The 'I' creates a homepage with external links that act as credentials to create a 'me' who is liked by the cyber brother/friend. Photos and other visuals are carefully chosen so as to create a profile that preserves only the best parts. Anything that is not liked is edited at the earliest. It is a self that is developed in interaction with the cyber society, much in the same way as our offline selves are developed (Robinson, 2007). It is now an extension of the offline self, rather than being distinct from it as was seen in the avatars of *MUD*ders (Rainie, 2004). The online self is being used to supplement and augment the offline identity. Or as Galvin (2003) very aptly puts it: cyberselfing is changing us from Homo Erectus to Homo Technicus (Galvin, 2003), an individual for whom the significance of the technology just cannot be ignored. And, it is often seen that people are not able to keep their online self or avatar totally segregated from their offline identities. A large mass of research is now pointing out that our online identities affect more salient aspects such as our attitudes and values (Peña, Hancock, & Merola, 2009; Yee, Bailenson, & Ducheneaut, 2009).

Why is it that virtual reality is taking such a hold over people? An important feature of virtual reality is the anonymity that it offers. As a result you can create your own profile and change it at will. This virtual reality is populated by a virtual public: people who are all virtual: what they are offline, we cannot even guess. One may feel that in this realm of virtual social interaction, there is no fear of judgments, there is no retribution, but in reality there is. Virtual communities and guilds are often formed and one is always careful of what one projects online. Being much more interactive than the *MUD* environment, today's virtual world has linkages to various types of online forums, as a result of which one tries to protect one's self, much as one does in front of an offline public. The paradox is that while we may have hundreds of friends online, we may have not even one offline.

5.9 Robots and Sociable Robots

Despite the widespread use of the term robot today, its root lies not in information technology but, surprisingly, in fiction and drama. The term is credited to Capek, a Czech playwright who wrote a play entitled, *Rossum's Universal Robots* in 1920, in which the robots turned against their human masters creating much of the fascination we have all held for these highly sophisticated computerized machines.

Since much before the turn of the last century, robots have helped man in manifold ways, from lifting heavy objects in factories to caring for our aging population. We have come a long way from the physical assistance that the first robots provided to providing psychological care and social interaction to interactive robots for psychological enrichment (Box 5.7). What started as mere toys such as the well-known

Box 5.7: Meet Pepper, The 'Love-Powered' Humanoid Robot That Knows How You're Feeling

'If the thought of a humanoid robot in your home makes your skin crawl, meet the friendly Pepper.

Pepper is a cute, wisecracking personal robot designed to bring joy to everyone, and Japanese mobile carrier SoftBank wants people to start buying it next year for the price of a high-end PC. The phone giant unveiled the autonomous, sophisticated machine on Thursday along with partners Aldebaran Robotics of France and China's Foxconn, the world's largest manufacturer of electronics.

Equipped with an array of audio, visual, and tactile sensors, Pepper is 120 cm tall and weighs about 28 kg. It has two arms and rolls around on a wheeled base, with a lithium-ion battery that can power it for at least 12 h. Its chest bears a 10.1-in. touch screen that can be used to communicate along with its voice and gestures. Its main function is to interact with people, according to SoftBank.

'We want to have a robot that will maximize people's joy and minimize their sadness,' SoftBank CEO Masayoshi Son told a press conference outside Tokyo.

The event began with a darkened stage and several minutes of theatrics as Son presented Pepper with a heart-shaped object. The robot then began interacting with him with a high-pitched voice and then introduced itself to journalists in Japanese. Pepper is the world's first personal robot that can read people's emotions, Son said, and it uses voice-recognition technology and proprietary algorithms to analyze people's feelings from their facial expressions and tone of voice. It will go on sale in Japan in February 2015 with a base price of ¥198,000 ($1929).

Pepper doesn't have 100 % recognition of what people say to it, Son admitted, adding it will improve with time. Its NAOqi operating system, a nod to Aldebaran's pint-sized NAO robot, has an 'emotion engine' as well as cloud-based artificial intelligence (AI) to help it understand people and respond to them.

'With cute robots, so cute that people want them at home, very easy to interact with and which are connected to the Internet, look at this potential we are opening,' said Aldebaran CEO Bruno Maisonnier. 'Many things can be done to improve education, healthcare, entertainment, flow management, you name it.'

(continued)

Box 5.7 (continued)

Robots will greatly change people's daily lives just like the PC, Internet, and mobile phones did in the past, added the head of Aldebaran, which is owned about 78 % by SoftBank. Pepper will debut at two SoftBank Mobile stores in Tokyo on Friday. From next year, it will initially be sold in SoftBank stores in Japan and through online channels. Overseas sales will follow at some point, Son said, and Pepper's emotional expressions will be customized to fit various cultures. So far, Pepper can speak English, Japanese, French, and Spanish. More languages are expected in the next few months.

The debut of Pepper is the realization of a 25-year dream for Son, as the CEO recounted how he was inspired by Astro Boy, a popular science fiction robot created by manga artist Osamu Tezuka in the early 1950s. The heroic machine became a template for friendly humanoid robots in Japan, both in fiction and reality. 'Pepper is a baby step in making robots with emotion,' Son said. 'Our vision is to create affectionate robots than understand people's feelings and then autonomously take action. So the joy of a family will become the joy of the robot.'

Source: Excerpted from Hornyak (2014)

Furby, and its later avatars including the AIBO and Kismet, sociable robots are also performing a host of other functions such as looking after young children and caring for the elderly with dementia. As the use of such **sociable robots** is on the rise, scientific studies have been conducted on the effects that they have on the human beings under their care (for good reviews, see Murdoch, Osterreicher, Guse, Roger, & Thompson, 2013; Bemelmans, Gelderblom, Jonker, & de Witte, 2012) and we are gradually seeing the evolution of a new discipline, **robotic psychology** (Box 5.8). No, it does not study the psychology of robots; rather, it focuses on the cognitive, emotional, and social effects that robots can and are having on human beings and how human-like characteristics can be introduced in robots.

Sociable robots*: robots that are capable of social interaction with the user*

Robotic psychology*: a part of psychology which focuses on the cognitive, emotional, and social effects that robots have on humans*

Do people prefer robots to humans? Believe it or not, research and experience is proving that many people do prefer robots to humans. A recent article in the *Science* entitled *The synthetic therapist*, suggests that "some people prefer to bare their souls to computers rather than to fellow humans" (Bohannon, 2015b, p. 250), and, in another article in the same issue, entitled, *Fears of an AI pioneer*, Bohannon (2015c) claims that Artificial Intelligence is as dangerous as nuclear weapons. The question that then arises is why is this so? How can lifeless machines (after all, a robot is nothing but a machine) be preferred to other humans by a species that by nature is gregarious? And Bohannon is not the only person who is now being dis-enamored by

Box 5.8: Robotic Psychology

LONDON: ERWIN is a great listener and responds only when required. He understands when we are unhappy. He is warm, responds to touch, and is the perfect companion for the elderly. ERWIN (Emotional Robot with Intelligent Network) is the world's friendliest robot, built by an Indian student in UK, capable of expressing five basic emotions while interacting with a human. Originally the brain child of Dr John Murray from the School of Computer Science, University of Lincoln, UK, ERWIN is helping scientists to understand how more realistic long-term relationships might be developed between humans and robots.

It is now being used as part of a PhD study to find out how some of the human-like thought biases in robot characteristics affect the human–robot relationship. It is hoped the research will not only help scientists to understand and develop better, more realistic relationships between humans and 'companion' robots, but that it could also help to inform how relationships are formed by children with autism, Asperger syndrome, or attachment disorder.

PhD student Mriganka Biswas said: "Cognitive biases make humans what they are, fashioning characteristics and personality, complete with errors and imperfections. Therefore, introducing cognitive biases in a robot's characteristics makes the robot imperfect by nature, but also more human-like."

"Based on human interactions and relationships, we will introduce 'characteristics' and 'personalities' to the robot. If we can explain how human-to-human long-term relationships begin and develop, then it would be easier to plan the human-robot relationship."

According to Mriganka, robots are increasingly being used in different fields, such as rescuing people from debris, in medical surgeries, elderly support, and as an aid for people who have autism. "For the latter two especially, robots need to be friendly and relatively more sympathetic and emotive to its users. A companion robot needs to be friendly and have the ability to recognize users' emotions and needs, and to act accordingly. So, for each category the robot needs to form a 'long-term' relationship with its users, which is possible by continuous interactions and the robot having its own personality and characteristics," she added.

Source: http://www.lincoln.ac.uk, **2014**

twenty-first century developments in computer sciences. The person who is regarded as the father of virtual reality, Jaron Lanier seems equally disillusioned. Through his books, *You are not a gadget* (2010) and *Who owns the future* (2013), he has voiced concerns about the undermining of the individual by collective forces over the internet. And, as science and technology go ahead with its development of more and more intelligent computers through insights into the areas of computational rationality (e.g., Gershman, Horvitz, & Tenenbaum, 2015) and machine learning (e.g., Jordan & Mitchell, 2015), the writing on the wall becomes even more clear.

5.9.1 Turkle's Second Self

Sherry Turkle at MIT has long been known for her pioneering work on the development of the virtual self, that is, the self, in the process of interacting in a virtual environment. In her book, *The Second Self* (1984), she has argued that computers are not mere tools; they become a part of our lives. Traditionally, artifacts such as computers were considered objects for helping us to perform various activities, but they are now guiding our thinking. It would be a mistake, she has argued further, to conclude that children are simply learning computers; they are, in fact, being influenced by such devices in their thinking, feeling, and behavior. In her later work, *Alone Together*, Turkle (2011a) contended that with recent advances in technology the relationship between humans and the computer is changing. An obvious corollary of this change is the expansion of ever-increasing expectations from computers. In essence, Turkle has emphasized the following consequences of the impact of technology:

1. The use of electronic devices is changing our identity; it may give us a sense of placelessness as we move from the real world to an imagined world.
2. The expansion of technology is resulting in the discovering of a new companionship, a situation in which there are no demands of reciprocity such as that in a real-world institutional setting.
3. Technology is acting like a broker who sets the direction and the limits for communication between two parties. By becoming a source of mediated communication, it reduces the chances of direct communication, say by posting a message.

Through a large number of experiments conducted on a variety of populations, ranging from very young children to the elderly, Turkle has attempted to gain insights into a variety of questions relating to the ways in which robots and other technology are changing not only our lives but also us. The answers are presented in research articles, books, and interviews. For one thing, you will have to agree with her when she writes that "objects do not simply do things for us, they do things to us, to our ways of seeing the world, ourselves and others" (Turkle, Taggart, Kidd, & Daste, 2006, p. 347). With the first exposure to computerized toys such as a Furby, who demand attention, who have to be fed, cleaned up, even amused, or else they start complaining, the toddler of today realizes that this toy is different from other toys, that it is more than a toy, it is almost alive. With the ever-increasing hassles of urban life, human neighbors may not even be aware of each other's existence, it is the Furby or the computer who becomes the nearest neighbor (*Alone Together*, Turkle, 2011a). Despite the fact that such sociable robots do very little for us, our response to them is relatively strong, in the hope that they will reciprocate in the same manner, forgetting that these are just machines that have been programmed to give certain responses. We not only start anthropomorphizing them, but we go to the extent of humanizing them.

There are many reasons for why we start treating robots like other human beings and come to prefer their company to that of humans. If one thinks about it, do not

> ***Nannybots***: *robots that can take care of infants and small children*

robots and interactions with them appear almost hassle free? Turkle's subjects reported that unlike humans, robots do not make untenable demands, they don't fake, they don't cheat, and they don't need to be understood. And, most of all, you can end your interaction with it whenever you want, by turning it off. How convenient it all seems? Today we hear about **nannybots** (robots acting as nanny for a small child) and interactive robots taking care of the elderly. In many of her experiments, Turkle and her team have left the robot with the subject and then came back later to observe and interview the subject. In most cases, the subject was very satisfied with the robotic relationship. Children interacted with the robot much as they would with real people, assigning to the robot, the same emotions and feelings (see Box 5.9).

> ***Robotic relationships***: *relationships that people form with robots*

While Turkle agrees that such **robotic relationships** are serving a very important purpose, acting as surrogate caregivers, she also questions whether we are not facing the danger of these same people disowning us completely? Started because robots "are better than nothing" we may be moving them to the idea that robots "are better than anything." This is why she feels that we are alone together: why we expect more from technology and less from each other (*Alone Together*, 2011). Are we not losing the "raw human part" of being with each other; those small gestures, that smile, those emotions that form the essence of

Box 5.9: Children and Sociable Robots

In most of the experiments that Turkle conducted in schools and nursing homes for the aged, subjects, either children or the aged were allowed to keep a sociable robot such as an AIBO or a My Real Baby for some time after which they were interviewed. We present excerpts from some of these very telling interviews.

> "**Melanie, a 10 year old**: Melanie believes that AIBO and My Real Baby are sentient and have emotions. She thinks that when the robotic dog and doll were brought to her school 'they were probably confused about who their mommies and daddies were because they were being handled by so many different people.' ……She sees her role with the robots as straightforward; it is maternal.
>
> One of Melanie's third-grade classmates is aggressive with My Real Baby and treats the doll like an object to explore (poking the doll's eyes, pinching its skin to test its "rubber-ness," and putting her fingers roughly inside its mouth). Observing this behavior, Melanie comes over to rescue the doll. She takes it in her arms and proceeds to play with it as though it were a baby, holding it close, whispering to it, and caressing its face. At home, Melanie has AIBO and My Real Baby sleep near her bed and believes they will be happiest on a silk pillow. …At home, she and a friend treat it like a sick animal that needs to be rescued. They give it 'veterinary care.' For Melanie, not only does My Real Baby have feelings, Melanie sees it as capable of complex, mixed emotions" (p. 351).

Source: excerpted from Turkle et al. (2006)

human interactions. We, and especially those who design robots, must remember that rather than replacing human care, robots are simply meant to augment human contact. We have to be careful to see that such human contact is not overly reduced or even completely removed from the nursing and healthcare arena, once again pointing to the fact that technology design and development without an understanding of its effects upon the user can be detrimental. Is this not an important role that can be played by psychology of technology?

> ***Moral HRI****: a dimension of human–robot interactions that attempts to study the ways in which people attribute morality to robots*

Is this what the designer of sociable robots had in mind? Are we becoming enablers of moving man away from man? Each one of us needs to think. The kinds of effects sociable robots are having on the people they are taking care of have important implications for current advancements in the field of robotics. Not only is it demanding developers and designers to think of the extent they want robots to replicate human behavior, but it is also leading to new disciplines being spawned. We have already mentioned the upcoming field of robotic psychology, but added to it is a newer dimension, that of **Moral HRI** (Human–Robot-Interactions) which is attempting to study the way in which people attribute morality to robots. Morality being such a crucial part of social interaction and codes of ethics preventing us from causing physical or even psychological harm to others, it seems but natural to expect robots, especially interactive sociable robots, to abide by ethical norms. Simultaneously, seeing the nature of human–robot interactions, social scientists have also started raising questions regarding ethicality in robotic design (e.g., Sharkey & Sharkey, 2012). A good example of the ways in which research is carried out in the area of Moral HRI can be gained from two online experiments reported by Malle, Scheutz, Arnold, Voiklis, and Cusimano (2015). Using the same sort of moral dilemma scenarios that Kohlberg (1976) based his studies on moral development in people, the authors concluded that as compared to their human counterparts, robots were expected to abide by utilitarian values to a greater extent and when their behavior was found to be otherwise, the chances of blaming them was greater than when humans behaved in the same manner. So, it seems that our attributions for morality change depending on whether we are interacting with a human or with a machine.

> ***Robotic pets****: animal-shaped robots who are treated as pets*

Even in Turkle's experiments, the question is not whether children will love their **robotic pets** more than their real life pets or even their parents, but asks Turkle, "what will loving come to mean?" Similarly, what will mortality or death come to mean to these children and elders who see the robots 'dying'? As one woman subject put it, "[AIBO] is better than a real dog … It won't do dangerous things, and it won't betray you … Also, it won't die suddenly and make you feel very sad." So, when such people see their robotic friends die, will this death not be different from the death of human beings? What will we make out of it? What will death come to mean to us? It may not be what death means to those of us who have not established relationships with robots. Box 5.10 helps us gain some insight into how people feel when the AIBO or some such robot is no longer working.

Box 5.10: Japan: Praying for the Departed Souls of Robot Dogs
Incense smoke wafts through the cold air of the centuries-old Buddhist temple as a priest chants a sutra, praying for the peaceful transition of the souls of the departed. It is a funeral like any other in Japan. Except that those being honored are robot dogs, lined up on the altar, each wearing a tag to show where they came from and which family they belonged to. The devices are "AIBOs," the world's first home-use entertainment robot equipped with Artificial Intelligence (AI) and capable of developing its own personality.

"I believe owners feel they have souls as long as they are with them," said Nobuyuki Narimatsu, 59, who heads an electronics repair company specializing in fixing vintage products.

Sony rolled out the first-generation AIBO in June 1999, with the initial batch of 3000 selling out in just 20 min, despite the hefty 250,000 yen (more than $2000) price tag. By 2006, Sony was in trouble because of the fierce competition from rivals in all fields and in March 2014, the 'AIBO clinic' which repaired the AIBOs finally had to go.

For Hideko Mori, 70, and many others, that nearly spelled disaster. Mori has had her AIBO for around 8 years. She enjoys the conversations she has with it and thinks it far more convenient than a real puppy. But in May last year her beloved AIBO, whose name is simply "Aibo," became immobile. She was then introduced to A FUN, a company that employs former Sony engineers, who fixed her machine in 2 months. "I was so happy to see him back to health and at home," she said.

The engineers at A FUN say that the AIBO owners see them more as doctors than as engineers and that their AIBO is not a robot but a family member. The problem therefore becomes one of restoring the 'health' of these robots and there are a large number waiting for 'transplants' from 'dead' robots after due respect has been paid to their 'departed souls.'

Source: AFP, Isumi (2015)

5.10 Moving On? From Avatars and Agents to Immersive Virtual Environment Technology

From time immemorial, humans have developed tools to help them to communicate. Starting with story-telling, we have gone on, to graphic arts, theatre, printed books, movies, radio, TV, and most recently to digital media. Each stage of development has enhanced our ability to travel between grounded and virtual realities. The most sophisticated aspect of this augmentation is what has come to be known as digital **Immersive Virtual Environment Technology**

Immersive virtual environment technology*: technology which creates a type of virtual reality in which the user gets absolutely immersed*

(IVET). What IVET has enabled us to do and which goes much beyond what would have been labeled as science fiction just a decade back is the ability to change the context of the relationship between man and machine.

But has this ability really been developed only during the latter part of the last century? Have we not always been able to travel back and forth in our mental time machine? As Blascovich and McCall (2010) put it,

> "Humans are clearly neuro-physiologically wired to travel mentally back and forth between grounded reality and virtual realities as well as among virtual realities themselves. Humans not only dream during sleep, they also dream while awake. Human minds wander often and effortlessly from grounded reality to somewhere else. Undoubtedly, mind wandering serves some adaptive function" (p. 286).

> ***Telepresence****: when the user can immerse as well as interact with simulations*

Probably, no other psychologist has argued in favor of the social nature of humans as vehemently as Batson et al. (1995). Based on his research studies, he has demonstrated the tendency of people to help other people, or altruism, which can be said to be a social orientation either with or without specific rewards. In other words, we are social animals who engage in a wide variety of communal activities. The extension of our self would afford us opportunities to display our inner persona via a virtual character replete with mental and physical features which can be tested through **telepresence**. If we can call our car "Mustang," ascribe human features to it and use endearments for it of the type we would use for a near and dear one, working in conjunction with an avatar would elicit a great deal of our own attributions since it is our own creation. Further, it would afford the opportunity to explore new vistas of human experience that would occur as the role of avatars reverse or change. For example, with change in the characteristics of a female, there would be an opportunity to experience the lifestyle and ethos of people categorized as gays, lesbians, or even transgendered identities.

While virtual reality has existed since the 1990s, when Jaron Lanier popularized the term, it has finally moved and has started impacting a variety of facets of both life and business. A recent issue of the *Time* magazine, (August, 2015) gives us an idea of the times to be as far as Immersive Virtual Reality Technology is concerned. Call it IVRT or AR or VR, the past few years have seen the dawn of a new era in the world of gaming in particular and computing in general. It has now become possible to project a virtual world onto the real physical world, and the beauty of it lies in the fact that the user of the technology becomes totally immersed in it. All it requires is the particular software and a pair of specially made headsets. When Google introduced *Google Glass*, this was the first step. While it was not very successful, it paved the way for *Oculus Rift* created by Oculus VR, a company founded by little known software developer Luckey Palmer. Mark Zuckerberg of Facebook was quick to see the opportunities created by this remarkable technology and bought Oculus VR in 2014, setting the stage to be followed by almost all the well-known gaming companies, each developing their own product, including names such as Google, Sony, HTC, and Samsung, Almost all of their VR products, such as the *Cardboard, Vive*, *Project Morpheus*, and *Gear VR* have already been released or will be released soon. Even Lanier, who at one time had started becoming pessimistic as far as VR is concerned, is of the view that this just had to happen, being as it is the next logical step, starting with the written language and progressing to the printing press, the photograph, the audio recording, and finally to film. Excitement reigns high because,

> "It (VR) can blur the distinction between you and the rest of the world. You have the option to map yourself to the clouds or the grass. When you move your body, all the clouds and animals can move in sync with you....and in about a year or two, nobody will find this hard to understand. This will become totally ordinary" (Lanier, 2015).

While at one time headsets costs were over $1000, they have become much more affordable with the commercial launch of Google's *Cardboard* and Samsung's *Gear VR* enabling people to use them for purposes other than complex gaming. New opportunities are being thrown up including the use of this technology in the field of industry and marketing (Besecker, 2015), such that inventories can be managed virtually, and a 3D virtual tour can be arranged for your prospective customers. Some companies have already started using it (e.g., Lowe's uses a *HoloRoom*), while hotel chain Marriot has a virtual honeymoon package, Volvo uses *Google Cardboard* for a virtual test drive while LandRover has a virtual showroom which promises 3D experiences.

> ***Cybersickness***: *nausea and dizziness much like in motion sickness but caused by using some types of virtual reality technology*

However, is there a downside to the use of these VR headsets? While companies are optimistic about the new VR technology, users have their issues, the greatest being issues related to health. With people showing signs of what has been named **cybersickness**, much akin to motion sickness, it seems clear that the future of this revolutionary technology lies in the extent to which companies are able to overcome these health issues (Lewis, 2015). Even the best known product, *Oculus Rift* is fraught with this problem and when Samsung initially released its *Gear VR*, it came with a health disclaimer warning people to stop using it if they felt nauseous or dizzy and stopped children under the age of 13 from using it.

The cause of cybersickness is still not very clear though it certainly seems related to brain functioning. While under real-world circumstances, too, the brain processes multimodal or multisensory data from the environment, these data are in sync with one another. For example, as a vehicle draws nearer to you (which your eyes can sense), the sounds made by the vehicle become louder (as sensed by your ear). Thus, the data from the eyes and that from the ears are in complete sync with one another. When we view virtual reality, the brain still expects such synchronicity, but it is not so, creating problems for further processing and interpretation by the brain (Lewis, 2015). It is only with the use of brain imaging and other such techniques that this quandary can be resolved which will then enable technology developers to overcome the health problems. Some advances have been made by changing the speed at which the headsets refreshes data. It has been seen that if the headset refreshes its data at a higher speed, these associated feelings of nausea and dizziness are not seen. Thus, as our knowledge of brain processes accompanying the use of VR technology increases, we will also be able to overcome most of the problems, pointing to the complex ways in which technology, biology, and psychology coevolve. There are signs of some improvement, with HTC promoting its product, *Vive*, by saying that this problem has been resolved by tweaking the technology. Even the Samsung *Gear VR* and the Google *Cardboard* have been improved so that the commonly seen nausea is not felt. Yet much more needs to be done to understand the psychological and especially the physiological concomitants of the use of VR if it is to take the world by storm as enthusiasts predict.

5.11 Applications of IVET and Other Virtual Reality Technologies

Given the fact that technology is affecting us in almost all walks of life, the applications of virtual reality technologies are enormous. From its usefulness in the educational setting to that in business, it has offered positive results. Here are a few examples of its application.

5.11.1 Social Psychology

The focus of social psychology, among other things, is on the interactions between individuals. The study of such interactions poses a challenge for laboratory research, adding not only complexity but also costs to experiments. For example, if a researcher wants to study the effects of group diversity on group performance, each observation requires not a single participant, but an entire group of participants. If a researcher wants to study a behavior that is evoked by a specific social interaction, it gets even more complicated: suppose a researcher thinks that individuals will share less information with an incompetent supervisor than with a competent supervisor. The experiment would require a supervisor who is either competent or incompetent for every experiment. Also, this supervisor should possibly exhibit the same interactions toward all experimental participants in one condition. While one way of doing this is by hiring an actor for the role of the supervisor, this is often not feasible.

In Schmid Mast's Laboratory they do such studies in a different way (e.g., see study by Bombari, Schmid-Mast, Cañadas, & Bachmann, 2015). They use virtual reality: a 3D virtual immersive environment. The experiment participant wears a head-mounted display (HMD) that gives one the impression of being in another world. In this virtual world, one can interact with avatars or virtual representations of individuals who are programmed by the experimenters in such a way that they exhibit certain interpersonal behavior, which is, of course, always constant and fully controllable by the experimenters. At the same time, the system logs data that is difficult to acquire in normal laboratory settings, such as interpersonal spatial distance between the participant and the avatars. In combination with verbal coding of the participant, one gets an extremely rich and reliable source for social interaction data.

5.11.2 Education

The U.S. Secretary of Education, Arne Duncan, in his draft on the National Technology Plan (Duncan, 2010) suggested that technology must be boosted in instruction in schools even if we have to achieve our goals by using videos or the avatar in the classroom. School teachers are realizing, more than ever before, the value of technology in teaching. A good example is the *Quest to Learn* program, funded with the support of local school budget, the MacArthur Foundation and the Gates Foundation, which involves digital games as a core idea to flourish the growth of a child. After such games were prepared for students,

> "A lesson doesn't look like a lesson anymore. It is now a quest. And while students at the school are put through the usual rigors of studying pre-algebra, basic physics, ancient civilizations and writing, they do it inside interdisciplinary classes with names such as *Codeworlds*—a hybrid of a math and an English class—where Quest blends skills from different subject areas. Students have been called upon to balance the budget and brainstorm business ideas for an imaginary community called *Creepytown*, for example, and to design architectural blueprints for a village of bumbling little creatures called the Troggles" (Corbett, *New York Times*, September 19, 2010).

Given the fact that children can download and create material on computers, the process of instruction and the concept of school are bound to change. Modern instruction is not limited to school hours any more. And the teachers would no more be "teachers" as the conventional gap between a student and teacher will blur with technological advances. A very well-established institution in human history, called school, is going through a radical change due to technology.

> ***Edutainment****: entertainment which has education as its goal*

It is well known that the virtual environment offers an opportunity to learn without being physically present in the school or college. The popularity of long-distance educational programs hardly needs to be emphasized. What educators find useful is that the characters and environments created in virtual settings offer a new challenge to students beyond what has been known as "**edutainment**." For example, Gee (2008, 2005) has been using this technique to enhance critical thinking among students, who face the challenge of having to look around through a character created by them and find options available for solving a problem. Besides this effort, referred to as a probing cycle, the second cycle, called telescoping, keeps the student focused in order to attain the desired goals. With this technique, Gee could engage students with enthusiasm for several hours.

In his book, *Rewired*, Larry Rosen (2010) cites the experience of Courteny, a high school student, who collects a lot of material from websites to prepare her article on Mayan culture, despite which her PowerPoint presentation appears boring to herself. However, when she used *Second Life*, to navigate around the pieces of information, she felt that it "offered a more engaging, compelling lesson about Mayan culture" (p.122). The virtual environment also tends to facilitate cooperative and collaborative learning.

5.11.3 Serious Games in Health and Education

Maxmen (2010) remarked that

> "The video-games that kids have been neglecting their science homework to play can now be harnessed to make science and scientific thinking fun again" (2010, p. 203).

> ***Serious games**: virtual games designed to promote learning and development of some type*

Often called serious games, these can be designed to train users on a particular skill set (Annetta, 2010), such as attentional skills (Wronska, Garcia-Zapirain, & Mendez-Zorrilla, 2015), spatial skills, problem-solving skills, etc. However, as had been the case with IQ testing, previous differences in video game experience are becoming increasingly important. It would be worth considering whether the introduction of serious games in the realm of education will weaken or magnify the already existing digital divide.

5.11.4 Teachable Agents and the Protégé Effect

> ***Teachable agent**: a computer character whom the child has to teach*

Through a series of experiments, Chase, Chin, Oppezzo, and Schwartz (2009) have demonstrated how by the use of immersive virtual reality technology, children are able to not only learn more but also develop prosocial habits. Using software called *Betty Brain*, the experiments use a computer agent whom the children have to teach, thus making it a **teachable agent**. They are actually not true agents but hybrids because though they reflect their owner's knowledge, they have a mind of their own, which is computer controlled. The experimental paradigm asks these children to teach these agents, such that the agent becomes dependent on the child. Interestingly enough, though the children had not been asked to learn, the experiments reveal that there were clear learning benefits in the sense that the children showed greater effort toward learning. Moreover, since this effect was more noticeable for low achieving students, it can be an effective way of getting such children to enhance their learning. Similar results have been obtained by others too, for example, Wagster, Tan,

Wu, Biswas, and Schwartz (2007), while Baylor (2007) found that the use of agents can be effective for giving **meta-cognitive** tips to children. There could be a variety of explanations for such **Protégé Effects**. Thus, Chase et al. (2009) conclude that the Teachable agent provides the child with an ego-protective buffer, since failures would be failures of the agent, though in fact it is because the child was unable to teach effectively. Another plausible reason could be that the child feels a sense of responsibility toward the success and failures of the agent, because of which the child tries harder and harder. These studies are just some examples of the innovative ways in which technology can be harnessed by educators.

> ***Meta-cognition**: knowledge about one's own cognitive processes*

> ***Protégé effect**: learning caused by teaching someone, even a virtual character*

Another example of how the much abused video game can be put to real use is the game "*Re-Mission*" designed for child cancer patients. A major problem with child cancer patients is getting them to adhere to the tight treatment regimen. This particular game has been designed keeping this problem in mind and allows the player to control a nanobot who shoots cancer cells, overcomes bacterial infections, and even manages signs of nausea and constipation. This game has now been distributed to over 200,000 patients and has been found very useful in teaching children how to adhere to treatment regimens (Granic et al., 2013).

More recently, Eric Carson (2015) has described how VR can be used for virtual field trips for children or to teach even children with special needs. Thus, a school in Ireland used *OpenSim* to recreate the ruins of an old monastery surrounded by a cemetery. While it did take weeks to build, they were able to use *Oculus* for the virtual trip.

5.11.5 Therapy

According to Botella, Garcia-Palacios, Banos, and Quero (2009), there are two forms of technologies that have shown impact in **cybertherapy**: virtual reality and **telepsychology**. The success of cybertherapy would depend on the effectiveness of simulated reality and creating an illusion of being out there in the virtual environment and if properly formulated, it could become a useful tool for Cognitive-Behavior Therapy. Telepsychology, on the other hand, is useful for those who can help themselves and for some reason, for example, disability or remote location, are unable to participate in the actual clinical setting. Botella and coworkers concluded that both technologies have yielded positive results, while Budman (2000) went to the extent of suggesting that the future of psychotherapy lies in computer-mediated communications.

> ***Cybertherapy**: use of virtual reality as a part of psychotherapy*

> ***Telepsychology**: delivering psychological help through the telephone or internet, in which the physical presence of the patient is not required*

Acrophobia: *fear of heights*

In the early 1990s, Dr Ralph Lamson at the *Kaiser Permanente Psychiatric Group* started exploring how he could use VRT to cure his own **acrophobia**. Later, tightly controlled studies at the same institute revealed that the success rate of the use of VRT to cure a variety of psychiatric conditions such as phobias and anxiety disorders was as high as 90 %. When patients are placed in a virtual world, they are exposed to visual and auditory inputs as per their treatment needs. For example in treating phobia, the feared object would appear less threatening. Along the same lines, psychiatrists at the University of Louisville use VR in cognitive behavior therapy to treat patients with social anxieties or phobias of things like flying, public speaking, or heights. The controlled environment allows doctors to expose their patients to simulations and direct them on how to cope with how they're feeling (cf. Dvorak, 2015). VRT is also being used to increase the effectiveness of psychoanalysis. Rosegrant (2011) reveals how technology can be used to understand patients better and describes how texting was used for a girl and game playing by a boy was used for understanding them better.

Takacs (2006) reported that such therapy has worked successfully in treating cases of depression, anger, phobia, pain, and navigation while Carlbring and Andersson (2006) review empirical work showing that self-help programs for treating panic disorders can be effectively delivered through the internet. One positive feature of this type of therapy is that the patient can set up his own schedule for therapy without waiting for an appointment with his/her physician. A virtual panorama with the placement of a live doctor and a nurse contributes to its authenticity made possible through complex work in virtual studios. Another study has been reported in the medical journal, *Frontiers in Neuroscience* (2014) on the use of virtual reality to treat the phantom limb pain of people who have lost limbs (Ortiz-Catalan, Hakansson, & Branemark, 2014). The therapy uses sensors that pick up on nerve inputs from the brain, and patients have to complete a game using a virtual limb. It helps them gain control: so if an amputee feels as though they've been clenching their fist, seeing a virtual limb that they control helps them learn how to relax the fist (Ortiz-Catalan et al., ibid). Virtual reality simulation has been used to train surgeons in the use of minimally invasive surgery such as that being used for heart bypass surgery today (Vapenstad & Buzink, 2013; Seymour, 2008). Katie Dvorak (2015) reports on how VR is being used by the Miami Children's Health system and how it has benefited hospitals while psychiatrist Albert "Skip" Rizzo of the USC, Institute for Creative Technology, created a war simulation entitled, *Full Spectrum Warrior* for treating PTSD among war returnees. As Rizzo pointed out at the American Psychological Association Convention at Washington DC in 2011, virtual reality has several advantages. These have been given in Box 5.11. However, Rizzo (2013) also cautions that "technology is really no more than a tool. The technology in and of itself, does not 'fix' anybody. Rather, these systems are designed to either train or extend the skills of a well-trained clinician, and in the case of *SimCoach*, to help a person to anonymously find the treatment they may benefit from with a live human provider" (Rizzo et al., 2013, p. 135). In March 2015, Rizzo

Box 5.11: Assets of Using Virtual Reality in the Psychology Laboratory

- Ecological validity
- Stimulus control and consistency ("the ultimate Skinner box")
- Repetitive and hierarchical stimulus delivery possible
- Cueing stimuli for "errorless learning"
- Real-time performance feedback
- Self-guided exploration and independent practice
- Stimulus and response modification contingent on user's impairments
- Complete naturalistic performance record
- Safe testing and training environment which minimizes risks due to errors
- Graduated, systematic exposure
- Distraction
- Gaming factors to enhance motivation
- Low cost functional environments that can be duplicated and distributed

Source: Rizzo (2011) Symposium on Innovative Technologies for Psychological Intervention, Consultation and Training, APA Annual Convention, Washington, DC

received the "Pioneer in Medicine Award" from the *Society for Brain Mapping and Therapeutics* (*SBMT*) and *Brain Mapping Foundation*. The award, presented at the society's annual meeting in 2015 recognized Rizzo for his role in the field of virtual reality medicine and his impact on treatment of patients with a range of clinical health conditions, including stroke, brain injury, autism, and posttraumatic stress disorder (Belman, 2015).

5.11.6 *Industry*

There are at least nine industries that are already using VR (Carson, 2015), namely, healthcare, entertainment, automotive industries, advertising, education, tourism, space, skilled trades, and military and law enforcement. Let us have a look at some of the very interesting and innovative uses that VR is being put to.

One industry sector that has gained tremendously from the use of VR is the automotive industry. From virtual showrooms such as those used by Ford in its *Immersion Labs* to help customers get a feel of their cars using the *Oculus to* Toyota which uses VR to train teenagers and parents about distracted driving and Audi which plans to use VR as part of their customization of cars as per customer demands, virtual reality technology has proved to be an important adjunct to the repertoirc of companies for enhancing customer experiences before the actual purchase takes place.

Advertising too has not been left untouched by this amazing technology. Branded VR experiences are taking on many shapes and digital marketing agencies are exploring how they might use VR to promote the brands of their clients. Merrell, an outdoor apparel brand, has in fact used VR to set up an experience where users could go trekking up and across treacherous mountain sides, while wearing their hiking shoes, of course. Is it not astonishing that even training in traditional skills such as welding is being affected? One immediate benefit is that using virtual reality training means money does not have to be spent on materials to practice on, and the trainees can repeat the task as many times as they need to. While it will never replace traditional training, it can make the process faster and cheaper.

Did you ever think of why companies hand out samples? The reason obviously is that unless you experience the product, hearing about it or seeing it or even being able to touch it is not sufficient. And, how many customers would be ready to spend money for a trial? And how does a tourism company distribute samples? They certainly cannot send prospective customers on an all paid holiday in the hope that with this experience they will go on one paid by themselves. The answer has finally been found. Virtual reality will enable some industries to give customers the hands-on experience. In December 2014, Destination British Columbia launched a VR experience called *The Wild Within* featuring two options: a boat ride and a hike in the mountains. The app was created to promote tourism to BC and it helps engage the traveler in an emotional conversation about why they should visit. Similarly, Marriott Hotels created a "teleporter" which lets users step into a booth, wear an *Oculus Rift* headset, and visit downtown London or a beach in Hawaii. The teleporter also caters to other senses, so users can feel wind in their hair and sun on their faces.

The potential of this new immersive technology appears unfathomable. A list provided by John Brandon (2015) is truly amazing and include tasks as varied as those mentioned as follows:

- Preview new office buildings
- Attend technology conferences
- Interview candidates
- All hands meeting
- Complex training sessions
- Employee and customer confrontation

5.11.7 *Limits of Virtual Reality*

> "If virtual reality becomes a part of people's day-to-day lives, more and more people may prefer to spend a majority of their time in virtual spaces. As the futurist Ray Kurzweil predicted, somewhat hyperbolically, in 2003, 'by the 2030s, virtual reality will be totally realistic and compelling and we will spend most of our time in virtual environments … We will all become virtual humans.' In theory, such escapism is nothing new—as critics of increased TV, Internet, and smart phone usage will tell you—but as VR technology continues to blossom, the worlds that they generate will become increasingly realistic, as Kurzweil explained, creating a greater potential for overuse. This technological paradigm shift brings a level of immersion unlike any that has come before it, and the handwringing has already begun. Early doomsday predictions aside, can virtual escapism can ever be used for good?"

This is what Carole Cadwalla wrote in the *Guardian* (2014).

One question that remains unanswered in the use of virtual environment is "is it real?" Is the avatar out there really me? Or, is it just a look-alike? Are multiple exposures of such avatar-type scenarios harmful in the sense that the gap between fact and fiction becomes blurred? Will children start having problems in distinguishing between the real and the simulated world? Will they become desensitized to behavior normally engaged in online (e.g., violence) but frowned upon offline?

Despite all the applications pointed out earlier, is the virtual experience an exact replica of the experience in the real world? One area of critical significance is efficiency in virtual environments ranging from limitations of human sensory and perceptual systems to movements. Navigation in a virtual environment would require constant updating of spatial information. Further, mental representation of real and imagined environments is not similar, for subjects generally tend to lose information much faster in the imagined surrounding than in the real environment. Stanney, Mourant, and Kennedy (1998) have shown how objects in a kitchen were poorly located in an imagined scenario, but with physical navigation in a real environment, there was a significant advantage. They attributed this difference to behavioral consequences. For example, when we move physically, an object has a chance to collide, but this is not going to happen in an imagined situation. Further, they found that actually touching an object outscored many other inputs that had also facilitated effects.

These issues are likely to become more complicated with the applications of robotic parts within our body. In his book, *The Singularity is near: When humans transcend biology*, Ray Kurzweil (2005) posited that with advancement in technology there would be fusion between artificial intelligence and human intelligence. With nanorobots implanted within our bodies, the dividing line between virtual reality and the real self would vanish, leading to the development of a new identity. With experiments in placing a neuro-circuit board in our brain to combat lost function, we may encounter a new self at the mercy of this new artifact that will flip back to malfunction upon withdrawal. We are entering an era of the exoself, where our new identity would be tagged to an artifact.

From time immemorial, a life lived outside regular society has generally been viewed as being dangerous and at the same time, unhealthy. In Japan, they use the term 'hikimori' to describe the large numbers of people who simply refuse to leave their homes, and according to most reports, these people display depressive and obsessive–compulsive tendencies. Closer to home, here in the Western world, there have been incidents of people losing themselves while immersed in the *World of Warcraft*. Former Warcraft player Ryan van Cleave felt that living inside WoW seemed preferable to the drudgery of everyday life and groups like WoWaholics Anonymous have been created to help players like Cleave.

While these are extreme examples, they certainly share roots with what has been termed escapism. And, looking at it from the viewpoint of Maslow's Hierarchy of Needs, discussed in Chap. 6, did he not place love and sense of belonging just above the physiological needs? We have already pointed out in an earlier section of how we are being estranged from our fellow human beings because of the onslaught of technology. Monika Kim (2015), in an article aptly entitled, *The good and the bad of escaping to Virtual Reality,* warns us of these and many more such dangers.

There are other concerns too. Several scholars have argued about the dangers of technology going viral and the cost of such recovery becoming unmanageable. Without proper control of such technology, the chances are that it would be abused (Rothbaum, Hodges, Smith, Lee, & Price, 2000).

One line of approach that empirically establishes the link between the real self and avatar stems from the measurement of personality orientations of both entities. Recently, McCreery, Krach, Schrader, and Boone (2012) investigated how one's real personality pattern was reflected in its avatar. Out of the *Big Five* personality factors (Openness, Conscientiousness, Extraversion, Agreeableness, and Neuroticism) only agreeableness could connect the real and fictional characters. Agreeableness refers to that aspect of one's personality on which is dependent our relationships with others ranging from compassion at one end to antagonism at the other. High scorers tend to be those who are trusting and forgiving, helpful, and straightforward (Kool, 2008). Of course, as the demands of software change in such a scenario, so does the outcome. However, the vital issue is: how would this connection influence our overall real-life behavior? This topic needs further analysis in future research.

When computers were first introduced, it was a source of stress for many workers as they were not ready to change their old ways of doing their job manually. It also created a digital divide, that is, those with financial inputs could afford it versus those who could not. This effect resulted in social exclusion of many people and technology became synonymous with wealthy lifestyles. However, as this chapter has attempted to discuss, both the users and the dynamics of these users have changed dramatically. Psychology has only recently started focusing on the psychological aspects of the use and abuse of technology. It is hoped that this chapter will inspire many a budding psychologist to delve into the mysteries of technology adoption.

Summary

That this is the age of machine-mediated reality would not be a misnomer, keeping in mind the rapid advancements in information and communication technology (ICT), smartphones, social networking sites, and the world of virtual reality. While our perceptions and cognition decide its initial adoption, it becomes imperative to understand what makes us slaves of such technology, the complex ways in which it changes all forms of social interaction and even our very personalities. Chapter 5 attempts to answer questions regarding technology adoption, in the light of current research in the use and abuse of communications technology and virtual reality. While dealing with topics such as texting, cyber bullying, problematic internet use, and internet addiction, the focus is on their effects on the psyche and the social life of the users. The second part of this chapter focuses on virtual reality and gaming, with the aim of understanding its psychological and social concomitants. The chapter ends with the ways in which the principles underlying virtual reality can be applied in realms such as education, therapy, and industry.

Suggestions for Further Reading

Gentile, D. A. (2009). Pathological video-game use among youth ages 8-18: A national study. *Psychological Science*, 20, 594–602.

Granic, I., Lobel, A., & Engels, R. C. M. E. (2013). The benefits of playing video games. *American Psychologist*, 69, 66–78.

Parisi, T. (2015). *Learning virtual reality*. Sebastopal, CA: O'Reilly Media.

Rosen, L. D. (2012). *iDisorder: Understanding Our Obsession with Technology and Overcoming Its Hold on Us*. New York, NY: St Martin's Press.

Chapter 6
Technology and Hedonism

Professor Semir Zeki from the University College, London has been oft regarded as the father of the upcoming field of neuroesthetics. Zeki has been delving into works of art not as an art historian or as an art connoisseur but to unravel the remarkable brain–behavior links with reference to the creation and viewing of works of art. Perhaps, the most amazing finding is that beauty perceived through the eyes, say a painting, and beauty you receive through the ears, say soulful music, do not activate different parts of the brain. Rather, both of them reward the same spot in the cortex (Ishizu & Zeki, 2011).

According to Zeki's findings, whenever we look at something beautiful, a certain area of the brain gets activated, thus going against the age-old adage that beauty lies in the eyes of the beholder. Nay! Beauty is not in the eye of the beholder: it is in the brain of the beholder, and in a very specific part of the brain, according to the paper mentioned earlier (Ishizu & Zeki, ibid). They have found, by examining fMRI images of their subjects' brains, that when people look at something they find beautiful, a portion in the front part of the brain called the medial orbito-frontal cortex "lights up." That is, there is increased blood flow in this area. Moreover, viewing art triggers squirts of dopamine, a feel-good chemical, in this orbito-frontal cortex of the brain, resulting in feelings of intense pleasure. Not only that, they say, "the degree of activity in the medial orbito-frontal cortex correlates very strongly to the degree to which you find a thing attractive." Is it not ironical that this is the same part of the brain that is involved in the cognitive processing of decision making.

Zeki believes that this is a near-universal response to beauty. But what is even more important is that we tend to seek beauty. The reason being, Zeki adds, is that the medial orbito-frontal cortex is a portion of the brain associated with pleasure and also reward. Thus, seeking beauty is actually seeking to reward the pleasure centers of the brain and it is interesting because it leads to a completely different definition of beauty. The lessons we have learnt from neuroesthetics have been detailed in Box 6.1.

The earlier findings are important for our discussion of psychology of technology. As mentioned in one of the earlier chapters, technology is not simply about

V.K. Kool, R. Agrawal, *Psychology of Technology*,
DOI 10.1007/978-3-319-45333-0_6

Box 6.1: What Lies Behind the Magnetic Charm of Beauty: Lessons from Neuroesthetics

Neuroesthetics is an amazing example of interdisciplinary research, with researchers in this new field combining principles from perceptual psychology, evolutionary biology, neurological deficits, and functional brain anatomy to address the evolutionary meaning of beauty.

What is at the basis of this understanding? Zeki (2008, 2009) proposes two supreme laws of the visual brain.

Constancy: this refers to the rare phenomenon that despite changes that occur when processing visual stimuli (distance, viewing angle, illumination, etc.), the brain has the unique ability to retain knowledge of constant and essential properties of an object and discard irrelevant dynamic properties. This applies not only to the ability to always see a banana as the color yellow but also the recognition of faces at varying angles.

Comparatively, a work of art captures the essence of an object. The creation of art itself may be modeled off of this primitive neural function. The process of painting, for example, involves distilling an object down to represent it as it really is, which generally differs from the way the eyes see it.

Abstraction: This process refers to the hierarchical coordination where a general representation can be applied to specifics thereby allowing the brain to efficiently process visual stimuli. The ability to abstract may have evolved as a necessity due to the limitations of memory. In a way, art externalizes the functions of abstraction in the brain, though cognitive neurobiology has yet to unravel how this is done.

Source: Zeki (2008, 2009)

being useful: it must, at the same time, appeal to the user. In this chapter, we will focus on how technology, which has been traditionally associated more with manufacturing output is becoming increasingly involved with our affective processes including feelings and emotions.

Can the same be applied to technology? Will beautiful technology act as a ploy for attracting more and more people toward it? One look at recent developments in the field of smartphones and even laptops makes it clear that to attract users, technology must not only be functional but it must also be beautiful. If this were not so, how does one explain the recent trend in technology toward a focus on characteristics such as shape, color, texture, and sheen? Hewlett Packard boasts about its latest laptop much along the same lines. Not to say that technological features are put aside, but as important are the looks of the product. Once we accept the fact that we seek beauty so as to reward the pleasure centers of the brain, the reason for the above does not sound very far-fetched.

Let's now look at beauty and especially pleasure from a more down-to-earth perspective. Just imagine: you see an advertisement of a new electronics store opening shortly quite a way from your home. You call a friend excitedly, telling her of this new store and how much fun it would be to visit it and what a whole lot of new products are to be available there. The day finally arrives when both of you, making sure that you are carrying not one but two credit cards (just in case one does not work, or even more, if your shopping exceeds the limit of one). You arrive at the store, all decked up, to welcome its new customers. The sales persons are extremely cordial, you are offered a steaming cup of coffee at the entrance and then you start on your shopping spree. Half an hour later, you are back at the exit, without even having visited the checkout point, with not an item in your cart. What happened? You shrug your shoulders at each other and say "what a waste of time, not even one interesting item; the same old stuff."

Does this bring in a sense of déjà vu? Yes, most of us have gone through such experiences when we went shopping not because we needed something but were looking for something interesting, something novel, and something about which we could boast about at our next get together. Simply adding more stuff of the type that we already have does not give us pleasure. Even four cars of the same make and same model does not bring as much pleasure as having three cars of three different makes and/or models. You want to change your laptop computer only if the new model is offering you something new, not just adding more memory to the old (you could do that with your old laptop by simply extending its memory through some device, without buying a new one). Remember the long validated Law of Diminishing Returns so oft quoted by economics: it holds true, be it for food, clothes, cell phones, cars, or any other object. More of the same stuff has lesser value for you than the original stuff. But change the food: bring in the dessert and everyone who had been complaining that they have eaten too much is ready for the new item, the dessert. The moment companies see sales dipping, they launch a new model and once again, the interest of the consumer is enhanced. Every time Apple introduces a new model of its phone, it leaves people gasping and gazing with an infinite urge to own it. Why is this so?

This is so because for most objects we have some sort of emotional reaction. As a matter of fact, our attitude toward technology, too, is hardly ever neutral. Tools are not mere tools that we use and then forget about them till we need them again. Technology cannot be considered to be simply a means to help us achieve our desired ends. Some objects give you pleasure, others frustrate you. And, the pleasure/displeasure seems to vary from person to person. What do you think about a vacuum cleaner or the garbage dispenser in your home? Not much, I guess. These are mere tools that we use as and when we need them. On the other hand, if you have grown up with that vacuum cleaner, or if the vacuum cleaner was given to you by your mother on a special day, say when you bought your first home, you would find yourself feeling sentimental about this gadget. You arc cmotionally attached to that humble vacuum cleaner. If that vacuum cleaner is one that you never liked because you found it difficult to move around, your attitude toward it would be different. The wise old man of yore was not wrong when he said that "one man's meat may be another man's poison." Each one of us is unique, with differences on so many aspects, not just demographic factors such as age and gender but also on personality, intelligence, interest, experience, to name just a few.

Affordances: *the limits to which any technology can be used*

In his book, *Things that make you Smart*, Donald Norman (1993) discusses why technology is not neutral. The reason is that for every technology there is a set limit to which it can be used. Norman calls these limits **affordances**. The important idea here is that it is we, nay, our perception which decides these limits or affordances. Objects have salient features and we use them on the basis of our perception of these features. Let us think of a container such as the one we use in our homes to store groceries such as sugar. It has a shape, needs space to be kept, and must be placed in a vertical position to hold the stuff and avoid spilling it. Beyond this basic use, we may find various other uses for the container, such as using it as a support for another object, say a photo frame, or as an aquarium for newly bought fish for our kid. In short, technology develops objects for a certain use but with time, this very usage shapes our perception about its usage, leading to new affordances being developed. With a little bit of creativity, the range of such affordances could be infinite, and, with each developing affordance, our reactions would change. There is so much that you can do with technology that you thought had become obsolete. What you were thinking of as being useless and taking up space in an already overstuffed cupboard is suddenly seen in a new light and you decide not to throw it away. Today, there are a number of websites which show you how you can do this (e.g., www.pcmag.com and www.cnet.com). As you find new uses, new affordances are created through your changed perception of the object. Along with the change in perception, has not your emotional reaction to it changed?

At the same time, one must remember that every object has certain limits. A basic calculator usually does not have a correction key. So, if while adding several values, you press the wrong key, you would have to redo the whole function, leading to feelings of frustration and the wish for a better calculator. Similarly, when we get computers to perform many tasks simultaneously (multitasking), we are happy, but at the same time it comes with the enormous price of cognitive load. This is because even our brain has limits to which it can operate. Even while Google Glass was about to go on sale, it had already generated a fearful reaction (Box 6.2).

Box 6.2: The Google Glass, an Eye Wear with a Mini Computer and Camera: A Threat to Privacy?

The *New York Times* reported (May 17, 2013):

> "The group, for which Representative Joe Barton, Republican of Texas, is a co-chairman, asked questions including how Google would collect and store data from the devices, how it would ensure that it did not unintentionally collect private data, how Google would protect the privacy of people not using Glass when they are with people using it and whether the device would have facial recognition technology."

6.1 Economics and Hedonomics

Economics deals with the issues of money as external stimuli. It attempts to explain how the accumulation of wealth will increase happiness, but as applied to human behavior this approach seems to have certain limits. A simple example is that at some stage additional wealth is less useful to a rich man than to a poor person (the **Law of Diminishing Returns**, again). Since money helps us buy tools, how many tools, for example, iPads or mobile phones, do we need? Or, how many iPads can we actually use? As in the case of money, after the initial lust for accumulating gadgets has dissipated, what would we do with the rest of products? In their article, "*Hedonics in consumer behavior*," Hsee and Tsai (2008) have contended that there are at least two questions that traditional economics fails to answer:

> ***Law of Diminishing Returns***: *a law in economics which states that the consumption of every unit produces a decrease in its value or utility*

1. Is happiness uniformly related to wealth or any other external stimuli? That is, as wealth increases, do we become happier? Is there a linear positive correlation between wealth and happiness?
2. Do consumers have complete knowledge of the extent to which external stimuli like money or gadgets would fulfill their needs? Do we know how much money, or how many cars, or how many luxury yachts will fulfill our need for money, cars, or yachts?

The fact is that, neither is happiness uniformly related to external stimuli such as money or gadgets, nor do consumers have full knowledge of how their needs would be fulfilled by this money or these gadgets. Research studies clearly show that income levels are related to happiness to only a limited extent and data from several countries support this conclusion (Diener & Biswas-Diener, 2002). Therefore, argue Hsee and Tsai, we must seek to investigate why individuals attempt to "optimize the relationship between external stimuli and happiness without having to increase the magnitude of the external stimuli per se" (p. 639). For example, given a similar amount of money or tools, why are some people happier than others. Such experiences are beyond the domain of economics and have led to the development of a different approach called **hedonomics**, made up of two Greek words, 'hedone' and 'nomos.' 'Hedone' means pleasure, a positive state of mental experience. 'Nomos' refers to principles or scientific study of a subject.

> ***Hedonomics***: *the scientific study of the principles of pleasure*

A major difference between the two concepts, **economics and hedonomics** is that while the former tries to help the individual to maximize wealth from a fixed source of resources, the latter tries to help by showing how to maximize happiness from a fixed amount of wealth, in the hope that happiness maximization will bring more satisfaction for individuals.

> ***Difference between economics and hedonomics***: *former helps individual to maximize wealth while latter aims at maximizing happiness*

One can also breakup the term, hedonomics, to mean the **economics of pleasure**, that is, how pleasurable objects mean better economics than the mere accumulation of things. Thus, if a product is designed to give pleasure along with its practical usability, more people will like the product and therefore more people would be ready to buy the product. In this chapter, we will focus on how our feelings and emotions are associated with the tools and technologies that we use and the ways in which they shape our behavior for the use for which they have specifically been marketed and also for uses beyond its basic use. An attempt will be made to analyze the principles underlying such feelings, how they act as the key to technology attractiveness along with the limits of these principles.

Economics of pleasure: *pleasurable objects mean better economics*

6.2 Cognition Versus Emotion

From the time of Skinner and the other Behaviorists, it has been contended that only observable behavior should come under the rubric of scientific psychology. With the cognitive revolution and the advent of the man–machine model and work on memory and other higher mental processes, the focus shifted to brain functioning. However, even cognitivists have generally felt that emotion should not only be considered independently of cognition but should also be seen as a deterrent to efficient cognition. Ever since then, the role of emotion has been given a back seat and has been seen as the black sheep that spoils rational decisions. Also, clear thinking would be possible only if emotions are eliminated, or at the least, kept under control. Much like the scientist, the lay person also thinks that decision making is a purely rational process, whether it is decisions regarding people, situations, or even gadgets, trying to brush away or even rationalize emotional reactions even if they try to force themselves into the decision-making situation.

Economists have been no better, and have taken man to be a rational person, with every decision being made on the basis of "hard" information, where the head rules over the heart. But as Simon was soon to point out with his **Theory of Bounded Rationality** (Simon, 1991), decisions can never be completely rational because that would be possible only if we have complete information about the problem. And, sad to say, we hardly ever have complete information on all aspects of the problem. Thus, Simon contended that rather than being totally rational, we "**satisfice**" or take the best possible decision under the circumstances and on the basis of the then available information, which hindsight may even prove wrong (because the information available later could be more complete).

Theory of bounded rationality: *a theory which proposes that since we base our decisions on available information and not on complete information, we can never be completely rational*

Satisfice: *take the best possible decision under the circumstances on the basis of the then available information*

However, is satisficing itself really and truly rational? Does it only depend on the amount of information we have? If this were so, every object should be considered only because of its practical utility. We should buy and use gadgets only because they help us to fulfill some goal. But, even when we have one mobile phone, we buy another one, or long for another. We have a car but we yearn for the latest model. The purpose served by our new car would be the same as that served by the old: it would transport us from one place to another. But as the Volkswagen advertisements put it, the others are simple cars; this one is a "caaaar"; it is "das auto." The Apple Mac-Air notebook is wanted because its sleekness excites us; the mobile phone with changeable shells is supposed to cater to every changing mood of the person; an expensive music system and black leather interiors in a car helps create feelings of upward social mobility. This excitement, this mood, this feeling goes hand in glove with the cognitive appraisal of the object's practicality, clarifying that decision making is based on both cognitive and affective aspects. The two are, in fact, intertwined to the extent that it may not be possible for the scientist, leave alone the lay person, to tease out differences between the two.

Emotional response*: our feeling of pleasure or displeasure regarding the object*

Cognitive response*: the meaning of and knowledge about the object*

Generally speaking, to every object, there would be both an **emotional response** as well as a **cognitive response** (Khalid &Helander model, Fig. 6.1). Whenever we see, hear, or even smell something, not

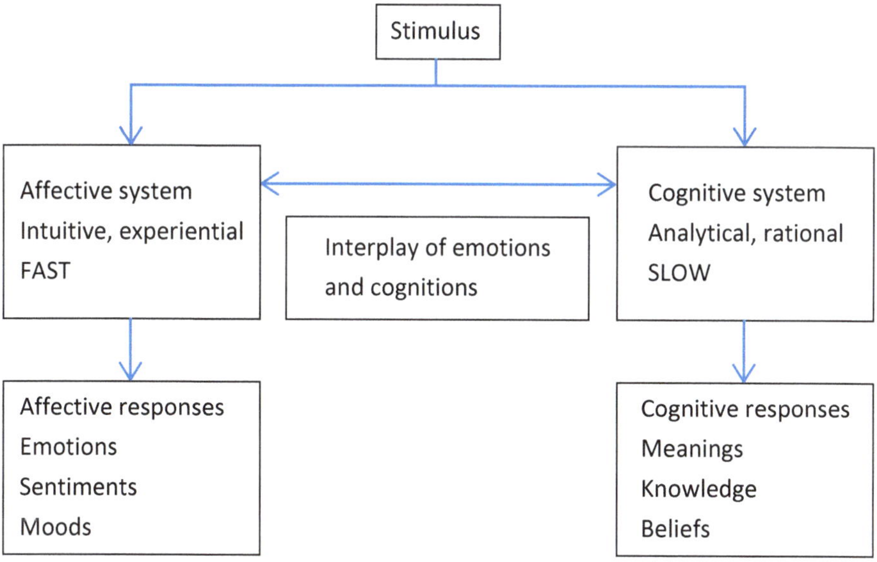

Fig. 6.1 Interplay between cognition and emotion (adapted from Khalid and Helander (2006))

only do we try to give it some meaning (what are we seeing, hearing, or smelling), but at the same time we decide whether we like it or not. In fact, is not our first reaction, "do we like it?" And, if by chance we do not, we may not even bother to go into its features and other details. It is filtered out by the system stopping it from being processed any further. This is because the emotional system is much faster than the cognitive system. Very similar to the earlier model is a model described in one of the earlier chapters, an economic model by Kahneman, who talks of two systems, **System I** and **System II**, the first being akin to the fast, affective system mentioned earlier, while the second is similar to the slow, cognitive system (*Thinking, fast and slow,* Kahneman, 2011). As Minsky (2007) explains in his book, *Emotion Machine: common sense thinking, artificial intelligence and the future of the human mind*, the traditional idea is that emotions tend to contaminate thinking and should therefore be controlled. However, we now know that emotions cannot be separated from thinking. Minsky even provides details of fMRI studies which demonstrate that emotion and cognition are so closely interlinked in the brain that excitation in areas related to one often leads to excitation in brain areas responsible for the other (Minsky, ibid).

***System I**: a fast affective system, based on feelings*

***System II**: a slower cognitive system based on thinking and rationality*

Current studies in neurobiology have also shown that the theory regarding the functional specificity of brain areas seems to have lost considerable ground. There are several reasons for the above. First and foremost, brain areas viewed as affective areas are seen to be involved in cognitive behavior and vice versa. At the same time, a variety of neurophysiological studies clarify that cognitive and affective areas of the brain are highly integrated (Phelps, 2006; Dolan, 2003; Davis & Whalen, 2001). As very aptly stated by Pessoa (2008), "complex cognitive-emotional behavior has their basis in dynamic coalitions of networks of brain areas, none of which should be conceptualized as specifically affective or cognitive" (p. 148). In fact, there seem to be what can be called hubs which control and mediate the flow and integration of information between brain areas. So great is the combined cognitive–emotional interaction at the level of the brain, that "at some point of processing, functional specialization is lost and emotions and cognitions conjointly contribute to the control of thought and behavior" (Gray, Braver, & Raichle, 2002, p. 4115). A recent study by John and his colleagues show that for behavior to be flexible (and no one doubts that it is not), cognitive and emotive areas must act in concert, lacking which, as in the case of pathological absence of emotions, decision making becomes impaired (John, Bullock, Zikopoulos, & Barbas, 2013). The above-mentioned studies, therefore, indicate that there is enough physiological data to support the contention that there is not only an interplay between emotions and cognition but also an integration between the two.

As far as technology is concerned, its adoption, use, and rejection are also part of our behavior. As such, should we not expect the same degree of integration between the cognitive and affective aspects in our interactions with technology as would be seen for other forms of behavior?

In other words, is it not imperative to study the ways in which emotions color not only our cognitive decisions to use technology, but also our post usage emotions, which will decide further use or abandonment of that technology, and changes in emotional reactions with continuous use of the technology over time?

People in the discipline of marketing management have also realized that emotions have a very important role to play. This is clear from the fact that emotional responses during product trials have a significant impact on both subsequent attitude formation toward the product (Kempf, 1999) and actual consumption of the product in the future (Menon & Kahn, 2002). Nowhere is this more true than when we consider video game playing. A multitude of studies have addressed player emotion as a fundamental part of the gaming experience (e.g., Ravaja, Turpeinen, Saari, Puttonen, & Keltikangas-Järvinen, 2008; Mandryk & Atkins, 2007; Sherry, Lucas, Greenberg, & Lachlan, 2006). In fact, the main reason for choosing, buying, and playing a digital game is the strong sense of enjoyment invoked by the game, as contended by Poels and his colleagues (2012), in a thought-provoking article titled, "*Pleasure to play, arousal to stay.*"

6.3 From Ergonomics to Hedonomics

As discussed in Chap. 2, technologies may be developed by scientists and engineers but their actual usage would depend on the extent to which the tool takes human physical and mental limits into account. This consideration has led to the development of many important disciplines that have contributed to the designing of technology. One of them is **ergonomics** or human factors engineering. The *International Ergonomics Association* has defined ergonomics as follows:

> ***Ergonomics**: the study of how tools and equipment can be designed taking human capabilities and their limits into account*

> "Ergonomics (or human factors) is the scientific discipline concerned with the understanding of interactions among humans and other elements of a system, and the profession that applies theory, principles, data and methods to design in order to optimize human well-being and overall system performance" (International Ergonomics Association, 2016).

The ways in which anthropometrics and the related disciplines of ergonomics have contributed to the designing of tools, both small and large, have been discussed in Chap. 2.

There are at least **three important targets of ergonomic design**, namely, safety, productivity, and operator satisfaction (Helander, 1997). However, we must also realize that with the evolution of technology, the targets for ergonomic design are constantly evolving and what Helander stated more than a decade back may have become outdated. Neither ergonomics nor economics can explain consumer behavior today, especially, when we see the speed with which materialism is on the increase even though to not much avail as far as happiness is considered.

> ***3 goals of ergonomic design**: safety, productivity and operator satisfaction*

> ***Affective design****: design which generates pleasure*

> ***Engineering aesthetics****: how to use aesthetics in the engineering design process*

> ***Consumer loyalty****: loyalty towards a particular brand of goods*

Another strong argument for thinking beyond anthropometrics and ergonomically designed equipment is the great range of diversity we see among users. They differ on demographic factors such as age, gender, and socioeconomic status and also on less obvious factors such as religion, political affiliation, culture generally and organizational culture, skill level, literacy, disability, etc. (Schneiderman, 2000). Since anthropometric measures are based mainly on race, gender, and age differences, ergonomically designed equipment may not satisfy every prospective user. This last objective, that is, the satisfaction of a variety of people, would be served better by hedonomics. Or as Helander and Tham (2003) put it, hedonomics is not about how to evaluate the user (as done in ergonomics), rather, it is about how the user evaluates the product. Just give ergonomics a moment of thought. Ergonomics is certainly important but is it enough? By designing a product based on ergonomic principles the designer is reducing discomfort and thereby displeasure. But does the removal of such displeasure mean the generating of pleasure? This is what **affective design** is all about. It refers to what Liu (2003) has called '**engineering esthetics**,' that is, how to use esthetics in the engineering design process, without compromising on engineering aspects. Affective design goes beyond ergonomic design in the sense that rather than focusing on negative emotions, it focuses on positive emotions. For far too long the emphasis has been on getting rid of negative emotions. Maybe the reason is similar to why the good, disciplined child in the classroom does not get the desired attention while the naughty child gets it all, simply because the latter is disruptive. Along the same vein, negative emotions are disruptive and so have gained all the attention, whether it be of psychologists, engineers, neuroscientists, or even market researchers. But, as Kotler, Kartajaya, and Setiawan (2010) point out, it is high time that we focused on positive affect which is what brings **consumer loyalty** and repeat purchase.

An important landmark is research conducted by Jordan (1998) who attempted to differentiate between what produces pleasure and displeasure as far as products are concerned. First and foremost, he stated categorically that if there are **emotional benefits** accruing from the use of the product, the product will lead to pleasure. On the other hand, if there are **penalties** associated with the use of the product, displeasure results. In an ingenious study, Jordan went on to analyze a whole host of factors associated with pleasure and displeasure and in the process isolated the exact factors that are salient in this context. These were found to be usability, performance, reliability and last, but not the least, esthetics. From the study it became clear that,

> "If human factors specialists concern themselves with only usability issues, then they will not—in all probability—be fully addressing user needs. In order that this can be achieved, human factors specialists must address issues, such as aesthetics, that they would previously not have become involved with (p. 32)…it may also mean that human factors specialists will have to evaluate a wider range of issues than they have traditionally" (p. 33).

Hierarchical model of human needs: *a theory by Maslow which proposes that human have different sets of needs which are arranged in a hierarchy*

We can also understand the difference between ergonomics and hedonomics by using Maslow's theory of motivation as an analogy. In his seminal work on human motives, Abraham Maslow (1943) presented a **hierarchical model of human needs** in which physical needs, such as hunger and thirst, and safety needs are placed at the base and therefore categorized as deficit needs. At the higher level, we seek self-expression and recognition and ultimately, display whatever is best in us in terms of our abilities to seek higher order goals. Reaching to the top of the pyramid would mean the development of self-actualization in an individual, a stage in which, according to Maslow, she would attempt to reach to the zenith of her potential and even search untapped avenues so as to develop herself to the fullest (Fig. 6.2).

Now, the question that begs an answer is: If technology serves our needs, including our hedonistic orientation, how does it fit into Maslow's theory? Simply put, technology is a tool that serves our basic physiological and safety needs. We store our food in the refrigerator to keep it from spoiling and construct houses for safe living. However, when a 3-year-old baby says that my phone is my buddy because it helps me to talk to my mom, technology is serving our love and affiliation needs. How about our growth needs? Needless to say, it is hard to conceive of any endeavor consisting of one's full expression of human abilities and pursuit of perfection in the absence of technology. In fact, technology has become an integral part of almost all human enterprise including the preparation of this book.

Based on Maslow's model, Hancock, Pepe and Murphy (2005) clarified the relationship between ergonomics and hedonomics, and this has been presented in Fig. 6.3.

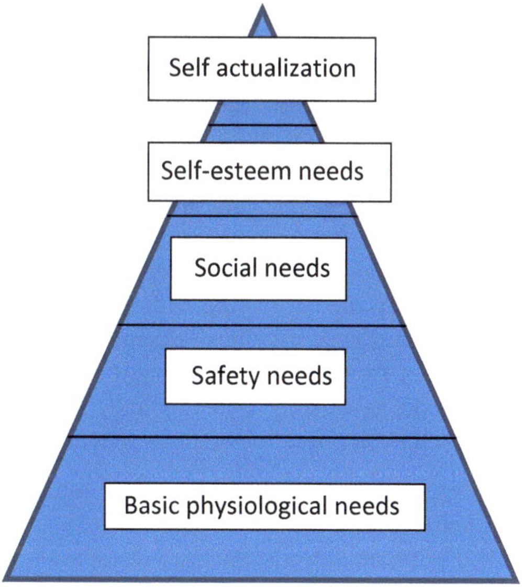

Fig. 6.2 Maslow's hierarchy of needs

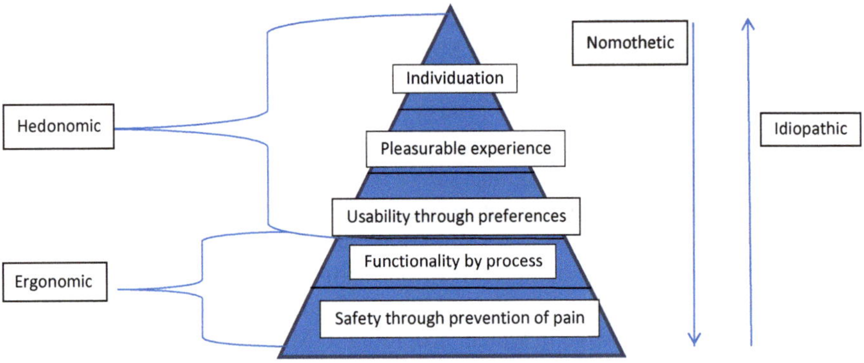

Fig. 6.3 Model representing relationship between ergonomics and hedonomics (adapted from Murphy (2005))

The emerging conclusion is that not only should technology serve certain ends, it should serve different ends as it evolves. Let us take some examples. When a tool is first developed, that is, during the first generation of the tool, it aims at the fulfillment of certain basic needs of the human being. Thus, a hammer was developed to insert nails into wood while a mobile phone was developed to enable people on the move to communicate with others. Thereafter, further evolution of the tool takes place and ergonomic features are incorporated and the tool is now found to be more usable and comfortable. The bulky hammer has become lighter while the mobile phone has become sleek and easy to hold. It is only after this stage of technology is reached that the aspect of pleasure is normally considered. Now, we have reached the stage in communication technology where factors such as the color, shape, and new affordances can be incorporated so as to enhance the self-esteem of the user or even allow her to develop her capacities to the fullest. A reflection on developments in the field of computers will help us to understand the earlier mentioned even better. When first developed some 60 years back, it was large, bulky, and served very few ends. Over the first decade or so, the lowly computer evolved, many more functions were added to it and this brought it to the second level. One can just think of the first main frame computer which needed a huge room to house it. But, soon, the size started decreasing and what we now know as the desktop computer was designed. Was the designer satisfied with that? Did he stop innovating once the desktop computer came into being? No! And, what is the result? We now have laptop computers that are vying with each other not only over technical features but also regarding size, weight, sleekness, and even color!

As seen from the diagram (Fig. 6.3), the needs served by technology can be understood as existing in a hierarchy consisting of five levels, similar to what Maslow had proposed for needs in general. According to the authors of the model,

(a) "Ergonomic needs, such as safety, functionality and usability, would be at the bottom of the pyramid", and
(b) "Hedonomic needs, such as pleasurable experience and personal perfection, would be at the top of the pyramid" (p. 70).

Nomothetic: based on norms for everyone

Idiopathic: based on individuals

As technology evolves, it moves from serving pure utilitarian needs common to a large number of people to serving the specific needs of a few. A recent advertisement of a telecom service provider is a good example. It shows how an individual can customize her service plan, depending on her differential usage of messaging, primary phone calls, internet, etc. So, all this while the service provider had been thinking of how to provide the most efficient services to the maximum number of people. Now that this compulsion has been satisfied, it can start thinking of other aspects, namely, the hedonic aspects. We can say that the service provider has moved from a **nomothetic** approach to an **idiopathic** approach. At the same time, it has moved from pure ergonomic principles to hedonomic principles in designing its product.

6.4 Hedonomics and Its Importance

Funology: principles of introducing fun into HCI

The basic postulate of hedonomics is that pleasure and usability should go hand in hand and that satisfaction with the product is maximal when there is an interplay between the usability of the tool and the extent to which it is considered beautiful (Hassenzahl, 2004). This trend is seen even in HCI systems (Human Computer Interactive Systems) in what has been called "**funology**." With reference to other technologies, too, we seem to have moved on, from mere ergonomics to hedonomics (Oron-Gilad & Hancock, 2009). So salient is user emotion that we see its effect even in complex systems such as nuclear power plants (Savioja, Liinsuo, & Koskinen, 2013), in the design of health care systems for enhanced patient engagement and adherence to medical regimens (e.g., digital health care, Lupton, 2013) and in the design of learning systems (D'Mello, 2013).

While ergonomics focused on the reduction of pain in the use of the tool and making it more comfortable to use, to reduce fatigue and thus enhance productivity, hedonomics focuses on pleasure derived from beauty and attractiveness (Khalid & Helander, 2006). Some of the effects of pleasure gained through the use of technology are now being detailed later.

6.4.1 *Time Perception*

Time perception: our subjective experience of the passage of time

Almost everyone would have had at least one common experience with technological devices: they give us joy. We feel happy when using a wide variety of gadgets and we often have the feeling of being literally lost in its use. You might have read the story of the "Pizza" mom who was so addicted to the internet that she continuously ordered pizza for 3 days to feed her kids! As we take an interest in the gadgets around us, we find that time is literally flying. Does this subjective experience of "time flying" differ from our objective measure of time? Sackett and coworkers (2010) studied their subjects' report of scenarios in which they were interacting with technology and the authors found that the subjects experienced that time had passed so quickly that they did not even realize how much time they had spent on it. While engaging in a task such as listening to enjoyable songs has been known to be a common example of when time passes very quickly, such distortion in the perception of time can be attributed to a metacognitive cue originating from our enjoyment in performing any task. Extending the application of their research, they further argued that "people often neglect the duration of events when judging hedonic value" (p. 116).

Do phone calls or emails cause excitement—"**a dopamine squirt**" among people and boredom in its absence? In fact, people have become so addicted to email or WhatsApp that they open their accounts or look at it as many as 25 times a day. Involvement of this magnitude of information obviously leads to a higher level of input that our brain has to process and respond to (Richtel, 2010). Compared to 1960, it is estimated that people are processing as much as three times more information now. What has this change done to us?

> ***Dopamine squirt:*** *a sudden surge in dopamine, a neurotransmitter which tends to get people excited*

Does this heightened level of activity cause any change in our cerebral processes? Research on this issue is mired in controversy. While it is claimed that prolonged stimulation by computer use might interfere with the development of brain of children for whom natural interaction with the environment would be more beneficial for brain development, critics argue that such activities help our brain to react appropriately and prepare us for selecting information in the midst of a vast array of information. The bottom line is that our brain has an enormous capacity to adapt, as aptly put by Steven Yantis of Johns Hopkins University. However, he had also cautioned that only after considerable research would we will be able to determine the consequences of such a continuous rewiring of the brain (Yantis & Serences, 2003).

6.4.2 Broaden and Build Theory

The importance of considering cognitive and affective factors in technology design is also clear if one thinks of the Broaden and Build Theory by Barbara Fredrickson (2008, 2001). She has classified **five positive emotions**: joy, interest, love, contentment, and pride and her research clearly shows that when people experience positive emotions, they develop an upward cognitive spiral which helps and encourages them to seek productive solutions to their problems (Box 6.3). Seeing the ever-increasing role of technology in our lives, we can extrapolate from Fredrickson's theory that when people experience pleasure

> ***5 positive emotions:*** *joy, interest, love, contentment and pride*

Box 6.3: Positive Technology: Enhancing Positive Emotions Through Technology

> "Although positive emotions can occur in adverse circumstances, the typical context of positive emotions is not a life-threatening situation. As such, a psychological process that narrows a person's momentary thought-action repertoire to promote quick and decisive action may not be needed. Instead, the positive emotions of joy, interest, contentment, pride, and love appear to have a complementary effect: They broaden people's momentary thought-action repertoires, widening the array of the thoughts and actions that come to mind ..."

So writes Barbara Fredrickson, author of the *Broaden and Build Theory* (Fredrickson, 2001, p. 219).

> "Moreover, they do so not simply within the present, pleasant moment but over the long term as well. The take-home message is that positive emotions are worth cultivating, not just as end states in themselves but also as a means to achieving psychological growth and improved well-being over time" (p. 218).

If this is the importance of positive emotions in our lives, it is imperative that technology does not lag behind. And it certainly isn't. An emerging field in this direction is that of positive technology which is based on the theoretical underpinnings of positive psychology propounded by Seligman and Csikszentmihalyi (2000), and claims that technology can not only foster learning and learning experiences but also promote psychological, emotional, and social well-being. The aim of the field is to investigate how Information and Communication Technology (ICT) can be used to improve the quality of personal life. It is believed that through what are termed serious games, one can fulfill truly serious aims, namely, that of fostering positive emotions, promoting engagement, and at the same time enhancing social integration and connectedness. Serious games are growing rapidly both as industry and in research. While MOOCs have been developed for a variety of courses ranging from online courses to cloud-based classes, current research in this up-and-coming area is throwing up a totally new perspective to computer gaming, which traditionally has focused mainly on its negative effects and links with violence.

and joy in the use of tools and technology, they will be far more productive in all their life endeavors than when they find a tool useful, but nonetheless, not very pleasurable.

Research in psychology also shows that when we analyze the range of our emotions, we find that negative emotions, such as anger, outscore positive emotions. In addition to having consequences for our survival, negative emotions receive greater importance than positive emotions, for reasons pointed out earlier and also analyzed by Kool (2008) and Agrawal (2001). It is okay to follow a message of peace in a holy place, but when attacked outside, most people would retaliate violently in a threatening situation. Finally, human beings tend to attribute differently in a negative mode as compared to when they are in a positive mode. For example, when our team wins, we say, "we won even without modern tools," but when we lose, we move to a negative mode causing a change in our expression, "we lost because we did not have the right tools." This shift in our perception of technology from positive to negative is a common illustration of differences in a wide range of attribution in not only sport settings but in various other aspects of day-to-day living.

6.4.3 Marketing of Products

Whether we think of the eternal sin committed by Adam and Eve, the negative libidinal energy talked about by Freud, or the positive psychology emphasis on positive emotions, pleasure and happiness have always been crucial for mankind, and technology is after all, only a means to an end, a tool that can produce either pleasure or displeasure. Over the years, more and more designers are becoming aware of these aspects of pleasure and companies are willing to spend more on the designing of beautiful products rather than just utilitarian ones. You may use any knife to cut vegetables but if it a beautiful one, in fancy colors, your desire for owning it and using it increases manifold times. The ever-increasing speed of advances in technology that we see today has been paralleled with increasing competition among manufacturers of technology and each company is attempting to outrace the others, using a variety of strategies. Consumer needs, and in fact, consumer psychology is becoming more important, sometimes even more important than the actual functional features of the product. A feature is said to be successful only to the extent that it is able to attract the consumer who is the user of the technology. As pointed out earlier also, hedonomics is an important aspect in the degree of satisfaction felt by the user (Demirbilek & Sener, 2003; Desmet, 2002; Hassenzahl, 2004). According to Naeini and Mostowfi (2015), customer satisfaction is related to customer purchasing behavior, feelings, and impressions. Moreover, the customer's purchase decision is dependent on the satisfaction of two types of needs, namely, functional needs and emotional needs. It is therefore imperative that distinct positive emotion arousing elements be introduced into the product to make it preferred by the customer.

Marketing strategies and the understanding of consumer behavior have also evolved in line with the above and with the current emphasis on the building of customer loyalty and relationships, marketing gurus such as Kotler, too, have been forced to revise their old models to include the role of customer emotions (Kotler et al., 2010). A whole new set of factors, hitherto not attended to, such as customer conflict, customer ambivalence, customer emotional intelligence, and their role during the process of purchasing, is now being emphasized (Taylor, 2009), while Kotler and his coworkers (2010) are of the view that we need to move to strategies that appeal to the whole person. Hedonomics helps in a big way to meet these helping designers to create what have been variously called hedonic design, esthetic design, or affective design.

6.5 Hedonomics and Hedonic Design

One may wonder what the term, **hedonic design**, refers to. Khalid and Helander (2006) define it as "design devoted to the pleasurable human-product interaction." Going back to the pyramidical structure given by Murphy earlier, it is clear that tools evolve much like living beings and as far as tool designing is concerned, there is a progression from concerns for safety and functionality to that of usability and pleasure.

***Hedonic design**: design devoted to pleasurable human-product interaction*

***Affective design**: design that is intentionally created to catch the user's attention and trigger an emotional response*

Another term used in hedonomics is affective design. According to van Gorp (2006), **affective design** is,

> "Design that's intentionally created to capture the user's attention, triggering an emotional response that will increase the likelihood of performing a certain behavior. The emotional response can be conscious or unconscious. For example, a brightly colored button will attract users' attention unconsciously by affecting the degree of arousal (i.e. physical stimulation). And the behavior could be any action, from clicking a button or signing up for a newsletter, to making a purchase online" (van Gorp, 2006).

Further, we must remember that the design does not always have to cause pleasure. It can cause annoyance (e.g., the beeping in a car to fasten the seat belt) or it could even cause fear (e.g., the sounding of a fire alarm). The simple idea is that it should draw your attention in a very big way and that is most easily done by arousing some sort of emotion, whether it is joy, curiosity, fear, or anger.

6.6 Achieving Hedonic Design

From the preceding discussion, it is very clear that technology can bring us a lot of happiness. In general, people feel good when they buy gadgets for their use, be it a car or an electronic device. While psychology has long focused on the performance of a user, that is, his capabilities to master the operation of a tool, current research has shifted to the responses of the user who evaluates a gadget. So, when I begin to use my iPad, how do I feel about it? Is it simple to use, strong enough not to break easily, is it attractive? While all this may not seem new to the reader, the fact is that until Mitsuo Nagamachi, in Japan, worked on Kanzei Engineering, Donald Norman demonstrated the emotional aspect of designs, or Picard wrote on affective computing, our scientific knowledge of this topic was almost negligible.

Let us start by trying to understand the varying types of pleasure one can obtain from technology. Khalid and Helander (2006) have developed a simple categorization consisting of five types of such experiences. These have been described as follows:

1. **Physical pleasure**: With reference to our body, the technological product can give us physical pleasure, such as when we get pleasure on touching the object. A soft surface of a cell phone, such as one with a silicon cover gives more pleasure than a cell phone with a hard, slippery surface or one with a rough surface.
2. **Social pleasure**: Through technology, our social status and relationships can be enhanced. A clear example of this is when our neighbors come and admire our newly purchased smart TV, with 3D, internet and what have you.
3. **Psychological pleasure**: Engaging in tasks through the medium of technology can cater to our interests and even results in its prolonged use, for example, when we listen to music on our CD player or listen to downloaded music on our computer while we work.
4. **Reflective pleasure**: we obtain reflective pleasure, for example, when we look for the best sound system or the finest musical instrument to enhance our experience.
5. **Normative pleasure**: We can use technology to enhance social values or for following social norms, for example, for rain harvesting, pollution control, and other environmental issues and derive pleasure in the process.

Physical pleasure: *pleasure gained through our sensations*

Social pleasure: *pleasure gained because it enhances social status*

Psychological pleasure: *pleasure gained because it caters to one's interests*

Reflective pleasure: *pleasure gained it caters to our thoughts about the object*

Normative pleasure: *pleasure gained because it helps us fulfill social norms*

The question in front of the designer, now, is to decide what sort of pleasure she wants the user of the technology to have and how to design the product accordingly. It is also clear that the first two dimensions of pleasure are superficial and the pleasure might not last as long as when the product leads to psychological or even reflective pleasure.

The problem with a simplistic classification of pleasure such as the above is that it fails to take into account the fact that our experiences of pleasure are very personal and their interpretation often varies across cultures. What may give social pleasure in one part of the world may provoke displeasure in another part. After a gunman killed several Amish children near Lancaster, PA, this community did not show fear or anger, by asking for police protection or a surveillance fence, believing that their children would be resting in peace in heaven. On the other hand, they showed extreme compassion and forgiveness for the perpetrators of the crime by raising several thousand dollars to help the widow of the assassin who was killed by the police (Kool, 2008). Conversely, it is also a fact that devoid of the social context, a lot of emotions are difficult to understand. If I laugh alone, will it make sense to those who are watching me? But once they know the context, they, too, might laugh. In short, emotions are contagious in nature because they tend to be interactive.

The expression of pleasure also seems to follow cultural patterns. After winning a game, to what extent do I express my pleasure? After buying my new tablet or Xbox, do I show it to everyone I meet, like the Greek scholar, Archimedes, who ran down the streets of Athens following his new discovery? Each culture sets some limits on the expression of pleasure. While in the West, a victory celebration over a rival in a competitive scenario is considered natural, in Stoic, Buddhist, and Gandhian traditions, emotional displays are replaced by a sense of contentment without hurting the loser (Kool & Agrawal, 2012).

As far as technology usage is concerned, culture decides to a very great extent the hedonic characteristics of the tool. Aligning tool design to cultural norms is an important aspect of hedonic design. An example is that for the same function, that is to turn on a light, light switches go up in the US, while in Europe, they go down. A technology designer who does not keep this very simple principle in mind would find it hard to find customers for a technology that could, otherwise, prove to be extremely helpful. He may have designed a beautiful switch but if it does not conform to cultural norms it will have no takers.

6.6.1 The Technology Acceptance Model

One model that has been developed for achieving hedonic design is TAM or the Technology Acceptance Model. Basically, the positive relationship between technology and its use can be examined in two ways:

(a) To what extent would the technological device enhance human performance?
(b) To what extent would it be possible to free oneself from effort while trying to master the technology?

In other words, the first domain of the relationship deals with the utilitarian aspect, while the second represents the hedonic nature of the relationship. Ayyagari (2006) has explained this further by proposing a two-dimensional relationship between technology and hedonism. First, if people perceive that the use of technology would improve their performance, it would lead to the fulfillment of one's extrinsic motivation. This type of motivation would diminish as the utility of the gadget reduces. My old computer is of no value to me and, therefore, I do not use it and neither do I care too much about it. On the other hand, intrinsic motivation based on perceived enjoyment of a gadget or effortless usage leads to its prolonged use. Why did Windows Vista not succeed? This was because, after getting used to one method, people probably did not want to change. Even though Windows Vista may be more efficient than Windows 7, changing from one technology to another often involves more physical and mental energy than using an older version. We form habits that take a tremendous amount of mental energy to overcome. Further, it is habits that bring stability in our behavior. As stated over a century ago, by the founding father of modern psychology, William James, without habits, the operation of the human society would become dysfunctional (James, 1890). It is this dependence on habits which narrows the gap between us and machines and keeps us from discarding gadgets even when they become obsolete. What leads to an intrinsic motivation to cling to the old tool is, often, because of the fewer mental resources needed to operate it.

An important element in psychological research is the identification of the features of the gadget in use (identified usually as an independent variable) and the response we give to it, that is, the dependent variable, common examples of which are measures of time and error. In order to assess affective or emotional human responses to a gadget, Helander and Tham (2003) initiated the following questions in research:

- "How can one measure affective design?
- How can one predict affective design?
- How can one predict user and customer needs for affect?" (p. 1269)

Anthropomorphizing: a tendency to ascribe human features to other living beings and even lifeless objects

Helander and Tham argued further that unless the information obtained from the earlier questions is interpreted in a broad theoretical framework, our understanding of the affective factor would remain poor or pseudoscientific. Let us take the example of people's habit of talking to their computer. It is commonly observed that a number of computer users not only talk to their computers but treat them as if they are their pals or living beings. In technical terms this phenomenon is called **anthropomorphizing** and refers to a tendency to ascribe human features to other living beings and even lifeless objects. Research on talkers to their computers showed that by and large, they tended to perceive their device positively (Luczak, Roetting, & Schmidt, 2003).

Another very common example relates to our satisfaction with mobile phones. Many people are so dependent and attached to this device that they would go crazy without it. How do we measure our affective relationship with the mobile phone? When asked about what consumers liked about their mobile phones, their responses varied from its colorfulness to its attractiveness and delicacy. Further, Yun and coworkers (2003) focused on variables such as curvature and texture as related to the design of the mobile phone. They and other researchers (e.g., Borsci, Kuljis, Barnett, & Pecchia, 2016 and Khalighy, Green, Scheepers, & Whittet, 2015) have concluded that what applies to consumer satisfaction of a product and its design variables could easily be, in most cases, found useful in the study of similar gadgets.

6.6.2 The Role of Arousal

Another set of factors have been pointed out by van Gorp and has been elucidated through Fig. 6.4 and its explanation. According to Trevor van Gorp (ibid) of *Affective Design Inc.*, this relationship between products and people is based on three factors, namely,

- **Pleasantness of products**: its attractiveness and ability to make people feel good.
- **Usefulness of products**: the degree to which it serves our need such that it demonstrates its worth over a period of time. The range of needs may vary from material to spiritual in content.
- **Usability of products**: the degree to which it is easy to use and its operations are easily understood.

> ***Priming**: activating thought processes by bringing back information to the conscious level*

On the basis of the earlier factors, one can intentionally attract the attention of our prospective customers and thereby enhance their arousal level. This enhanced arousal level will, in turn, lead to other psychological processes such as thinking and emotions and will decide how the customer reacts to the product. Arousal can therefore be seen as the main source of our hedonistic relationship with the product. Very often, we may be bombarded with a wide variety of information and may not even realize that we are attracted to a product. Even when we are not consciously aware of it, our brain continues to process the information at a covert level. Once there is an activation of the neural circuits in the brain, through what has been called **priming**, we perform better in bringing back information to our conscious level. Experiments on priming have demonstrated that after exposure to pictures consisting of people, animals, and other things, subjects did better in solving incomplete words than in a control condition wherein there was no such exposure.

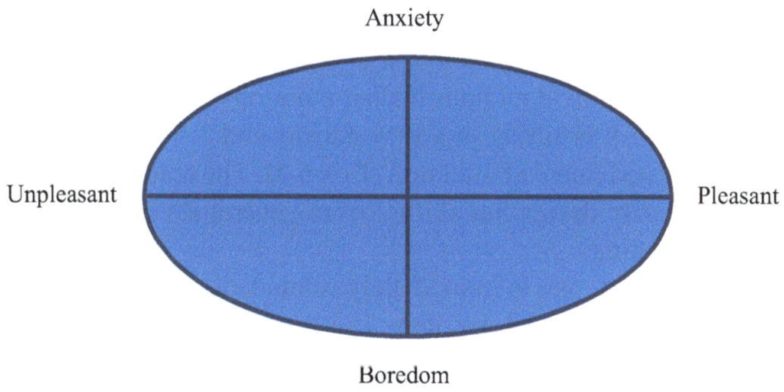

Fig. 6.4 2×2 matrix to explain relationship between a product and the individual (based on van Gorp (2006))

When emotions are described in terms of arousal levels, they can range from low to high arousal or from boredom to anxiety. With the dimension of pleasantness/unpleasantness added, a 2×2 matrix is formed, which can be used to describe the relationship between a product and an individual (van Gorp, 2006; Fig. 6.4).

Let us think of technology and the way people react to its changing characteristics. Most of us find music to be very pleasant, but imagine the music in a discotheque, with laser lights flashing around, young girls and boys swirling around to the music, and an older person may well find the same music to be unpleasant. Now, if this discotheque is next to your home, you soon get used to the unpleasant music and do not react to it. In other words, the unpleasant music fails to arouse any interest in you and your attention is not drawn by it. It is unpleasant as well as boring. Suddenly, the sounds change. Maybe a new electronic music instrument has been introduced by the band or it is playing a different type of music. Though still unpleasant, your attention is drawn to it, or, in terms of arousal, your arousal level becomes high. You have now moved to another area of the earlier figure, namely, that part that shows that though you find the sound unpleasant, your attention is aroused by the change. We get bored by the ringtone of our mobile phone and so keep changing it so that it attracts your attention every time it rings. Yet, if you are not careful, the changed tune may not sound very pleasant. You have to choose one that is not only novel, so that your arousal level is high but also pleasant to the ear. It is clear that the perception of the same stimulus may change as circumstances change leading to different ways in which our brain reacts to it. Designers of technology keep both dimensions in mind: the end product should not only attract the person's attention but should also be perceived as being pleasant, to the ear, to the eye, or to touch, as the case might be.

6.6.3 Designing for Flow

An extreme example of full attention is **flow**, as described in the work of Csikszentmihalyi (1990), which requires intense focusing on a task to achieve a target. In doing so, an individual finds his/her emotions enmeshed with the task resulting in a very narrow and extremely concentrated range of attention (Box 6.4). The reader will notice that this discussion on attractive design has shifted from general attention to flow which is, in itself, very captivating.

> ***Flow**: the experience of being completely absorbed in whatever one is doing*

In the earlier context, one is immediately reminded of the famous **Yerkes–Dodson Law** of the 1960s which proposed that optimal performance is achieved at moderate levels of arousal. Getting flow follows a similar pattern. Too little arousal causes boredom and so does

> ***Yerkes-Dodson Law**: optimal performance is achieved at moderate levels of arousal*

Box 6.4: The Psychology of Flow

During the early 1970s psychologist, Csikszentmihalyi was fascinated by the way in which artists would get so absorbed in their work so much so that they would forget everything else, including the need for food, water, and even sleep. During the 1980s, he started conducting research on this phenomenon and when he asked people to describe this experience of theirs, they said that they had the feeling of being carried away by a water current. It is this which led to the phenomenon being named 'flow.' Though not new: similar states have been mentioned in Eastern philosophies such as Buddhism, Hinduism, and Taoism, referring to it as 'doing without doing' or 'action of inaction,' it was the first time that such a construct came to be considered in modern psychology

Explaining the mechanism behind flow, Csikszentmihalyi says that normally people are able to decide what they want to attend to. However, when they are in a flow state, they are completely engrossed with the task at hand, and surprisingly, even without making any conscious decision to do so, lose awareness of all other things, people, time, and even distractions and bodily needs.

Recent years have seen the phenomenon being applied to a variety of spheres, including education, medicine, music, and sport. The most recent entrant is that of technology and of trying to discover ways of creating flow through appropriate user-focused technology. Generally three conditions have to be met. These are as follows:

- The activity must have a clear set of goals.
- The task must have clear and immediate feedback.
- There must be a good balance between the perceived challenges of the task and one's own abilities.

These principles have been used for designing intrinsically motivated computer systems. One way is to combine the Technology Acceptance Model (TAM) with Hedonic-Motivation Systems, which can then be used to fulfill user's intrinsic motivations in online gaming, virtual worlds, online shopping, digital music libraries, social networking, to name just a few. The interesting aspect of such designs is that they are able to include advanced cognitive absorption constructs such as those of joy, control, and focused immersion in the computerized experience.

not lead to flow. Similarly overarousal causes anxiety which is again a distraction to our cognitive system and so reduces our interest and intensity of attention to the task at hand. Achieving flow in design is not difficult and in most cases designers do keep it in mind even though they may not be actually thinking of why it is so effective. There are at least three important factors for achieving flow. First, reducing distractions leads to greater flow (just think of how advertisements in the margins on websites distract you and lead to lower levels of concentration); second, immediate feedback has been found to be an effective factor (it is used in so many computer and video games causing a literal squeezing of all your attention and energy toward that particular game). Another factor that game designers keep in mind is the level of challenge. You are always attracted toward something that you feel you will be able to achieve. In other words, there would be greater flow for tasks of moderate difficulty, forcing you to give greater and greater attention simply because you feel that the goal is reachable.

Advanced learning Technologies*: computerized technologies which focus on introducing academics through serious games*

A good example of how flow can be achieved is the case of **ALT** (**Advanced Learning Technology**). We will all agree that learning cannot take place if the learner is bored and nowhere is the proverb, 'you can take a horse to water but cannot make it drink,' more applicable than in the learning situation. The knowledge level of the teacher is important but even more important is the pedagogy. Of course, if the learning material itself is made interesting, there is nothing like it. From early research by Csikszentmihalyi (1975) to research in the current decade (e.g., Pekrun & Linnenbrink-Garcia, 2014; Daschmann, Goetz, & Stupnisky, 2011; Pekrun, Goetz, Daniels, Stupnisky, & Perry, 2010), it has been shown that there are certain conditions under which the learner fails to learn. For example, if the learning material does not appeal to the student, she may get bored. Moreover, if the student does not have any choice as far as the learning task is concerned, she may start feeling helpless. And most importantly, when there is a gap between skills and the task at hand, either because the task demands more than the skills of the learner or vice versa, the end result may be frustration or even despair and anxiety. But research also clarifies that the learning situation can very easily arouse positive emotions, such as curiosity and interest (Silvia, 2009), delight (D'Mello & Graesser, 2011), and even flow like states (Csikszentmihalyi, 1990). This is where technology can prove to be of great advantage.

Affect Sensitive ALT*: ALT which coordinates emotions and cognitions of the learner*

The last two decades have witnessed an educational revolution in the form of ALTs or Advanced Learning Technologies which use intelligent tutoring systems, animation, simulation, and immersive educational games, and in the process are able to achieve flow-like states in the learner. One can go a step further and introduce **Affect Sensitive ALTs** (e.g., Baker, Gowda, Wixon, Kalka, Wagner, Salvi et al., 2012) which not only detect the students affect but are also able to respond to it and so increase engagement. More details on how this can be achieved have been presented in Box 6.5. A meta-analysis of research (D'Mello, 2013) suggests that the next-generation ALTs should not only incorporate the analysis of affective states of learners but would also do well if they could handle negative affect as well.

Box 6.5: Affect Sensitive Advanced Learning Technologies: How to Hook the Student into Deeper Learning

To date, the multibillion dollar gaming industry has not been able to market many games that incorporate important academic content. The reason has been brilliantly stated by Graesser, D'Mello and Strain in a chapter in a recent book entitled, *International Handbook of Emotions in Education* (Pekrun & Linnenbrink-Garcia, 2014). They write:

> "A serious game could be engineered that brilliantly manages the tradeoff between fun and work. However, one would need to be very clever in designing such a game because young learners are skeptical of games that have any semblance of academic content. The game designer would need to smuggle in the academic subject under the student's attitudinal radar" (p. 473).

They are convinced that research in the social sciences, especially psychology, can come to the rescue of the technology developer and help technology coordinate emotions and cognitions in the context of learning. The following principles have been enunciated by the authors:

1. Create a state of flow (active engagement) to the point where fatigue and boredom disappears.
2. Provide an engaging story narrative that not only sustains the learner's interest but is also well integrated with the academic subject.
3. Reward the student with points to introduce extrinsic motivation and introduce the game-like environment.
4. Allow active student control so as to increase feelings of autonomy.
5. Interact with the student through turn-by-turn conversations to promote interactivity and social presence.
6. Give the insecure student material that he can master so as to increase his self-confidence.
7. Provide timely feedback.
8. Provide feedback and guidance regarding student's emotions. A discouraged student may need an explanation that difficult material is sometimes confusing and even boring.

So, go ahead, develop a serious game and hook the student!

6.6.4 Levels of Designing

Donald Norman is, undoubtedly, one leading scholar, who has been able to demonstrate through his research the ways in which designs influence our behavior. When Norman (1990) first published his book, *The Design of Everyday Things*, he did not have much to say about the role of emotions in design. It is in his subsequent publications (Norman, 2004), that he concentrated on how "everything we do, everything we think is tinged with emotion, much of it subconscious" (p. 7). Citing the work of Noam Tractinsky, he argued that individuals tend to prefer attractive designs and view them as being superior to those designs that lack such appeal and referring to Masaaki Kurosu and Kaori Kasimura, who had shown that while attractive and unattractive ATMs were operationally similar, the more attractive ones were rated as being superior in performance (Norman, 2004). It may be of course assumed that the Kurosu and Kasimura samples consisted only of Japanese subjects and therefore this bias may be a Japanese trait. However, further research has shown that such behavior is universal and tends to manifest itself across cultures. In his next volume, *Emotional design: why we love (or hate) everyday things* (2005), Norman focuses completely on this all-important facet of emotions in design.

Norman's strongest argument is that many of our decisions have been made even before we begin to process the information at the conscious level. He offers a very interesting example of walking on a plank that is placed on the ground or 3 m above the ground or 10 m above the ground. We will readily walk on the plank placed on the ground, but may hesitate at its 3 m height and without any thinking simply refuse to walk at the 10 m height. When a design is ugly, we instantly take a decision not to buy the product and it takes a herculean effort to convince ourselves that the product could indeed be useful for our purpose.

Along the same vein, Norman has argued that, contrary to our common belief, people do not always make rational decisions even though they believe that they had operated in a rational manner. Illustrating this point further, he invites us to think about our decisions regarding the ordering of food in a restaurant or the purchasing of a dress. A number of our decisions are based simply on affective and emotional factors such as appealing colors, novelty of objects, and its general attractiveness, giving a clear indication that subconscious factors may, in many cases, overpower our rational mind. Just think of the countless number of times when you have bought something on the impulse. In all probability it was guided by affective factors such as your mood at that time or other emotional factors, or you were simply attracted to it by its color or design, and shop owners would be sadly affected were it not for this phenomenon of impulse buying.

Analyzing the aforementioned phenomenon, Norman (2004) has differentiated between three aspects of design that have important bearing for things we use in our life.

***Visceral level:** based on physical features such as shape, color, look, feel and sound*

The Visceral level: Our brain's response to a stimulus such as temperature, a smiling baby face, or sweet taste is likely to arouse an instant positive affect. On the contrary, heights, darkness, and bitter taste generally elicit negative affect. In between these two extremes, people engage in behaviors that elicit joy as well as fear such as rock climbing and bungee jumping (adventure sport which gives a thrill only because of the risk involved). The latter forms of behavior are opted for as per our choice involving higher levels of functioning in which our cognition and emotion form an even more complex relationship.

Box 6.6: Is Our Taste Influenced by the Crockery We Use?
In the simplest form, our visceral response to a gadget is based on its appearance—the cool color, pattern, or shape. The following quote from Norman illustrates this point so well:

> "At the visceral level, physical features—look, feel, and sound—dominate. Thus, a master chef concentrates on presentation, arranging food artfully on the plate. Here good graphics, cleanliness, and beauty play a role. Make the car door feel firm and produce a pleasant chunking sound as it closes. Make the exhaust sound of the Harley Davidson motorcycle have a unique, powerful rumble.....Yes, we love sensuous curves, sleek surfaces, and solid, sturdy objects" (Norman, 2004, p. 67).

At the visceral level, people's reaction is based on looks alone. Design experts know how to exploit this feature so as to boost sales or save a dying product, as the following examples clarify:

- Sales of Apple computers skyrocketed as new colors were introduced. In contrast, IBM's computer looked stale and stodgy.
- PT Cruiser, Audi TT, Chevy HH, and Volkswagen Beetle were able to attract the attention of consumers and helped change the image of the company.
- The looks and appeal of the Microsoft Surface Pro Tablet versus that of the iPad.

Citing evidence from evolutionary development across species (known as phylogeny), Norman contended that when animals respond in set prewired ways, called routines, they act at the visceral level. A spider's attack on a fly, insects reacting to light, and avoiding noxious food are common examples to illustrate visceral functioning. Visceral responses are based on the sources that link an individual to its environment, for example, a fear response of an infant to thunder, or crying on seeing the milk bottle when he is hungry. Nature provides various colors and sounds that attract living beings to improve their adaptability and effective coexistence. For example, symmetrical fruits and flowers appear attractive and tempt insects, animals, and human beings alike. On the other hand, when two halves of a face are not symmetrical, we do not find it attractive. In short, nature is full of its own designs that try to attract us. This is the beginning point for visceral responses (Box 6.6).

> ***Behavioral level:*** *focus on the function and utility of the object after we like the look of it*

The Behavioral level: Whereas a product based on looks raises the temptation to possess it, we soon begin to think about its use. As Norman (2004) stated, "In most behavioral design, function comes first and foremost; what does a product do, what function does it perform?" (p. 70). When we look at a machine such as a tractor, a thresher, or at some other agricultural machinery, what are we thinking about? Most people would say that they are thinking about the specific purpose for which they can be used. The same is true about automobiles, horse-driven carts, and so on, but as the functional properties of these mechanical devices come to be known and established, the focus shifts to their visceral properties. Given two machines similar in their output, it is likely that we would prefer the one with decent colors and an overall better look (visceral). But first things should come first. What would we do with a machine with a cool appearance if it does not function to serve our purpose?

In a mall, a friend of the author saw a logo. It looked like a donkey oriented toward the left. Then he looked further up and read, STAR TRAC Spinner NXT. Being familiar with this company which manufactures machines for gymnastic purposes, he looked back at the logo. Now it looked very different, a guy on a bike. I guess the company overdid its effort to make a design attractive (visceral) by somewhat compromising on its functional appeal. The important consideration for a behavioral design is to convince a consumer about its usefulness. Unlike a visceral design that elicits a sudden, natural response, the task of a designer of a product is not that simple.

According to Norman, there are two considerations that dictate product development for its usage:

(a) Innovation, and
(b) Enhancement

When common gadgets such as the typewriter and computer were developed, people knew about their specific applications. As they explored and experienced a newly discovered product, its functional aspect was clear. On the other hand, enhancement of the same product would require an effort to show that the product is useful beyond its original purpose. When mobile phones arrived in the market, most people thought that it was meant for business, but now it has replaced the traditional phone. In places like Iceland, the landline phone has disappeared and in developing and Third World countries, the use of mobile phone is associated with the status of the owner.

The bottom line of a behavior-centered approach is that after knowing about a product, we should be able to use it smoothly. It requires knowledge about the operation of a gadget, memorizing the steps in its operation, and getting the desired results. Most innovators know that for their product to succeed in the market, it is important to pretest it with potential users through what has been called pilot testing, and, then carefully apply the findings. This feedback is a key issue in the usability of a product (Norman, 2004). Further, there is no substitute for practical experience gained through continuous experience in the optimal development of a product. Again, armed with such feedback, carpenters, machinists, and athletic professionals have known for decades and centuries what makes a tool genuinely functional, though there are times when we fail in our efforts, as we will elaborate in the next chapter.

> ***Reflective level:*** *focus on what the gadget means to the user, apart from its functional use*

The Reflective level: This type of experience with design relates to our higher levels of functioning. Does this product enhance my self-image? Does this product suitably represent my cultural identity? Is this product

Box 6.7: A Popular Bollywood Actor's Obsession with Watches
Saif Ali Pataudi, a popular Bollywood actor married to another diva from the same industry, has a massive collection of luxury watches. He says,

> "They all have their places… Sometimes just sitting at home, I change my watch three times a day. I like to open the box, clean my watches and admire them. I have a few, chunky gold ones that go well with long hair, a dark sun tan and white pair of shorts. But usually, I like very simple watches with leather straps," he says.

Source: *The Times of India, July 4, 2013*

meant for kids and how would an adult look if he used it? The key issue that dominates at the reflective level is the continuous monitoring of our association with the product and linking it to our self-awareness. Rolex is a watch that shows time, but when you ask people if it indicates something more, they would say it is more than a watch. The website of Rolex mentions that it is an ornament. Norman (2004) similarly cites the example of the Swatch Company that added another dimension to its traditional approach of time display with fashion. Why don't we wear the same tie or the same pair of shoes every day? Can this logic apply to the wearing of a watch? It can, if we find a way to link it to the emotional needs of the individual. If I wear a Casio sport watch, does it reflect my athletic preference more than when I wear a Rolex? Most people would say, of course, it does. Go to any town or city in India and you will find a lemon and a green pepper, tied around a thread, hanging in front of many cars and trucks. It is considered customary/auspicious over there. A car decorated with flowers to transport a newly wed couple leaving for their honeymoon does not appear to be a regular car. In a very emphatic way, Norman argues that technological products soon find their way to our cognitive system—the ways in which we perceive, attend, and value things around us—along with our affect and emotions such as likes or dislikes, pleasantness or unpleasantness and the two decide our behavior of whether we will approach or avoid the product. Can you imagine people becoming obsessed with watches? One such example is given in Box 6.7.

6.7 On Designing the All-in-One Product

In the previous section, we have discussed a wide range of variables that need to be considered to create emotional appeal in a product. The question now is, how do we combine all these factors, how do we create a technological marvel that will attract each and every person who looks at it? Simply put, an ideal situation for effective marketing and sales would be to make a product look pretty (visceral), easy to use (behavioral), and valuable (reflective). In his paper presented at a conference in Italy, Norman (2003) argued that there are two perspectives of a technological product. First, there is the designer perspective which is guided by the appearance of a product and its utility. Second, there is the user perspective which would, in turn, involve:

(a) Perceptually induced reactions (visceral)
(b) Expectation-induced reaction (behavioral), and
(c) Intellectually induced reaction (reflective)

While it is possible for the designer to manipulate visceral and behavioral reactions relatively successfully, reflective reactions are not as easy to predict. Such reactions are dependent on a variety of factors. As stated earlier, the Microsoft Surface Pro has suffered so far in the market owing to the ambivalence of the consumers regarding its reflective value: Is it a tablet (too heavy?) or a laptop (not really a laptop?) or cool (appears stodgy when compared to an iPad)?

6.7.1 The Affective-User Design Model

We now describe a model that can be used to create a tool or a technology that will satisfy all the conditions mentioned earlier. The model is based on two sets of variables, namely, design variables and user variables, and the relationship between these two. The exact relationship between the design variables and the user variables has been envisaged by Khalid (2006) in the model described later and presented diagrammatically in Fig. 6.5.

First, we have to be aware of the fact that we have, here, two systems which are distinct and yet the designer will have to interrelate the two in order to create a gadget or a tool or even a technology that serves its purpose and at the same time appeals to the consumer. The first system, namely, the designer's environment consists of three factors:

- **Artifact**: affective quality such as appealing to the visceral, behavioral or reflective aspects, as pointed out by Norman earlier.
- **Context of use**: the goal of the object or how and for what it is going to be used.
- **Society trends**: norms and fashion, in terms of size, color, shape, and features.

The second system is that of the affective user which has, as discussed earlier, two highly interrelated systems of cognition and emotions, each being influenced by a whole host of factors such as culture, gender, age, curiosity, esthetics, to name just a few.

The task of the designer is to match the previously mentioned and create something that will satisfy all the aforementioned factors. In order to do this, customer

Fig. 6.5 The affective-user design model (based on Khalid (2006))

needs are studied through detailed market research processes. Yet, by the time the product is launched, customer needs may change, often because of some competing technology or product. With the speed at which new technology is being developed, this has become the major problem for both manufacturers and sellers.

While measuring emotional reactions, it is necessary to focus on negative as well as positive affect, since the aim is to maximize positive affect and simultaneously minimize negative affect. As pointed out in an earlier section, benefits of using the tool produce positive affect while penalties associated with its use lead to negative affect. It is the latter which causes user anxiety and fear, frustration and anger, confusion and annoyance, while positive affect leads to pleasure, excitement, and even awe. Khalid (2006) has put it very succinctly:

> "A poorly designed application with complex user interface and controls impedes natural interaction, inducing negative emotions, a desire to quit, moodiness, and sometimes swearing at the machine, or even kicking it. ….Pleasurable interaction may be derived by integrating adaptability into designs and providing design features that enhance user control" (p. 416).

6.8 On Measuring Affective Responses to Objects

The starting point of affective design would therefore be the measurement of emotions and emotional reactions to objects. There are various techniques that have been evolved and these have been described very briefly, including two from outside the western hemisphere.

6.8.1 The Semantic Differential Technique

A very useful technique for research on affective factors related to the use of things is the semantic differential technique (Kool & Agrawal, 2006; Oskamp & Schultz, 1998). If you are asked to rate your mobile device on several factors such as newness, social status, complexity, pleasantness, and speed, it would not be difficult to rate your choices. Based on a list of adjectives that describe such preferences, the semantic differential technique classifies responses into three domains which can then be used by the designer.

6.8.2 Semistructured Interviews

Jordan (1998) has used a series of semistructured interviews to arrive at the hedonic considerations of certain products. In a study described by him, an interview consisting of three parts was used. The first part was based on finding products that gives pleasure to the subject, followed by questions going into a detailed analysis of why they found this product pleasurable and the nature of the felt pleasure, namely, excitement, security, confidence, etc. The second part of the interview asked the subjects to focus on products they considered displeasurable and their characteristics, along with the nature of the displeasure, that is, whether they were feeling cheated, annoyed, anxious, frustrated, angry, etc. The last part of the interview consisted of certain general questions referring to how they are affected by the associated pleasure or displeasure and when did they become aware of these feelings. By a thorough analysis of the responses obtained, the author of the study was able to isolate those product features that led to pleasure and displeasure and even more so, he was able to obtain information on how these emotions affected the person. The data collected through this method would, thus, be of help to both manufacturers and marketers of the products.

6.8.3 Kansei/Kenzei Engineering

Founded by Mitsuo Nagamachi, Professor Emeritus of Hiroshima University and CEO, International Kansei Design, the term Kansei is a Japanese word meaning "psychological feelings and images generated in a certain product" and is one of the methodologies developed by companies, yearning to stop dropping sales, for affective or emotional design. According to Shimizu and colleagues (2004), the goal of Kansei engineering is the identification of product properties and to assess the correlation between those properties and design characteristics, so as to design a product which gives pleasure to the consumer. There are three focal points of Kansei design, namely:

- How to understand customer Kansei?
- How to translate this into product design?
- How to create a system and organization for Kansei-oriented design?

Schutte and coworkers (2004) have even formulated a general model incorporating five steps, starting with the choice of the product domain and ending with model formulation and validation. Seeing the popularity of Kansei design, along with its complexity, special software for Kansei engineering has been designed (Box 6.8).

Box 6.8: Kansei Designing and Manufacturing

When Professor Mitsuo Nagamachi created Kansei Engineering (KE) little did he realize that what he had created would be used for products ranging from the lowly showerhead to a car. The purpose of Kansei engineering is to transform people's emotional needs into concrete elements for detailed design elements. For nearly four decades, KE has been developed as a productive

(continued)

Box 6.8 (continued)

research discipline, highly connected to the industrial world, claiming numerous innovations and market successes. Mazda Miata (also known as MX-5) is the first and the most famous project to be developed to be followed by so many car making companies—Nissan, Ford, Saab, and Volvo to name just a few. And, are car makers the only ones to use Kansei? Certainly not! Today, we see it being used by a wide variety of industries such as textiles (e.g., by Wacoal & Goldwin), food (e.g., Nestle), electronics and home appliances (e.g., Sharp, Panasonic, Samsung), and even cosmetics (Shiseido).

But what is the essence of Kansei Technology? Though still not very well defined, Levy (2013) writes:

> "Embracing this ambiguity, and based on the synthesis of a larger set of definition attempts, a compound description of kansei has been proposed (Lévy, Lee, & Yamanaka, 2007):
>
> - Kansei process gathers all the sensory qualities functions, including interactions between them (Clark, 1993).
> - Kansei means are all the senses and other factors such as personality, mood, and memory.
> - Kansei result is the fruit of kansei process.
>
> It appears to be a unified perception providing a qualitative meaning and value of one's environment. It is the synthesis of sensory qualities" (Levy, 2013, p. 85).

6.8.4 *Citarasa Engineering*

This is another methodology developed for hedonic designing and is based on a Malay word, meaning emotional intent and aspirations. In Citarasa engineering the starting point is a description of customer needs, rather than a description of product properties as in Kansei. This, in fact, is the basic difference between Kansei and Citarasa. It is based on the assumption that the customer understands her needs and also knows very well the extent to which a certain product can fulfill those needs. Khalid (2005) has developed the CATER project, that is, the Computerized Automotive Technology Reconfiguration system which has attempted to use the Citarasa method in automobile design.

6.9 Choosing Technology: Problems with Affective Design

The relationship between a design and the choices made by an individual are far more complex than one can envisage. One may design an object that has all the above-mentioned characteristics and yet, when it enters the market, we find no

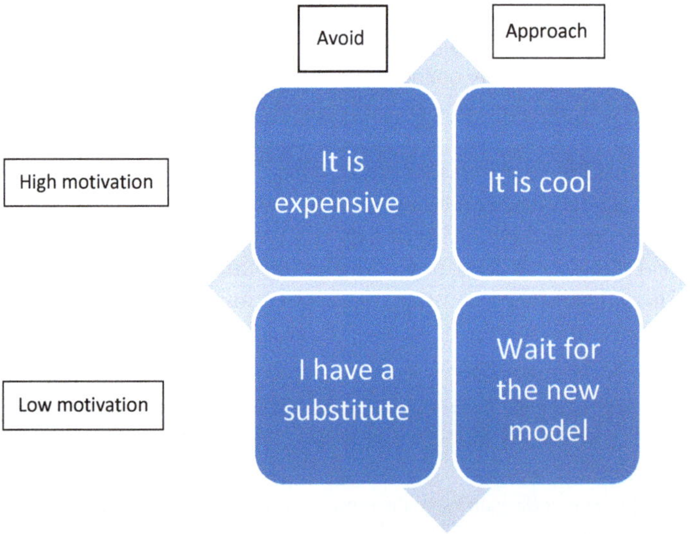

Fig. 6.6 On buying an expensive electronic gadget (adapted from van Gorp (2006))

takers for it. Why is this so? The reason is that besides emotional aspects that attract us to the object there are a whole host of factors that determine our final choice.

In general, people prefer, or tend to approach, simple, familiar gadgets and dislike and avoid their unfamiliar, complex counterparts. Approach behavior is associated with pleasure while avoidance is caused by pain or fear. If you are used to driving a car with automatic transmission in your country, would you approach a car with a manual shift drive in your car abroad? Probably not. However, if you are highly motivated to save money in places where automatic rental cars are ridiculously expensive, you might settle for a stick shift. How does this relationship between our approach–avoidance tendencies interact with our motivation to prefer a design? In its basic form, it is explained in Fig. 6.6 (van Gorp, 2006).

Through the use of marketing research techniques, consumer behavior experts attempt to forecast the preferences of consumers—what they would buy or avoid. In the section on choices later in the chapter, we will return to this topic again.

Focusing more intensely on our appraisal of products and how our emotions get involved, Desmet (2002) has identified five types of emotions that products could arouse in us:

1. Instrumental: things that help achieve our goal
2. Esthetic: pleasing to our senses
3. Social: gadgets that enhance our status
4. Surprise: a very special and different type of gadget and
5. Interest: things that cater to our hobby

Desmet has argued that several antecedent factors such as the symbolic significance of products and culture also influence product experience. In their later work,

Desmet and Hekkert (2007) looked beyond emotional factors to add esthetic and meaningful experiences in explaining the relationship between the user and the product. They wrote,

> Through cognitive processes like interpretation, memory retrieval and associations, we are able to recognize metaphors, assign personality or other expressive characteristics, and assess the personal or symbolic significance of products. (p. 60)

In defining the interaction between the user and the product, we need to look beyond the instrumental (utility) function of the product. Whether the individual cares about a product or not and why s/he cares is also important.

6.9.1 Personality Aspect of Designs

As we interact with people we form judgments regarding their personality and tend to think that they are of that particular type, even though, they may actually be of a very different type. Thus, a person may be perceived as being arrogant because of not talking too much when actually that person does not want to talk because of a deep seated inferiority complex. But, the reality is that it is this perception of the person that will decide our future interactions, not her actual personality. In the same way, as we use objects, or as we interact with certain companies through their products, we tend to assign certain personality traits to those objects or those companies. Although we know that electronic devices are not animate, we still respond to them in many ways that appear to resemble our responses to our fellow humans. We often tend to anthropomorphize gadgets and other objects that we use in our day-to-day life. People often remark, "I do not like this product. It is sort of cold and unfriendly." Or, at other times we say, "It appears very dominating." Thus, a black flower vase may appear to be sad. Add to that visual features such as darkness, a straight and/or angular shape, a heavy base, and it starts giving the impression of dominance. On the contrary, a flower vase having a golden color, with soft, round contours evokes feelings of submission. If the technology is easy to use, we call it "friendly" technology. It has been suggested by researchers such as van Gorp (2006) that products should be viewed as living objects with which people have relationships and it is these relationships that will decide our attraction or repulsion to those objects. We do not want to be seen with an object of a brand that makes us feel ashamed of ourselves, and, therefore try to seek products that will make us feel proud and which will raise our self-esteem, much in the same way as we do not want to be seen with people who have been marginalized by society and rally around those that have been afforded a high status. Thus, hedonic design also attempts to design objects that are perceived as being friendly, and submissive, rather than being unfriendly or dominant. Shapes (such as rounded versus angular), colors (e.g., pastel versus dark colors), and typography (italic versus block letters) all lead to producing a design that attracts people because they are perceived as having a positive personality.

At the same time, it should be noted that there are two distinct issues in understanding the effect of designs on our emotions:

(a) The nature of the design, for example, its visual or graphic appearance. What do we feel when we look at it?
(b) Our interaction with a design that leads to pleasure and gratification.

When a developer works on a product, s/he creates it with the users in mind. Having this in mind, Donald Norman referred to what we now know as "user experience." It refers to aspects of a product that are beyond its basic functional nature. For example, users may be interested in

Is the product easy to use (yes, they know it is functional)?
Is the product intuitive (like other web browsing sites)?

When people begin to look at products in terms of its value—pleasantness, approach, and motivation—as stated earlier, van Gorp (ibid) proposed the emergence of the following relationship between products and individuals (see Fig. 6.7).

From the 2×2×2 matrix earlier, it is clear that our choices will depend on all **three** factors, the pleasant–unpleasant dimension (i.e., the nature of the emotion evoked), the motivational aspect (i.e., the degree to which it is useful for us), and the arousal dimension (i.e., the extent to which it presents something novel and attracts us). When are we most likely to purchase the gadget? Obviously, when the product fits box A, because it conjures pleasant experiences, it seems useful and at the same time, it does not bore us. We would definitely reject it if the product falls in box denoted as H in Fig. 6.7 as it is being perceived as being unpleasant, not of any use and boring to the hilt. The problems faced by the designer of any technology are clearly more than what he had asked for when he had conceptualized it.

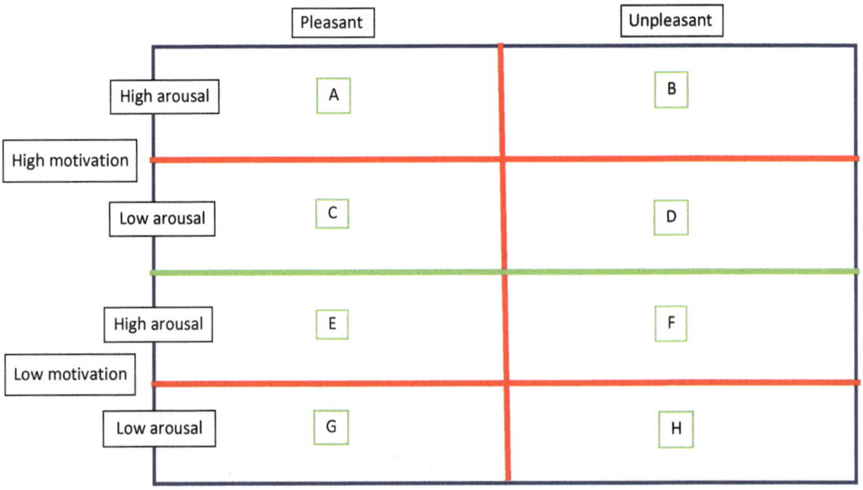

Fig. 6.7 Diagram to explain the relationship between products and individuals (based on van Gorp (2006))

6.9.2 More Choice or Less Choice

Thanks to technology, we are living in a world full of choices. With a plethora of cars, motor cycles, or computers, it becomes harder to select a product than when we were offered only a limited number of choices. This explosion in consumer choices has been studied by psychologists recently with a view to assess its impact on the behavior and mind of the users. In Chap. 2, we had mentioned that human beings have an inclination to choose from the bounty of products offered by nature, which in turn helps us to optimize our adjustment to the environment.

Having a choice implies a sense of control over the environment. In their work entitled, *Born to choose: the origins and value of the need to control*, referred to in earlier chapters, Leotti, Iyengar, and Ochsner (2010) contended that evidence from biological and clinical research indicates that exerting control is needed to produce results necessary for our survival. When people have a choice for food, mating, or housing, their well-being is enhanced. This capacity to make choices has remained a part of our phylogenetic evolution even hundreds of centuries later. Pradeep (2010) has illustrated further that human beings have options far exceeding even the wildest dreams of the cave man. When we go to buy a car, we do look for engine power, petrol consumption, tires, etc., but we also get attracted to new, shiny things (Pradeep, 2010). Imagine our choices for cars 30–40 years ago and compare them with what we find now—greater variety, shapes, and colors, all making rational choices far from easy. Yet, we keep craving for greater variation expecting them to arrive at faster and faster rates during our lifetime. Are such choices adaptive?

In his book, *The paradox of choice: why more is less*, Barry Schwartz (2004) has pointed out that as we go about maximizing our benefits in buying a product that has several varieties in the market, we end up getting frustrated, with the end result of not buying the product at all. A customer may want a BMW but it may not have the features that a Lexus provides and therefore he may want to wait for a BMW model loaded with Lexus features. According to Schwartz, maximizing in a world of multiple choices may delay purchase of a product or altogether diminish the appetite for it. In other words, too many choices may have a psychological cost and may lead to counterproductive behavior. Consumers may also feel depressed and, in hindsight, begin to ruminate about missed opportunities that would no longer be available. Exploring the connection between frustration and purchase of high-technology durable goods, Strebel, O'Donnel, and Myres (2004) concluded that the pace of technological change has a significant impact on buying behavior. Their findings are based on a measure having two dimensions: the ways in which frustration is processed by consumers and frustration with the pace of technological change. Their findings suggest that the relationship between buying behavior and complicated, frequently changing technological products is highly complex.

On the contrary, there are occasions when having choices may bring its own benefits. The positive side of enhanced choices is that it could lead to an increase in reaching an informed decision. After reviewing many products, a consumer is definitely more informed than the one who is restricted in the range of his samples.

With availability of choices and knowledge of products, people would be able to reduce the probability of making risky decisions by eliminating irrelevant features of a product. Finally, an increased number of choices could contribute to the development of self-determination. When a teenager is commended for his choice of a gadget, he is likely to say, "it was my choice!" This connection with technological products evokes positive emotions such as joy and empathy. When people have been able to do a market survey, they tend to have an idea of what they are exactly looking for and if they quickly settle on a product, they are displaying what Simon referred to as "satisficing" (choosing on the basis of the information we have). They end up probably being more content with the product as compared to those who keep on attempting to maximize their options.

Through their research, Iyengar and Lepper (2000) compared the buying behavior of consumers who had fewer options with those who had greater options and found that as far as sales are concerned, products offering fewer choices outscored those with a higher number of choices. They contended that the findings apply to a wide variety of settings ranging from selection of jams to options in retirement plans that are offered by employers.

Further, Botti and Iyengar (2004) found that consumers are less happy when someone else makes the choice for them regarding a product perceived to be unpleasant. When computers were introduced and offices had to replace old machines with new automation, the older and aged workers were unhappy but had no choice since the choice had been made by their employers. All they could do was to move on to adapt to the new realities at work. Another implication of this study addresses the issue of adjustment of older people with technology. For example, postal service for communication on social security matters has been recently abolished in the USA and all the beneficiaries will have their benefits transferred directly to their bank accounts. With this, it has become almost imperative for every beneficiary to learn some basic technology, no matter how averse they are to technology, in order to manage their finances. Under such circumstances where people are left with no choice, there emerges a deep sense of lack of control or mastery over the environment, with the development of a sense of helplessness. Yet, one cannot ignore the positive side of the earlier step, since it would reduce paper usage considerably, bringing contentment to both environmentalists and the Social Security Administration Department.

6.10 Hedonomics and Cognitive Biases

> ***Cognitive biases****: the ways in which our cognitions color our perceptions*

As technology is moving toward occupying a greater chunk of human functioning, it would affect us in at least one fundamental way: how do we select a technological product out of the several brands available in the market? As per Norman's model, when we are invited to select a product based on visceral, behavioral, and reflective attributes, logical reasoning would say that a product based on all three aspects would be the most valued choice (Lindstrom, 2011). However is this always so? Both theory and experience suggest otherwise.

We started the chapter by trying to explain how our cognitions and emotions form an integrated system with each affecting the other. Because of such interaction between and integration of the two systems, there are a number of ways in which cognitions determine whether we are attracted to the technological product or not and whether we will choose it or prefer not to have anything to do with it. Choices based on pure hedonomics are simply not there. We will focus on two theories to show how cognition is not purely rational. Rather, a number of biases tend to color our cognitions.

6.10.1 Prospect Theory

> ***Prospect theory****: a theory regarding the cognitive appraisals of the prospect of loss or gain*

> ***System I****: a part of the cognitive system that is fast and helps us to act immediately*

> ***System II****: that part of the cognitive system that is slower but based on detailed cost-benefit*

Prospect theory as advanced by Daniel Kahneman and coworkers (Kahneman & Tversky, 1979) emphasizes that no matter what the endeavor, people tend to weigh their prospects of loss and gain. At the same time, they are more averse to loss than they are attracted to gains and the former tends to raise more intense feelings than gains. According to this theory, human beings function at two levels (as discussed in previous chapters too). The first level is called **System 1** in which our lower centers of the brain actively mediate our responses and help us to act automatically and suddenly. When we need a fast response to deal with a threat, there is no time to think, and an instantaneous response is needed. The second system, namely, **System II** is rational and involves the higher centers of the brain. As such, if this system is called into play, responses become slower but are often better. We weigh various possibilities and perform a cost–benefit analysis for each, figuring out the consequences of our behavior and then decide. A loss is seen as a threat and as such we generally want to get over the threat as soon as possible. In order to react as quickly as possible, Kahneman is of the view that we rely on System I thinking and in the process fail to weigh the losses associated with the chosen behavior. So when we lose our cell phone, we get so anxious that we buy the first cell phone we see, failing to look carefully at the products offered. We would have gained considerably by making a more complete market survey but at that time it seems more important to get over the anxiety due to the stress of being without a phone.

Now take an example of a credit card which has become an integral part of our life. They are offered in various colors and are easy to carry in our pocket (visceral appeal). They are also very easy to use as we simply need to swipe them to transact a payment (behavioral appeal). And finally, it is indicative of our status to pay and affiliation with a corporate culture. The American Express Business card says it all (reflective). In terms of prospect theory, this behavior of using a credit card, mediated through our technological orientation, could bring to the forefront the positioning of our emotions and cognition in the above-mentioned two systems. For example, with an ease in swiping the card and an urge to buy a product while our dopamine flow is at its peak, impulsive buying may occur. On the other hand, with our rational System II intact, we may as well think that we would be able to always return the product. So, why don't we buy it instantly? Technology provides such affordance for both systems at levels that are unique to human experiences. Imagine what we were like (or sometimes still are) when we carried a large amount of cash to do a business transaction, but now we can perform the same transaction often with unknown people with a credit card for $5000 even though we may have only $700 in our bank account. Technology has precipitated our hedonic experience with the swipe of a card that empowers us for material gain in seconds, in contrast to a trader or farmer who had to struggle for centuries to raise capital to meet his needs. The downside of this privilege of a credit card is that an average American is loaded with a huge debt of thousands of dollars on it. Moreover, with an uncontrollable urge to buy a product because of System I analysis, it would be a greater challenge for a teenager than for a mature adult to refrain from using a credit card to purchase something that is just a swipe away.

Conversely, we might have come across several people who develop a distaste for credit cards as they make people overly vulnerable to spend and in the process lose their savings. The point that we are trying to bring home is that while similar behaviors, for example spending behavior, would have existed in the past in the absence of technology, it has initiated behavioral challenges that were hitherto unavailable to us.

6.10.2 Attribution Theory

In our day-to-day life we are constantly trying to evaluate the cause of certain actions and their results. Events without explanations tend to create a feeling of fuzziness, with cognitive overload as its end result. Since our cognitive system is of limited capacity and we are being bombarded by new information that is trying to make space for itself in this limited capacity system, finding ways of reducing the load is extremely important. One way of doing this is by using rules of the thumb or **heuristics**. One such heuristic is related to the attributions we make or the ways in which we relate

> ***Attribution Theory****: a social-psychological theory which explains how we evaluate actions and their results*

> ***Heuristics****: rules of the thumb used to simplify information so as to reduce cognitive load*

effects to causality. In our attempt to save our extremely important self-concept, we are generally disposed to using the **self-serving bias**, by which we attribute our own failures to external events while our successes are always due to our own efforts. So, if I fail an exam, I tried my very best but the examiners were negatively biased toward me. On the other hand if I do well, we generally say, "Oh, that's because I studied so hard." Contrary to this situation, the successes of others are explained as being due to external events while their failures are due to faults in the person per se. This is regarding the attribution of causes to my successes and failures vis-à-vis those of others, or the **fundamental attribution error**. But do we react in much the same way in our interactions with technology?

> ***Self-serving bias***: *a tendency to protect our self-esteem by explaining our success as being due to internal factors but failures due to external factors*

> ***Fundamental attribution error***: *difference in our attributions for our own success-failures and those of others*

Like any other form of behavior, our choices in terms of technology are also not free from cognitive biases about which we may or may not be consciously aware of. We normally believe that to be placed in a situation with a large number of choices is always good and that having more choices would be better than having only one choice. This is simply a case of false judgment. When we compare products, we become aware of their limitations as well. According to Schwartz (2004), we tend to carry several false attributions about our choices. In a study, Iyengar (2011) compared American and French families regarding their decision to take the life support off, an aversive heartbreaking decision on one side, but sustained by a hope of a miracle with the aid of technology. While French families did not have a choice as the decision was based on medical regulation in effect, their counterpart American families had the choice of removing or not removing the life support equipment. Those members of the American families who faced emotional overload due to continued support and later took the decision to take off the life support felt enormous guilt while French families were saddened but took the inevitable loss at ease. Why was this so? The difference lies in the fact that Americans felt the guilt because they themselves had taken the decision. The French, on the other hand, did not have any choice, being governed by law, and so there was no guilt.

What happens when an error occurs while operating a machine? Is the failure a human error or a machine dysfunction? In aviation industry, this is debated time and again, especially when an experienced pilot flies a thoroughly inspected plane. When we started writing this chapter, the CNN network carried a story about the Asiana Airlines crash landing at the San Francisco Airport (July 7, 2013). First, experts looked at the safety record of Boeing 777 and then examined the conditions of the navigation system as repair work was being done and the landing guide system which was not active when the plane landed. In other words, when there was nothing wrong with the pilot who was within few seconds of landing the plane, the failure of the machinery was under scrutiny first. This is an example of a **situational bias**. It would be considered

> ***Situational bias***: *tendency to put blame on situational factors rather than on the individual*

a **dispositional bias** when later investigation would show that this was the first time that the pilot, though experienced otherwise, was landing the 777 Boeing in the very busy air traffic of San Francisco. According to various news agencies covering the accident at that time, the cause of the accident remains unclear, especially regarding whether it was the pilot's inexperience, or, whether it was due to the malfunctioning of equipment at the airport or in the plane.

> ***Dispositional bias:*** *tendency to put the blame on the individual*

Do temporal factors change our attributions? Based on experiments conducted by Parducci and his coauthors (Parducci, 1965, 1995; Wedell & Parducci, 1988), we find that this is indeed the case. In a study on wine tasting, one section of Parducci's subjects tasted the inexpensive wines most of the time but the expensive one only occasionally. The other group of subjects was put in the reverse order of drinking, that is, they tasted the expensive wines frequently and the inexpensive wines rarely. A comparison of the two groups revealed that a better consumption experience was reported by the former group of subjects, that is, those who consumed expensive wines rarely. The implications of this study are enormous for our understanding of hedonic experiences with technology. With the rapidly changing scenario of technology, the gadgets that we are currently using are likely to become old very soon (much like inexpensive wines consumed most of the time) and the occasional use of a latest product would yield enormous pleasure. When our friend uses the latest iPhone and we get to handle it occasionally, we are attracted toward it. But once you buy the same phone, you start getting bored with it. If this was not the case with human nature, companies would never be able to continue to sell products after just a few cosmetic changes have been made.

6.10.3 Projection Bias

Generally speaking, each of us is in a certain state of arousal at any one time. As Hsee and Tsai (2008) contend, and as also elaborated in an earlier section of this chapter, we may be bored or we may be overexcited. We may be in a state of satiation or be undergoing a strong need. If we are satisfied with our car at the present, we fail to attend to good car deals and in the future may even repent not having taken a closer look at the deals, when our car suddenly starts giving trouble. The reason is that the present state of satisfaction is used to predict events in the future: we do not need a car today so we take it for granted that we will not require one in the future also and hence fail to pay due attention to ongoing deals. In other words, we have projected our present need state onto future needs and events and so lose out when the future brings a situation and a need other than what we had foreseen. Or as Hsee and Tsai (2008) put it, people tend to project their current state of judgment when making predictions about a future condition. A very popular preference of Mercedes sedan customers is its black color that is usually priced a bit higher than cars of other colors, but will

> **Projection bias:** *tendency to project one's present need state onto future events and needs*

this hold for its new GLE models? So companies, too, show a projection bias, predicting customer needs for one product on the basis of the needs for another.

Lay rationalism*: a bias based on what people consider to be rational as against what in reality is rational*

Citing several such cognitive biases that could have implications in the context of technology, Hsee and Tsai (2008) use the term, *lay scientism* to illustrate ***lay rationalism*** in the use of technology, a bias so common with our preference of gadgets. This kind of bias shows up when we try to base our choices on what we consider hard, objective attributes rather than soft, difficult to quantify attributes. The problem is that what we consider objective factors may not be objective at all, and it is this which makes it a bias. A recent study shows how this can operate. When consumers were given a choice to select either a powerful stereo system or one that was rich in sound, the choice depended to a very great extent on what had been framed as an objective, easy-to-quantify characteristic. Explaining the results of one of their studies, Hsee and Tsai write,

> "For half of the participants, power was described as an objective wattage rating and sound richness as a subjective experience. For the other half, power was described as a subjective experience and sound richness as can objective quantitative rating. When power was framed as being an objective attribute, more participants chose the more-powerful stereo than they predicted they would enjoy it more. When sound richness was framed as an objective attribute, more participants chose the richer-sounding stereo than they predicted they would enjoy it more. In other words, the objectivity/ subjectivity manipulation had a greater influence on choice than on predicted experience. This finding corroborates the notion that consumers base their choice not purely on predicted experience, but also on what they consider "rational" in this case, objective" (p. 650).

6.10.4 Medium Maximization

Medium maximization*: a bias which makes us prefer objects which serve as a medium for receiving a desired outcome rather than its immediate outcome*

Another cognitive bias that has deep implication as far as our interactions with technology are concerned is medium maximization. People generally do not pick up an object that offers an immediate reward as an outcome, but prefer that which serves as a medium for receiving the desired outcome (Hsee & Tsai, 2008; van Osselaer, Alba, & Manchanda, 2004). Instead of focusing on the outcomes people predict their choices in terms of how the choice will help them to maximize the medium. For instance, when people were given a choice of 60 points or 100 points for a shorter or longer task with outcomes of receiving vanilla or pistachio ice creams in various conditions of the experiment, they mostly chose to work on the nonpreferred longer task because it led to getting pistachio ice cream. Why? Though thcy actually preferred the shorter task, but this led to a reward which they did not want, namely, vanilla ice cream. On the other hand, the nonpreferred longer task was a via-media to a larger or better reward, pistachio ice cream and so they were willing to work on it. If the medium provides some kind of certainty, people might chose to work for longer periods of time and sacrifice their happiness. "Yet when people work," writes Hsee, "the immediate reward is not happiness, but a medium, money" (p. 652).

One implication of this finding is that people often find technology to be a medium for enhancing one's outcome. This scenario could cause an illusion of certainty among people who might start viewing their efforts as an advantage over another scenario in which they have no support from technology. Technology might push us into a number of such decision-making scenarios. Baby boomers often asked this question to themselves: whether it was right for them to learn how to use the internet to send a letter or should they have kept on writing letters as usual and sending them through the postal service? In short, what forms of behavior supported by technology would tend to have better predictability or certainty in terms of desired future outcomes? Unless and until we are able to sort out the level of certainty with which technology will help us in achieving our goals, dispositional attributions will continue to remain salient as far as our choice and use of technology is concerned. Do not be surprised if people say, "I knew which equipment I needed in order to succeed" rather than "I succeeded because of the equipment."

6.11 Applications of Hedonomics

This section will try to explain how the principles of hedonomics have been used in different sectors of technology. The focus is to bring to the fore the ways in which applications of hedonomics have made the technology more popular among its users and, even more importantly, to showcase how hedonomics is not a mere theoretical formulation but has already been applied to create technology that leads to greater satisfaction among users.

6.11.1 e-Consumer Behavior

With the advent of e-commerce, the number of internet shoppers has continually been on the rise. In fact, e-shopping has become a win–win setting for shoppers and sellers, the latter not having to maintain large physical inventories and mammoth staff for managing either inventory or the brick and mortar stores. At the same time, the former have the ease of making purchases from the comfort of their sofa without having to brave the plights of driving especially in not so pleasant climes. To be able to attract more and more people to their online sites, companies have been undertaking research using methods of ethnography to understand the needs, demographic characteristics, and motives of the user of online shopping sites.

Very interestingly, research findings in this area tend to validate the Hancock, Pepe and Murphy model mentioned earlier. The early e-shopper tended to focus mainly on functional and utilitarian aspects (Brown, Pope, & Voges, 2003). Also, they were different from the traditional shopper in that they tended to be younger, more educated, of higher socioeconomic status, and also were more likely to be males (as cited by Dennis, Jayawardhena, Merrilees, & Wright, 2009). At the same time, research, for example that by Jayawardhena and Wright (2009), has clarified that today's e-shopper and the traditional shopper are not only very similar but also that both tend to focus on social and recreational considerations as against being guided by pure utilitarian aspects. In other words, the e-consumer, though originally guided by utilitarian aspects now wants the sites to move on to the fulfillment of other needs.

The traditional Technology Acceptance Model (TAM, Davis, 1989), too, has been criticized for ignoring a variety of factors as a result of which this theory has been extended and now includes a new dimension, namely, enjoyment (Davis, Bagozzi, & Warshaw, 1992). Dennis and his colleagues (2009) go a step further and propose a new integrated model of e-consumer behavior which incorporates various hedonomic principles, including experiential factors such as enjoyment, and e-interactivity; situational factors such as convenience; and consumer traits such as education, income, age, and gender. This list is of special significance because empirical findings very clearly point to differences between male and female shoppers, with the former tending to be wanting to shop quickly while the latter prefer shopping for fun and enjoyment and enjoy spending time in the process. The authors of the model, thus, point out that e-shopping sites might well have different segments for male and female shoppers which would cater to the specific needs of each.

Marketers, too, are now interested in personal characteristics other than demographic variables, age or education, and income. As the internet is penetrating deeper and deeper across social strata, the online shopper is not just the young, educated, rich person. Using variables from shopping in general, five shopping orientations have been distinguished (Loevenich & Lingenfelder, 2004; Diehl, 2002), namely, experience, service, price, convenience, and brand orientation (Burkolter & Kluge, 2011). While the effects of the latter four are quite clear, it is the first that is of special interest from the hedonomic point of view. It is generally assumed that shopping in the physical world provides an experience, an adventure, a kick, which online shopping may not be able to provide (Zhou, Dai, & Zhang, 2007). Surprisingly, empirical data collected by Burkolter and Kluge (2011) in Germany show that online shoppers also got the kick that traditional shopping provides especially in some product categories such as apparel and furniture. Ways to increase the experience in online shopping have been suggested by Kim and Forsythe (2007), who feel that as technology advances, emotional and sensational stimulation can be increased as through rotated views, 3D views, or even virtual tryouts, as described in Box 6.9.

Box 6.9: 3D and Virtual Tryouts in Online Shopping

A website, "joy of clothes: your personal stylist" allows you to create your own model in three easy steps in less than 3 min. The steps are as follows:

Choose your body shape and color

Select your mouth, eyes, and nose

Choose your hairstyle and color

All you do now is to click on the 'save' button and start creating models. The site even has a video to help the not so tech-savvy!

Another site, www.trylive.com offers solutions for a myriad of virtual try-on experiences, ranging from eyeglasses to furniture to watches and jewelry. It says, "enchant your customers" and "rock your business" with cloud-based virtual try-on and product visualization solutions.

Even Rayban, the international eyeglass expert, has a virtual mirror, giving what it calls the "ultimate augmented reality experience which lets you virtually try-on the latest Rayban styles."

If you are having difficulties in deciding from the myriad options presented to you, there is another eyeglass vendor site, on which you can even save all your virtual try-on pictures and share them with your friends and family to help you take the decision.

Online shopping is definitely moving into the next era!

6.11.2 Online Gaming

In probably no other form of technology is the hedonomic aspect of such great importance. An individual "plays" a game, to enjoy, to have fun. Needless to add that there are times, as in professional games, where economic considerations far outweigh the hedonomic ones, but generally, they must contain the fun element.

Is online, video or computer gaming fun? In depth interviews (e.g., by Hussain & Griffiths, 2009) clarify that while all of the above are played for fun, or are based on the principles of 'funology,' Massive Multimedia Online Role Playing Games (MMORPGs) are successful only to the extent that they provide fun to the players. The intensity of fun provided is such that individuals reported being so "sucked into" the game that what seemed to be barely an hour was actually 4 or even 5 h (Hussain & Griffiths, 2009). At the same time, the interviewees reported that these games were potent enough to aid stress management and help them overcome negative feelings and alleviate moodiness, corroborating what Gardner had suggested way back in the 1990s (Gardner, 1991) that Nintendo games be used as an adjunct to psychotherapy. Twenty years later, technology seems to have provided us with even better games.

These games clearly show that it is not the more games there are, the better is the bargain that people get. Rather, it is the nature of the game that is important and game developers are constantly trying to upgrade their technology and change the games according to the changing needs of the players. Once again, we see a clear transition from one stage of need fulfillment to another. While people generally start playing just for fun and play those games that are easy to understand (functional level), they soon move to multiplayer formats and even form guilds of various types (social level) and finally to avatar games wherein they are free to experiment with a variety of identities that fulfill their personal needs (individual level).

Further ideas provided by Hussain and Griffiths (2009) is that MMORPGs could incorporate more of those aspects that lead to positive effects and decrease those that have negative effects. Thus, more rest breaks would be useful and the game could force the player to take a break by giving reward points for doing so. This would help lessen the burden of game addiction. The games can also be made more educational by inclusion of features that tax the players' analytical communication and decision-making skills rather than being simply repetitive in nature. The game developers are surely keeping in mind that hedonomics is more important than mere economics. Even more so, hedonomics seems to be making for better economics because it is the former that helps to rake in the moolah.

Emotioneering*: a combination of emotions and engineering*

Affective gaming: If one tries to analyze the typical game world (i.e., the virtual environment where the game takes place and in which the player is immersed), one notes that there is a player and a large number of nonplayer characters (NPCs). Thus in the game *Pac-man*, the game world is the labyrinth, the player is represented by Pac-man, a character that has to survive and collect all the dots in the labyrinth, and the ghosts who try to stop Pac-man are the NPCs. The challenge for game developers is that after a while the behavior of the NPCs becomes predictable and the player starts losing interest. Games would benefit if emotions could be incorporated in these NPCs. In that case, the software would recognize emotions in the player and consequently introduce tailored emotions in the NPCs, making them more like human opponents who are able to socialize. While some games do simulate a certain level of human like emotions, for example, *Creatures*, by *Cyberlife Technology* and *Sims* by *Electronic Arts*, there is still a long way to go. Even a decade back, there were at least 1500 known ways for evoking emotions in games, but to do so, one must draw from fields as varied as computer sciences, AI, psychology, and physiology (Freeman, 2004), leading to disciplines such as Affective Computing (Picard, 1997) and design skills such as **emotioneering** (Freeman, 2004). The latter, the name of which itself suggests the combination of emotion and engineering, entails 32 techniques for injecting emotions into games based on three dimensions. The first of these is **sound**, which not only augments the storyline and provides information but also increases engagement (Collins, Kapralos, & Kanev, 2014). The second dimension is **device**, for example, electrode covered T-shirts to gather player biometrics and convert them into emotional states (Dupire, Gal, & Topol, 2009), or EEG measurements (De Peuter, 2014). The last dimension is **interface**. Kromand (2007) identifies two types of avatar interfaces, namely, central avatars (which act like a shell in which the player steps into and the avatar becomes the embodiment of the player) and acentral

Box 6.10: *Gamygdala*: A Pluggable Emotional Engine for Gaming
Till recently, an emotional engine that could be used in a variety of games as a simple pluggable black box was not available. Since the Artificial Intelligence (AI) of different games would be different, an AI independent emotional engine was required so that emotions could be built into any game, making them not only socially complex but also affectively realistic. The breakthrough seems to have finally been made with Popescue and his colleagues managing to bring to the gaming world just such a pluggable black box (Popescue, Broekens, & van Somerenet, 2013). Moreover, the device has been very innovatively named: *Gamygdala*, taking inspiration from the cortical area, amygdala which is responsible for emotional behavior to a very great extent. *Gamygdala* supports 16 emotions and defines two types of emotions (internal emotions and social emotions directed at another NPC, that is, the Non Playing Character).

Source: Popescue et al. (2013)

avatars (in which rather than becoming the player's body, the player plays with the avatar). Once again taking the example of the game *Sims*, the character requires a build-up of sympathy through which the player connects: this is the acentral avatar. No matter how it is done, it is clear that "emotions sit at the heart of a game player's level of engagement ….and that understanding the role of emotions in creating truly immersive and believable environments is critical for game designers." In fact, much can be obtained by the interested game developer by drawing from the comprehensive conceptual system for building emotions in games formulated by de Byl (2015). A very innovative device for building emotions into games is *Gamygdala*, described in Box 6.10.

6.11.3 Online Music Services

How one technology fuels another is very well illustrated by the spurt of online music service providers. Advances in transport technology (both by surface and by air) helped man to move around the world with ease and comfort. Today one can travel to any part of the globe, including the Antarctica without too much hassle. As one started moving around for business and pleasure the need to be in touch also increased. This led to the evolution of communication technology, from the phone to the cell phone and to the email. As if this was not enough, one needed ways of relaxing on long haul trips very often undertaken alone. And, lo and behold! We now have a variety of gadgets that provide music to you without your having to carry audio cassettes, CDs, or even a pen drive. You can download music on the go, often for free and otherwise through payment. It is this that has led to the mushrooming of service providers for online music.

However, the irony is that the service is still being resisted by many, those who do not trust these sites because they fear the entry of malware into their computer or cell phone or tablet, necessitating research into how customers' acceptance and commitment can be got (Sanchez-Franco, Ramos, & Velicia, 2009). Especially in a sector where illegal providers thrive, one problem is to convince the customer that the company is a legal one. In other words, how can one get the customer to trust the company? One strategy is through relationship marketing as used by the company *Rhapsody* by enhancing customers' overall encounter, to build a sense of trust and thereby belongingness and thus retain their customers (Sanchez-Franco & Rondan-Cataluña, 2010).

6.11.4 Technology for the Differently Abled

> ***Ambient Intelligence:*** *identifies situations and then delivers appropriate services to assist the person*

> ***Emotion aware AmI:*** *AmI that can detect emotions of a person*

> ***Emotional neuro-fuzzy services:*** *services that can morph according to new environmental demands and human emotions*

A major hitch faced by differently abled people is the management of their life and day-to-day activities. While for some, moving to turn of the light may be painful or even impossible, for others, who are hearing challenged, the ringing of the doorbell is meaningless. With the ever-increasing advances in medical sciences, the life span of human beings is also on the rise, leading to a new group of older people who may require help. An innovative technology that would be helpful for all of the above is **Ambient Intelligence** (AmI) which uses technology to design a futuristic environment which "dips the human in the heart of technological development" (Acampora & Vitiello, 2013). By using an invisible and interconnected set of devices, the technology aims at identifying different situations that a person may face and then delivers an appropriate set of services for assisting him in these situations. These systems have been enhanced by the addition of systems that can detect the emotions of the person through **Emotion Aware AmI** which operate through sensors that can detect facial expressions, hand gestures, body movements, and speech. The end result is that we have a very intelligent system that can offer personalized services leading to greater user satisfaction. We even have what are called **Emotional neuro-fuzzy services** that can continuously morph in order to adjust to new environmental demands and human emotions.

6.12 A Caveat

Before we end this chapter, we would like to clarify that by focusing on emotional appeal, we in no way, are of the view that anything that gives pleasure to the individual should be designed. As time has gone by, we humans have been warned against excessive pleasure that may at times be even equated to lust. Alcohol and drugs are just two examples. Though they may bring pleasure to the individual, their manufacture and distribution has to be limited. Hedonomics, especially in the realm of technology could also use this warning. Too much of anything is bad and produces harmful effects for the user. Technology designers would well heed the warning given by Epicurus, who though often referred to as the father of hedonism, is actually a **Eudaimonic**, or a person who believes that happiness and well-being are not one and the same thing. Everything that brings happiness may not bring well-being in its wake and we can find ample examples of that. Video games may bring happiness to the child who is playing it, but may not be good for his eyes or even his wrist. Car speed leads to many advantages, not the least being optimal fuel consumption, but a car that accelerates very easily could lead to an accident. While we aim for hedonomic design and manufacturers of technology are trying their best to woo customers by designing gadgets that bring pleasure, they should also incorporate the Eudaimonic principle into design by seeing to it that technology not only remains a blessing for us but also promotes the well-being of the user.

> ***Eudaimonic**: a person who believes that happiness and well-being are not one and the same thing*

It is in this context that a completely new area is developing, that of **eco-hedonism** or **sustainable hedonism**. As the name suggests, the focus is on maintaining a balance between what we as humans want and like and yet maintain ecological balance.

> ***Eco-hedonism**: keeping a balance between beauty and ecology*

> ***Sustainable hedonism**: maintaining balance between what humans like and what is sustainable*

The need for making and using technology cautiously has been very succinctly stated by Braungart. Instead Braungart wants to apply good chemistry, which would be possible if we start mimicking nature. The example provided is that of the cherry tree, no part of which is ever wasted, whether it be the blossoms that turn into fruit, humus, or a new tree itself.

What are Braungart's plans? He has already established a company EPEA, whose trademark venture is ***Cradle-to-Cradle(C2C)*** and has been described in books such as, *Cradle to Cradle,* the first book coauthored by Braungart and McDonough, published in 2002 which received widespread support and the *The Upcycle*, in April 2013.

> ***Cradle-to-Cradle technology**: technology that helps in the recovery of raw material used for the product*

Braungart's German clients include personal care products giant Beiersdorf and lingerie maker Triumph, mail order company Otto, and cosmetics maker Aveda. Braungart advises Volkswagen, Unilever, and BMW. With his help, *HeidelbergCement* developed a special cement that purifies the air once it has been processed into concrete. And, in 2013, Puma introduced the first fully recyclable athletic clothing collection, which includes compostable shoes.

And it is not just Germany that has taken to this new approach to hedonism. The US is not far behind. Arnold Schwarzenegger, who became the governor of California soon afterward, declared his sunshine state to be a C2C project region and film stars like Meryl Streep, Cameron Diaz, and Susan Sarandon promoted the concept. Director Steven Spielberg is a C2C fan. After Hurricane Katrina, Brad Pitt teamed up with McDonough to have 90 houses designed in accordance with C2C criteria built in New Orleans. A major achievement was when McDonough planted sedum on the 100,000-square-meter roof at automaker Ford's Rouge River plant. The green roof cleans rainwater and saved Ford the cost of a $50-million wastewater treatment plant.

Soon to join were other countries across the globe including Denmark, islands in the North Sea and China. Did you know that even the famous sporting company Nike makes sneakers with C2C certification and China, home to the world's largest stroller manufacturer, sells special C2C certified strollers. Airports are also catching up: an area at Amsterdam's Schiphol Airport is being developed on the principles of C2C.

An even more futuristic idea is that goods should be leased instead of being purchased, with producers being required to take them back. The logic is that once manufacturers know that they will have to take their products back, they will start using high-quality material. The innovative ideas include, say, a company buying 7 years of sitting, instead of an office, or, an individual buying 20 years of looking through a window, instead of a window.

The earlier mentioned text provides numerous examples of the practicality of the idea of sustainable hedonism. One does not have to use ugly products or products that give rise to negative emotions. On the contrary, one can combine human interests with those of the community and the world at large. Technology would thus be an enabler not only for the individual user but would also help in the maintenance of scarce natural resources and ecological balance.

Summary

A psychological analysis of behavior focuses on three facets, namely, the cognitive (knowing), affective (feeling), and conative (doing) aspects. Along the same lines, the psychology of technology would be considered lacking without due attention being given to the affective aspects, that is, in terms of feelings and emotions. At the same time, Chap. 1 has pointed out very succinctly that the neutrality hypothesis of technology fails to carry much weight today, since every artifact changes and is changed by the user. In view of the earlier, a dimension that is important is hedonomics, which unlike ergonomics, does not evaluate the user; rather it is about how the user evaluates the artifact. While ergonomics and anthropometrics make the technology safe, functional, and usable, it is the inclusion of hedonomics that provides the individual pleasure and satisfaction. Call it hedonomic design, affective design, engineering esthetics, this is the only way in which twenty-first century technological products can face the bludgeons of competition. The focus of Chap. 6 is on hedonomics and is divided into five sections, the first, introducing hedonomics and explaining its need. The second section discusses the rubrics of affective design and ways of achieving it. Since the basis of hedonomics is emotion, a third section provides an overview of methods used by psychologists for the measurement of emotions. Emotional responses are, however, often guided by our cognitions and this is discussed in the fourth section in terms of cognitive biases affecting emotional responses. The last section provides a brief overview of applied hedonomics, followed by a word of caution and a section on sustainable hedonism.

Suggestions for Further Reading

Csikszentmihalyi, M. (1990). *Flow – The psychology of optimal experience.* New York: Harper Perennial.

Norman, D. A. (2004). *Emotional design – Why we love (or hate) everyday things*. New York: Basic Books.

Oron-Gilad, T., & Hancock, P. A. (2009). From ergonomics to hedonomics: Trends in human factors and technology. In Y. Amichai-Hamburger (Ed.), *Technology and psychological well-being* (pp. 131–147). New York: Cambridge University Press.

Turner, P., & Turner, S. (2013). Emotional and aesthetic attachment to digital artifacts. *Cognition and Work, 15*, 403–414.

Chapter 7
Psychology of Technology in the Twenty-First Century

"I fear the day that technology will surpass our human interaction. The world will have a generation of idiots"

—Albert Einstein.

In the previous chapter, our focus had been on the hedonic or the pleasurable aspect of technology. Let us now consider the other side of the technological coin: its negative role, as has been perceived in the course of human history and has remained, in a psychohistorical sense, a part of the human psyche. Mayans believed that after living in a world perceived as a failure, we would undergo some sort of transformation in the fourth era, known as the *Balktun* period. They even set December 21, 2012 as the doomsday, when the world would finally come to an end, as a consequence of our own doing.

In ancient Hindu mythology too, we find a similar expression regarding the end of human existence—the *Kalyug* or *Kali yug*, as it is called. Hindu mythology describes the history of mankind in stages, similar to that of the Mayans, namely, in terms of four stages but differing in the number of years and content of these stages (see Box 7.1). According to the Hindu view, we would reach Judgment Day when we enter the fourth and final stage of our existence. This era has, unfortunately, already begun. It seems that both the Mayan culture and Hinduism had foreseen the severe psychosocial consequences of technology, and especially, the impact of the age of machines on human nature, tormented and torn as it would be by a wide variety of dilemmas. With moral forces failing to act as forces of social control, the machine age could very easily turn such dilemmas into chaos, greed, and violence. While the length of the period in each of the four stages of Hindu mythology might not sound very meaningful to modern science, the ability to forecast the consequences of the machine age and its impact on human nature is evident. On the one hand, we see the phenomenal rise and success of technology, but on the other hand, we stand witness to the increasing ambivalence of human nature and the subsequent entrapment of moral and ethical conduct.

V.K. Kool, R. Agrawal, *Psychology of Technology*,
DOI 10.1007/978-3-319-45333-0_7

Box 7.1: The Age of the Machine

In Hindu mythology, the following four eras of life on earth have been specifically categorized with reference to changes in behavior—the last bearing the brunt of technology:

1. **Satya yug**: This is a simple, peaceful phase of humankind. Any wrong-doing would make the entire kingdom suffer.
2. **Tretayug**: The focus is, now, on the local community. Any wrong-doing would hurt the community at large.
3. **Dwaparyug**: It goes down one step further, from community to family, with the family having to suffer and absorb wrong-doings.
4. **Kalyug**: The age in which the individual would be the focus and machines would have an enormous effect on the behavior and thinking of humankind.

It is not only during ancient times that human beings held such pessimistic views regarding their survival. It has remained salient during almost the entire history of humankind. The fear of a doomsday was very common during the Spanish conquest in the 1500s and Europe's second Great Flood. And, it was in the sixteenth century that Nostradamus had predicted the end of the world and had set a date for it (is it not surprising that this is again 2012, the same as what had been predicted by the Mayans!). Even as we have progressed, having at our disposal a plethora of tools and a variety of medicines to enhance our survival, the central question of our existence remains gripped in some type of fear. In addition to natural calamities threatening our survival, the inability to manage our own creations has become problematic despite the fact that we claim sophistication through technology. This issue was, probably, best highlighted by novelist George Orwell in his seminal book, *1984*. It dealt with the various problems caused by unrestricted government encroachment on the privacy of its people and clarified how surveillance would threaten human existence. When Edward Snowden, consultant to the US National Security Agency, leaked secrets regarding American surveillance on several countries and on its own people, Peter Foster of the *Financial Post* wrote,

> "Orwell's main theme—the psychology of power, the use of propaganda, the dumbing down of language to restrict the scope of thought, the pursuit of history, using children as propagandists and "spies"—are still relevant" (Foster, 2013, http://business.financialpost.com).

This is, indeed, an interesting phenomenon as far as psychology is concerned, and it would be salient to examine the ways in which a threat posed by surveillance technology has led to an increase in our level of self-awareness. After the Edward Snowden incident, bookstores around the globe reported a record increase in the sale of Orwell's book, *1984*, and many have run out of stock of even other books written by him. The complex ways in which technology is driving us towards situations, similar to those described in *1984*, is presented in Box 7.2. Similarly, the

Box 7.2: Alan Zegas (2013): We're Closer to George Orwell than We May Think

"People today, for personal convenience, use technology that obliterates their right to privacy. Electronic devices and social media, such as Facebook, provide the means for government to know where we are, who we are, what we look like, who we associate with, our political and religious beliefs, purchases, finances, the names of our friends and family, our likes, dislikes and more."

"In Orwell's world, the 'Party,' sapping people of their humanity, permitted people to think only those thoughts it dictated. Individuality was outlawed. Should a person think a prohibited thought, he would disappear, usually at night, and so too all records of his existence. History was doctored by the 'Ministry of Truth,' which dictated 'facts' and used revisionist control of the past to justify future goals of the Party."

"The psychological independence of the Party's subjects was destroyed; people blindly followed instructions. Slogans of the state—'war is peace,' 'freedom is slavery' and 'ignorance is strength'—were plastered everywhere, so that words were robbed of their meaning and the Party's subjects remained confused, rendering them vulnerable to government control."

Source: *Star-Ledger, July 5, 2013*

authors of, *In* the *Wake of 9/11: The psychology of terror*, Pyszczysnki, Sheldon, and Greenberg (2003) reported that nationalistic feelings among Americans became stronger as they began to deliberate fear generated by the danger of terrorism which has been on the rise supported, as it is, by destructive technology. On the other hand, Muslims in America began to display American flags on their automobiles very prominently and frequently so as to avoid retaliation and the consequent fear of alienation. A study conducted by the *Pew Research Center* found that almost one-third of people in America also believe that a flying body in the universe might hit the earth by 2050 (*Pew Research Center*, 2013). The ensuing threats from technology, especially those related to nuclear technology, make us extremely vulnerable to fear and prime our memory regarding a doomsday. Further, with each technological happening, we may find it increasingly difficult to fight off frightening thoughts related to the negative aspects of technology and its power to end the world.

7.1 Technology as an Intervention

In the context of the four goals of psychology of technology (referred to in Chap. 1), from learning in classrooms to monitoring criminal behavior, it is clear that technology will have consequences for not only understanding behavior, but also for its application in a wide variety of issues ranging from human development at the micro level to fabrication of a web-based society at a macro level. To begin with, the first and foremost question is: why do we want, say, *X* type of technology? Can we manage without it? What are its short- and long-term consequences? It is not long back when we were still debating the issue of allowing calculators for the teaching of mathematics or the use of computers in the writing of papers. Just think of it: if we had continued with the same old-fashioned approach, we would have appeared antiquated by now. A lot of technology seems to have become a part of our life—it tends to simply walk into our lives, albeit on some occasions in a more fanciful way such as in driving a car for the first time. Most people would not even recall when they stopped writing letters and realized that they have almost entirely switched to emails! When did we realize that writing a letter on paper and sending it after affixing a stamp became a nuisance for us? Or, when did we stop walking to the neighborhood grocery store and began, instead, to drive there in our automobile? The day is not far off, when our refrigerator will display that we are out of milk or that we are running low on protein products. The point, here, is not to claim that we were not techno-oriented in the previous century, but to impress upon the reader that the demands of life would be such that without technology, our adjustment to the changing world would become increasingly difficult, if not impossible. Sensing the magnitude of this problem, Marsden and Hollnagel wrote (1996),

> "The spread of information technology, however, means that there are many situations where users interact with information technology systems because they have to do it rather than because they want to do it. The possibility of doing it in another way has simply disappeared. Examples include finding a book in the library, personal financial transactions, the health sector, traffic and transportation, process control, etc. A trivial example is the typing of a letter, since many offices no longer have a type-writer.…conventional modes of interaction disappear, often in the name of efficiency!" (p. 2).

> ***Intervention****: an organized program geared to assist the client in some way*

> ***Teaching machine****: a technological device to help students learn through the application of appropriate reinforcements*

Specifically, interventions based on technology are far more complex than what we can generally envisage and the time has come when it has become mandatory to start thinking about this in a more serious manner. Let us start with what is an intervention. An **intervention** is an organized program geared to assist a client in a systematic way. A good example is *Head Start*, a program widely acclaimed to be the largest program in the history of the USA undertaken to accelerate the cognitive development of children. In the context of technology, when Skinner developed his **Teaching Machine** to demonstrate how correct responses can be learned through reinforcement of behavior, he was essentially chalking out an intervention to augment human learning (Chap. 1). Later, his ideas gave rise to concepts such as Token Economy and were used in various programs to help people learn desired responses.

In the absence of an intervention, our behavior may become chaotic. Imagine what would happen if we did not have a program for learning how to type on a QWERTY keyboard: most of us would still be hitting letters like a chicken pecking for its meal! As far as psychology of technology is concerned, salient interventions would be those in which technology has helped to augment and support human potential and the results of which are reflected in some enriched form of behavior. Having said this, let us consider a few examples illustrating our dilemmas in the use of technological interventions, albeit well-programmed ones:

- As a responsible parent, at what age would you allow your child to watch television?
- As a responsible parent, at what age would you allow your child to use a computer?
- As a responsible parent, what would be the age of the child at which you would lift parental controls on television?

> ***Neurogenesis****: study of cell development and replacement of cells in the brain*

First of all, we bet that parents would give a wide variety of answers regarding the time when they would give permissions for any of the above. Earlier (Chap. 6), we have mentioned that the American Pediatric Association has strongly recommended that parents should avoid television viewing by their children until after the first 2 years of life and thereafter, restrict it to only about 2 h per day. In fact, they recommend that children should spend more time with interactive tools, so that their brain has a greater opportunity to develop. Recent research on **neurogenesis**, which is the study of cell development and replacement of cells in the brain, suggests that infants who are exposed to sensory overstimulation, such as through TV viewing, at an age as young as 3 months, might manifest delayed cognitive milestones with poor attention, delayed language development, and abnormal behavior as the end result (Nada, Khater, & Saeed, 2014).

> ***Technological intervention****: an intervention which uses technology to assist the user in day-to-day activities*

Given the state in which we are at present, the challenge before us is: what is a good, acceptable test for evaluating any **technological intervention**? We are of the view that one of the major challenges in the application of technology-based programs would be to realize, in the first place, *the extent to which a user wants it or does not want it* (in other words, the motivational level of the user). When programs are mandated by an agency, such as being told to bring a computer by a school or a college, a client is faced with a "*you have to do it*" situation. In this particular case, the action of the user is likely to be based on prior experience and hence it is employable, but is it immune to scrutiny in terms of optimal human performance? On the other hand, what happens when *free-market type of products* keep flooding the market and often tease clients into wanting to try them? For example, have you ever thought about the merits and demerits of aerobic exercise machines placed in a fitness center and widely used by the members of a club? The usefulness of a technological product, such as an exercise machine, appears convincing, especially when placed in an appropriate health enhancement context such as a gymnastic club. In such an appropriate, concrete setting, it looks trustworthy, as most of us would tend to remark. The question is: is it really trustworthy? The immense growth of technology is bound to flood the free enterprise market with a variety of gadgets that would not only burden our decision-making, as discussed in Chap. 6, but would also bring consequences of unforeseen magnitude. Did we ever anticipate that computers, when overused, could cause carpal tunnel syndrome?

> ***Narcissism**: the human tendency of self-love*

> ***False consensus bias**: the tendency to overestimate the degree to which others agree with him*

Technology is also being used for exploiting the basic human tendency of self-love (**narcissism**). There are countless products in the market that promote our narcissism, ranging from plastic surgery to hair removal and even artificial hair implantation. We invite readers to have a look at the technology promoted by the product, "*NO! NO!*" supposed to be used for the removal of unwanted hair. From customer reviews, we have come to learn that it works well for some, not so well for others, and not at all for still others. A common risk reported by many customers is burning sensations in and even actual burning of the skin, especially if the user is not properly trained in its application. Our curiosity led us to find out more about its application and modus operandi as provided in its manual. What we found astonished us. First, this company does not conduct a program to train its clients, and second, operating the instrument is not as easy as it sounds in the manual, as a result of which, the risk of burning the upper surface of the skin is always there. The same may be said for many other products and we are sure that the reader would have often felt likewise: that operating a certain gadget or using what appears as a simple product is not as simple as the manufacturer makes it sound. Why is this so? The fact is that technologists often suffer from what is commonly known as the **false consensus bias**, which means that a person has a tendency to overestimate the degree to which others agree with him or her. When engineers create their products and provide information about it, they simply begin to believe that most people would also understand its operation, and, in the same way as they did. This false consensus belief is so powerful that it gets reinforced even when someone truly believes that her opinion is supported by only a few members. What feeds this tendency of overestimation is the pervasive self-esteem of the individual that emerges out of the innovation of a product. On the other hand, if "making it so easy to use" is a commercial ploy used to boost the sale of a product, then it is all the more reason that the manual should be written with reasonably good illustrations. Those who frequently buy prefabricated products, and assemble them at home, would have experienced the consequences of such a false consensus effect, for they very soon realize how simplistic the manuals are and often feel that the manuals should have been prepared more carefully, in a way that would have made it easier for them to assemble the product.

> **Pluralistic ignorance**: *a notion based on public support but of which most people disapprove*

An effect which is the opposite of the false consensus effect is **pluralistic ignorance**, a notion based on public support to the extent that it has become a social norm, but of which most people tend to disapprove. In other words, an individual believes that he/she disagrees with the majority of people, though, in fact, he/she agrees. If you go back to the example of the aerobic machine placed in the appropriate context of a gym, does it not hurt you as you do your workout? But do you say so? Certainly not! So, even when someone says that the machine hurts, and you also feel that it hurts, you remark otherwise, simply because it is the in-thing to go to a gym for a workout. It has become a social norm and we dare not go against the norm, social beings as we are by nature. And, when everyone starts behaving in this fashion, it leads to what can be called pluralistic ignorance, or a general societal ignorance regarding the negative effects of the technology, feigned though it may be.

After having discussed the ways in which the false consensus bias and pluralistic ignorance operate, it is only logical that we focus on the user of technology in the next section of this chapter.

7.2 Where Is the User in User Technology?

If you open the *Pew Research Center* web site, you will find a survey,

"What kind of user are you?"

One of us went to this website but did not answer any question on the first page of this survey, and instead, moved to the next page. Guess what we got,

> "If you are Tech Indifferent, you are not a heavy internet user and although you probably have a cell phone you don't like its intrusiveness. You could easily do without modern gadgets and services."

The purpose of the above-mentioned exercise is to find out, not only the effects of technology on our lives, but also, to focus on the pattern and pace of adjustment to technology in life. If we take the slice of time between the years, 1900 and 1999 and catalogue the changes in our day-to-day living during this period, we would be amazed. Let us, now, consider the next time frame, that of 2000–2099. We guess that the *Pew Research Center*, if it would survive until late in the century, would then ask questions regarding our preference for making a trip to the Moon or to the Mars, about the comfort level of the space capsule's hotel room, and so on.

Although the above is hypothetical, the message for different fields of knowledge, including psychology, is clear and this is to take into account unexpected avenues, and, to prepare the human self for changes as far as we can envision. "The mind is understood," wrote Michael Glassman (2012), "as a 'user-friendly' device, where 'executive functions' and 'black box' technology efficiently processes information and solves problems" (p. 309). The day is not far off, when we will see an era of web-based societies, with technology partnering the human brain and engaging in a coevolution based on thinking in "webs" and "web trails." Indeed, we are on the brink, with many such technologies already underway, as has been discussed in the previous chapters. Many more are in the making. In other words, we would, then, be operating beyond our biological limits and could be viewing computers as an extension of our brain and thinking. In an earlier chapter, namely, Chap. 6, we

Box 7.3: The Techno-self

What will this techno-self be like? What and who are we? Are we becoming or have we already become "soft selves?" And, what would this soft self be like, what would be its boundaries? Philosopher, Daniel Dennett gives us a formula for assessing the boundaries of the self. According to him, "I am the sum total of the parts I control directly" (Dennett, 1991). As such, our bio-technological selves will be with reference to all that we can directly control. This may sound easy enough, but do you not see an identity crisis looming large in front of you? With technologies moving at a pace faster than we can adapt to them, how much will we be able to control? We are brought back to thinking beyond "being" to "becoming." The inherent dynamism, characteristic of the self, becomes even more so in an environment wrought through technological hybridization.

have already focused on how "selfing" has morphed with recent advances in virtual reality technology. Accordingly, the role of psychology in the twenty-first century would also have to change and would be aiming at seeking answers to such extensions of the techno-self (see Box 7.3).

While technology grows alongside advances in the biological sciences in matching brain activity by mapping the interconnectedness of neurons and plasticity of the brain and thereby creating an artifact analogous to the extension of our own self, it is the user who must still remain the center of attention. This would involve not only laypersons, who would be the consumers of such psycho-bio-tech types of products, but also developers who would have to envision this product without showing biases such as the false consensus effect discussed above. In fact, Glassman's concern, and also that of others such as Dror and Harnad (2008) who have used concepts similar to the "cognitive commons," (regarding extended artificial computer social networks), is thought-provoking, to say the least. It may even become one of dread, similar to Orwell's community, despite the fact that technologists of the mind claim that in such a network "information will be treated with equal value" (Glassman, p. 316). The questions are many and the answers are still awaited. We are back to square one and are forced to raise the question once again, where is the user in this technocracy?

Saariluoma and Oulasvirta (2010) argued that humans should be treated as actors and not as factors (as in human factors engineering), but unfortunately, the course of research, as stated in Glassman's quote given above, has shifted from its original psychological orientation to that of technological coevolution. We present, below, a very strong statement made by them in their article, *User Psychology: Re-assessing the boundaries of a discipline*,

> "Ultimately, in order to avoid being a slave to engineering, psychologists must start defining their work according to psychological themes rather than in respect to technological boundaries" (p. 318).

Continuing further, they write,

> "This means re-thinking how we answer such questions as
> 'What is the subject of psychological research in this area,'
> 'What are psychological theories about,'
> 'What kind of theory is useful,' and
> 'What is the role of psychological research in constructive effort like design' " (p. 318).

***MMI**: Man–Machine Interface*

***HCI**: Human–Computer Interface*

Indeed, their argument sounds meaningful, for many research articles in the area of human factors engineering remain devoid of any significant application or amplification of a psychological theory. If we remain confined to the conventional thinking that has dominated the past four decades as far as **MMI** (man–machine interface) or **HCI** (human–computer interface) is concerned, the contribution of psychology will definitely shrink. The need of the hour is for psychology to expand its theories, leading to the possibility of its usefulness in a wide variety of specific activities ranging from aviation to ground traffic, and from environmental architecture to product designs. Imagine the scenario later in this century. Our calculated guess is that it will be all about the pervasive role of technology in our lives. At the same time, have we given enough thought to our own behavior? What would happen to our feelings and emotions in a predominantly virtual environment? Are we thinking along those lines? Or, are we planning to think when that day dawns upon us? As of now, we can only engage in guesswork. Maybe, movies like *Avatar* would provide the much-needed guide for our cognition and mental frames.

***User-centered**: putting the user first*

Coming back to the question of remaining "**user-centered**," it is important to realize that technology plays, or at least, should play, only a supporting role, despite it being imbedded into almost everything we do in our lives. Given this premise, the unit of study in technological behavior is activity which is intentional. Along these lines, activity theorists have shown the limits of the information processing approaches in human–machine interaction and have been able to offer a relatively better alliance with psychology (Noe, 2009). Since psychology, as an empirical body of knowledge is predominantly directed at the study of individual behavior, while the focus in activity-centered analysis is on the interaction of the individual with societal forces, psychology has not been able to find its place in the activity-centered approach. We will return to this issue later in the chapter.

In the conventional "usability" approach, technology has always received precedence. It was only after the invention of a certain tool or technology that its applications were sought and the product was tested and adjusted as per trials on human beings. However, such an approach fails to support the notion of "putting the user first." One can find examples of such an approach in the countless gadgets that are part of our everyday living but which could have served us better. A classic example is the computer keyboard (Box 7.4). While several versions have been created and tested, no effective replacement seems to be in the offing and so it seems that we will have to probably live with its problems forever.

Box 7.4: On Using Keyboards

Keyboards are broadly defined as arrangements of buttons or keys which act as mechanical levers. In English-speaking countries, the standard keyboard is known as QWERTY, referring to the letters beginning with Q on the board. Other variations are the DVORAK and the AZERTY (EUROPEAN). It is well known that on the QWERTY keyboard, the letter A is on the extreme left and is usually pressed by the left hand's little finger. As we know that letter A is typed a lot, is it helpful to the user in terms of the strain on the little finger, which is probably the weakest finger in terms of muscle strength? As keyboards are shrinking in size, have we thought of its optimal size? Are curved arrangements and placing it in parts better suited to the user? There are no simple answers. There is a greater emphasis on the tactile experience now than ever before, and we are still wondering what impact such variations or additions would have in the context of our physiological limits to our capacity of using the fingers (or other parts of body under special circumstances).

So, what is the solution from the user's perspective? We advocate that the use of a **proactive approach**, in which the user is put first. We need a paradigm shift from:

Proactive approach: *an approach in which problems are foreseen and solutions developed thereby preventing the problem*

Product first ⇒ People, to
People first ⇒ Products

Possibly, such a user-centered approach would not only retain, but also focus on the significance of cognitive attributes, would take into account the range and dimensions of human responses and reflect any or all of the affective and emotional characteristics of individuals. For engineers, this would probably be difficult to understand because the majority of them do not begin with this premise and even if they do, the development of a product, per se, is their primary mission. In the case of physical disabilities, such parameters are concretely visible to the engineers, but it is not so easy with mental properties. When they work along with an entrepreneur, they face further constraints in terms of marketing, profit-making, and government controls. The user is the last to be considered. Therefore, it is of utmost importance that psychology develops its own platform to reverse the common practice of such a top-down approach. One recent development that has brought tremendous hope for user psychology is advances in a relatively new field of research, **neuroergonomics**, which is the study of the brain and behavior at work (Parasuraman, 2011; Parasuraman & Rizzo, 2007). Elucidating this upcoming field of study, Hancock and Szalma (2003) had written in their article, *The future of neuroergonomics*,

Neuroergonomics: *study of brain and behavior at work*

"It is our view that the state of current human interaction with technology is more akin to the activities of the naïve beginner as compared to the polished response of the expert. This

> is largely because current human interaction with technology has to proceed through highly limited input-output capabilities curtailed by the qualities of perceptual processes and the response capacities of the motor system…. Eyes, ears and skin for input and fingers, toes and voice for output are very limited ways for the brain to convert imagination to reality" (pp. 238–239).

It is encouraging to find that this field has been growing steadily and the interested reader will find substantially useful information in *Handbook of human factors and ergonomics*, edited by Salvendy (2012).

We are of the view that the field of psychology will be able to enhance its status, in this very important realm of practical knowledge, only when it has a clear focus on user psychology, engaged in and utilizing its theories and practices, rather than focusing on computers and other artifacts as its central base. With overwhelming developments in technology influencing our behavior in myriad ways on the one hand and the user placed in a central position and viewed first and foremost from a psychological perspective on the other hand, inventors, entrepreneurs, traders, designers, and salespersons, too, will begin to examine the dynamics of the users—their cognition, motivation, and emotions as reflected in technologically oriented behavior. It will bring psychology to the forefront rather than keeping it in the back seat. It is apparent that this development is long overdue as far as the growth of psychology is concerned. Given the significance of the user and the current status of research in psychology and allied disciplines, we believe that psychology of technology can and will benefit from user psychology in the twenty-first century in a number of ways. We will address a few of them in our discussion below.

First, user psychology offers a **psychology-based approach**, rather than one which is conceived on a technology platform. With research comes evidence which is the corner stone of not only conceptual knowledge but also applied use. In fact, such an endeavor would benefit the advancement of technology, as well, because growth in technology becomes possible only through human thinking and operation. Technology helps the user as much as the user helps technology. A symbiotic relationship between technology and user psychology would create a win–win solution for both and history stands witness to the advantage of such a relationship. For example, when computer programming began, the computer programmers benefitted from psychology in a number of ways. Thus, they found the concept of chunking in human memory, that is, our capacity to group information into categories, say of seven units each, very useful for the assembling of information, an aspect which is of utmost importance for their work.

> ***Psychology-based approach****: an approach which focuses on the psychology of the user*

Second, it is apparent that the motivational differences between users of technology in terms of primary and secondary intentions, for example, their ability to differentiate between need and want, will dominate the growth of psychology of technology during this century. With advancement of technology, often in ways that astonish even the creator of that technology, users would manifest varying intentions for the use of technology. Are you using technology because you prefer it? Or, are you using it because it is the in-thing to do? Or, are you using it because there is no other way out? It could also involve the use of different forms of technology for differing purposes. The answer will involve intraindividual differences for different aspects of technology. I may choose to email a routine letter, but send a birthday greeting via a traditional mail through the post office to make it look highly personal. On the other hand, if I am tech-savvy, as assessed through the *Pew Research Center* survey, I may look for a fancy technological way to impress the recipient, conveying that I really meant to make my greeting special even through technology! The choices that individuals make are rooted in their experience of culture and traditions and, with the user placed in a key position, technology will be able to cater to such intraindividual choices. Conversely, when technology dictates choice, such as dealing with internet banks with all transactions managed electronically, the psychological factors of fear and trust determine the relationship between the individual and the technology. Intentions vary in our interaction with technology, and in this process create contextual relations of experience that connect us with the world. Subscribing to this line of thinking, Martin Heidegger and other scholars from the phenomenological school of thought have been warning us against employing a simplistic explanation based on mechanical explanations of the human mind for the relationship between human beings and things around us (see Chap. 1). Whereas the focus of MMI is on the activities that connect the individual with technology in the outside environment, for Heidegger, it is the intentions of the individual that are central for establishing connections with the external world. Because intentions give direction to action, the critics argue, the user's experience and what it means to him/her is relevant for, and should be the primary focus of, psychology of technology.

The **third** issue in user psychology is the appreciation of individual differences in any type of technology. As mentioned in Chap. 2, applications based on anthropometry tend to undermine individual differences in terms of percentiles or other statistical outcomes. As a result, we have products that fit many but not all. Further, as mentioned in Chap. 2, with aging, the size of the human body shrinks and cognitive abilities decline, but in the predominantly technological environment, these factors receive hardly any attention. Go to any public place and you see the creation and accommodation of things for "average" people only. We witness such limitations everywhere—from boarding planes to using a Blackberry phone with a miniaturized keyboard. Critics may argue that it is a policy issue, most likely dictated by financial constraints, but they should also know that technology does not operate outside human endeavor and "brings the thing-like instrumentalization of human nature" (Kompridis, 2009, p. 25). The availability of a product makes a difference as has been so aptly presented by a well-known social psychologist, Leonard Berkowitz (1993): that the finger pulls the trigger is as true as the gun attracts the finger. The availability of weapons around us tends to prime our thoughts for their use and the affordances they offer.

The **fourth** issue relevant for user psychology is linked to the development of theory in psychology. The strength of any science lies in the robustness of its theories, and a weaker science, such as psychology, needs to project the ways in which its theories prove to be relevant in various domains of life. For example, for a long time, the applications of social psychology in the field of technology were almost unknown even though technology has been with us for centuries. An edited volume, *People's Reaction to Technology*, made its appearance during the 1990s (Oskamp & Spacaman, 1990). Similar publications have appeared sporadically, but a concerted effort on this or other technology-related subjects have remained, by and large, loosely covered under the subdiscipline of industrial psychology. Progress in psychology of technology surfaced with publications in journals on human factors engineering and ergonomics but even this is a relatively new development as compared to the growth in other domains of psychology. Neither did such growth lay due stress on psychological concepts nor was a mainstream psychological movement ready to support this neglect. When we browsed through books on applied social psychology at the beginning of this century, we were shocked to find that there were no chapters on psychology of technology in books on applied social psychology (Kool & Agrawal, 2006). Engineers work with people in a social setting. The artifacts they produce are subject to attributions and other psychological features. Despite this, while social psychology was attempting to address many aspects of life—religion, education, law, sport, recreation, and business, to name just a few—technology remained a distant cousin for scholars in psychology being considered only for its productive value in the form of *technology in psychology* as opposed to *psychology in technology* (see Chap. 1). A major constraint could have been the need for an interdisciplinary approach in order to address problems such as those that have been raised in this volume. We believe that during this millennium both aspects will grow closer, albeit with some reservations (discussed below).

While it would be good news for both psychology and technology to straddle the boundary of each discipline and blur the lines either way—psychology in technology or technology in psychology and thereby strengthening their causes, humans viewed as a simple, static prostheses in human–computer interaction might well jeopardize the alliance between the two approaches. Glassman (2012) has argued that unless such a prosthetic-oriented approach is viewed as being dynamic and malleable, we will continue to have two viewpoints:

- **Engineers' psychology of technology (EPT)**
- **Humanists' psychology of technology (HPT)**

***EPT**: Engineers' Psychology of Technology*

***HPT**: Humanists' Psychology of Technology*

Further, based on the revolution in information technology, that is so rapidly changing the world, Glassman argues,

> "The co-evolution of human technique (the way we determine the appropriate form of the computer) and technology (the ways the changes in computing technology push technique forward) and how this interactive association has impacted conceptualizations of human cognition and the workings of brain, bringing psychology to a critical new juncture in examining human relationships with and to information" (p. 309).

Lastly, as more scholars, now than ever before, are engaging in research on psychology of technology, engineers have begun to understand and appreciate how psychological knowledge can be used in a variety of meaningful ways. Gone are those days of ignorance and the random, casual consideration of psychology in the world of engineers. On the other hand, the hesitation on the part of psychologists in approaching technology has been mostly due to their own ignorance. Not many have ever received training in technology and even if they did, they did not gain much encouragement due to the preoccupation of psychology with traditional areas that have dominated teaching and research in psychology departments around the globe. We expect a major shift in this scenario, in fact a paradigmatic shift; or else, psychology will continue to move with one wheel of its wagon locked. Cyber-therapy, cyber-counseling, distance learning, and simulation are just a few trends that demonstrate how the scenario of psychology will keep changing in the near future.

7.2.1 Opaque and Transparent Technologies

In terms of EPT and HPT mentioned above, we can distinguish between technologies on the basis of the extent to which they mesh with the human user. On the one hand, we have technologies that can be classed as **opaque technologies** which are normally the result of EPT. Calling them opaque does not mean that they are difficult to understand. Rather, it is a technology that keeps coming in the way, requiring skills and capacities that do not come naturally and thus take up much of our limited attentional resources even during routine tasks. In contrast, a **transparent technology** is one that is based on HPT and is fitted to and integrated with not only our lives but also our biological capacities. When using opaque tools, the user constantly distinguishes between herself and the tool. While, when it becomes transparent, the user literally looks through the tool. In the course of technology use, many such tools become transparent either because of simply using them or because of practice. One good example is the spectacles we wear. After a period of time, we become aware of them only by their absence. Other examples include the blind man's cane, sporting gear, and even musical instruments.

> ***Opaque technologies:*** *technologies that require skills and capabilities which do not come naturally to the user*

> ***Transparent technologies:*** *technologies that mesh in with the user's natural skills and capabilities*

The point that we want to make is that psychology of technology will become increasingly important for developing tools that are transparent. Rather than retro-fitting them would it not be much better and even more economical to design them such that they are more or less transparent, if not completely transparent? Donald Norman (1999), cognitive scientist, puts it as a historical transition from technology-centered products to those that are human centered. In earlier chapters, we have mentioned products such as HULK and TALOS, exoskeletons created to provide almost superhuman powers to army personnel. The reader will recall how earlier versions had to be discarded because they proved to be far too heavy with the user expending more energy in carrying the apparatus than in actually using it. In contrast, transparent products would be those that "wear their functionality on their sleeve" and exploit the natural strengths of human brains and bodies. The user almost never needs to open the manual. Yet, is it not surprising that this genre of technology is slowly but surely vanishing from the scene. We can think of countless opaque products but how many of them are transparent in the true sense of the term? Norman also gives us the reason for this situation. In the words of Andy Clark,

> "At first, creating a product that can DO THE JOB is hard enough, let alone aiming for products nicely fitted to brains like ours. As time goes by, however, the vendors must seek to extend their market beyond the gung ho early adopters and technophiles. They will need to sell to the average user who simply wants a cheap, reliable, and easy-to-use tool. The technological product then comes under cultural-evolutionary pressure to increase its fitness by better conforming to the physical and cognitive strengths and weaknesses of biological bodies and brains. In quasi-evolutionary terms, the product is now poised to enter into a kind of symbiotic relationship with its biological users. It requires widespread adoption by users if its technological lineage is to continue, and one good way to achieve this is to provide clear benefits at low cognitive and economic costs" (Clark, 2003, p. 39).

We foresee that psychology of technology with its focus on user psychology would help, in more ways than one, in the creation of such transparent tools and technology.

7.3 The Interdisciplinary Focus of Psychology of Technology

We expect the reader to have become familiar by now with the fact that the nature and scope of growth of psychology of technology depends on how it relates to other sciences. From the material presented in the chapters of this book, it is obvious that various branches of biology—from evolutionary biology to anatomy and physiology, and developments in technology ranging from information technology to biotechnology, would continue to chalk the growth of psychology of technology as a subdiscipline of psychology. In the context of medicine, human factors and ergonomics and psychology, Karwoski (2012) has provided a comprehensive description (to name just a few of the categories mentioned by him, under medicine: cardiology, psychiatry, neuroscience, community medicine; under human factors—HCI, affective ergonomics, nanoergonomics, participatory ergonomics, ergonomics of aging; and under psychology, cognitive psychology, social psychology, educational psychology).

In the next section of the chapter, we will focus on three areas: sociology of technology, convergent technology (the latter including biotechnology and information technology), and the Internet of Things (IoT).

7.3.1 Sociology of Technology and Psychology of Technology

An important avenue of growth of psychology of technology would relate to the emergence of sociology of technology as a significant source for understanding human behavior. Sociologist Ogburn (1922) had pointed out, almost a century back, that with the support of social factors leading to the acceptance of technology, available technological components would naturally lead to further inventions. Raising a question, "with boat and steam engine available, would you not see a steamboat in the near future?" Ogburn (1922) opined that for a wide variety of significant inventions, many inventors worked separately and had absolutely no interaction at the personal level. In other words, these inventions cannot be said to be the monopoly of a single person, who is credited with the final creation of the invention. Thus, the notion of creativity was viewed in terms of (a) pieces of information, (b) availability of a concept for invention, and (c) permission to build a new product. Hughes (1989), a historian of technology, concluded that an invention was not something that emerged suddenly. Rather, it has generally been a gradual change of existing technology. For example, there was nothing so revolutionary about aircraft propulsion in the substantive context of old water and gas turbines. In their concept design of 2013, the scientists at the Imperial College, London concluded that the project for landing on Mars would not involve any new fundamental discovery (Box 7.5). Additionally, technology involves a system. For example, to have missile power, we need to have a warhead, propulsion, and launch pad and control units working in a logical order. Modern technology is mostly integrated, for it comes in bits and pieces. Therefore, various units of technology, its controllers and suppliers and the overall system managers form a team to implement the inventive effort.

Hughes argued, further, that as technology becomes deeply rooted in our lives, it begins to be viewed in a social context. Take, for example, electricity and its supply—it is technology, but it is also imbedded in economic, political, ecological,

Box 7.5: Mission to Land on Mars, 2013

Scientists at Imperial College, London have designed a concept mission to land astronauts on Mars. The plan envisages a three-person crew journeying to Mars aboard a small two-part craft. The craft would rotate to generate artificial gravity and use a heat shield to protect itself against solar flares.

The crew would then return to the Martian orbit in a present craft fuelled with ice from beneath the planet's surface. The concept, developed in conjunction with the BBC, is intended to spark further debate about the technical obstacles and risks that would have to be overcome in order to put humans on Mars.

"Every part of this mission scenario has been demonstrated one way or the other, including the in situ propellant production on the surface of Mars," said Prof Tom Pike, who led the Imperial design team.

"There are big, big jumps between a demonstration at one level and putting together the engineering systems for a mission, but they are engineering challenges. They are not fundamentally about making new discoveries."

Source: Poeter (2013)

and cultural factors, in terms of its cost, pollution levels in its production, electrical grids around the habitat, and so on. Finally, when technology is the key source in making judgments about changes at social and individual levels, it is said to constitute technological determinism. The fear of the individual is of not finding the self in the midst of advances of such great normative significance. Our very identity, the roles we play and the ways in which we transact our daily business is affected in more ways than one and often makes us feel that even frequently experienced forms of behavior are interventions, determined and guided by technology. When *PayPal* credited billions of dollars in an individual's account recently by mistake (Wills, 2013), we began to wonder what would happen to our accounts with this company. Rodney Brooks (2001), of MIT's *Humanoid Robotics Group*, stated that we are heading towards morphing "ourselves into machines." An interesting outcome of such technological determinism is that effects caused by technology are often not perceived as immediate sources of change. As a result of which, people may not be able to accommodate these changes at the cognitive level with problems of adjusting and adapting to them as the end result. This produces, what has been referred to, as technological lag in a society. On the other hand, certain slow appearing effects of technology take a sudden turn and "events develop at a furious pace," wrote Kurzweil (1999) in *The Age of Spiritual Machines,* "and that is what we will experience as we enter the twenty-first century" (p. 6). Norman (2004) has cited how mobile phones, once thought of for use primarily in business, have been selling around the globe at a phenomenal rate. They have become almost priceless in those remote areas of the world where traditional phone lines were never laid. Just a few decades back, while going out we would make sure that we had our wallet, but now, don't we make sure that we have our mobile phone, too? If we miss it or lose it, it is as if we are missing a part of our life. Compare this scenario to the statement made by Martin Cooper, pioneer of wireless communications, in 1981: "Cellular phones will absolutely not replace local wire systems."

The growth of psychology, as a science with the focus on behavior and mental processes, is, therefore, subject to developments in the sociology of technology, and at an extreme level, will witness what is known as "technological determinism." When the world becomes techno-centered, with or without our choice, change in behavior would be considered as an outcome necessary for adapting to the new conditions. Scott (2012) has offered several examples of the individuals' relationship with technology in the context of modern social psychological concepts. For example, a norm of reciprocity involves helping those who have helped us. In the context of technology, the norm of reciprocity would mean that the technology should return the favor of the cost of obtaining that technology or its services. However, will there be a guarantee that such a cost would lead to a reasonable substitute in terms of the norm of reciprocity? Will, whatever we have paid in terms of the cost of obtaining that technology, be repaid by the convenience or even the joy of using the technology? Human behavior is always viewed in a context and using technology in such normative behavior scenarios may help to create an image of status (Osborne, 2011). Taking such a perspective further, Scott refers to a study by Agarwal and Prasad (1998) in the context of *Personal Innovativeness with*

Informational Technology (***PIIT***). It is defined as "the willingness of an individual to try out any new informational technology" with an assumption that it is a trait of those who find value in technology and are willing to invest in it (p. 206). As expected, the results clarified that those who scored high on their test were willing to invest more in technology. With technological determinism around the corner, like McClelland's high achievers, we would not be surprised to find the number and intensity of PIITers rising during the current millennium. Scott concluded that the acceptance and continuance of technology is likely to grow in personal, as well as, in relational terms.

PIIT*: the willingness of an individual to try out any new informational technology*

Probably, there will be no area that will be able to glue together the two disciplines, psychology and sociology, as much as communication technology. Research in the context of group settings as well as web and mobile technologies have offered ways of collaboration between individuals on the one hand and novel forms of communication, on the other. Regarding research at the group level, relatively older studies in social psychology have shown that heterogeneous groups, albeit slower, offered a greater variety of solutions than their counterpart homogeneous groups. Applied to technology, groups performed slower in information search, but offered a small but significant accuracy (approximately up to 15 %) compared to individual searches (Forlines, Vogel, & Balakrishnan, 2006). At the individual level, technology has radically changed human communication. Our writings and expressions are loaded with technological expressions such as "boot your brain" (see Rosen, 2010, 2012).

7.3.2 Convergent Technologies and Psychology of Technology

Based on the proceedings of a workshop under the aegis of the National Science Foundation, Rocco and Bainbridge (2002) prepared a report on, *Converging Technologies for Improving Human performance*. The goal was to garner support for an interdisciplinary effort between four primary areas, namely,

Convergent technologies*: technologies which cut across discipline borders and are based on advances made in various areas*

- **Nanotechnology**
- **Biotechnology**
- **Information Technology**
- **Cognitive Science**

This NBIC project, also named "**nano-bio-info-cogno**," was undertaken so as to create a vision for meeting the ever growing pressure of unprecedented changes in technology, with much more to follow with the passage of time, on human performance (Figs. 7.1 and 7.2). While the applied nature of a discipline

Nano-bio-info-cogno*: referring to the convergence of nanotechnology, biotechnology, information technology, and cognitive science*

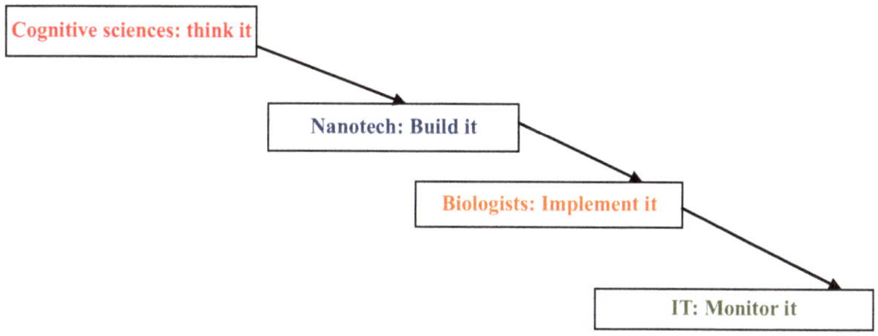

Fig. 7.1 Convergent technologies NBIC (Nano-Bio-Info-Cog)

Fig. 7.2 The NBIC Tetrahedron

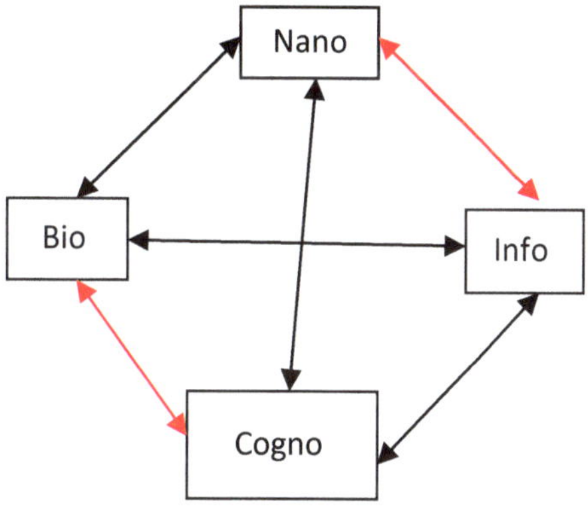

normally connects with another discipline in the form of an obligation (Saariluoma & Oulasvirta, 2010), this report focused on bringing together sciences and technologies, in the true sense of the word. In as much as psychology of technology is concerned, the principal interest is on how this model will

> "Allow us for the first time to understand the natural world and cognition in terms of *complex, hierarchical systems*" (Rocco & Bainbridge, 2002, p. 3).

More specifically, the focus of this report was on

> "All too often we communicate poorly with each other, and groups fail to achieve their desired goals. Our tools are difficult to handle, rather than being natural extensions of our capacities. In the coming decades, however, converging technologies promise to increase significantly our level of understanding, transform human sensory and physical capacities and improve interactions between mind and tool, individual and team"…. "Each scientific and engineering field has much to contribute to enhancing human abilities, to solving the pressing problems faced by our society in the twenty-first century; but combined, their potential contribution is vast" (Rocco & Bainbridge, 2002, p. 4).

The frustrating part of this report, as applied to psychology of technology, is that even with such a prestigious NSF-sponsored enterprise, the capacity to offer hope to psychology as a science with cognition as its mainstay, extends to only a limited time frame, namely, a "time frame of 10–20 years." At the same time, a participant in the NBIC workshop offered hope to the scholars engaged in this interdisciplinary work, by writing:

> "If the cognitive scientist can think it
> The Nano people can build it
> The Bio people can implement it, and
> The IT people can monitor it" (Rocco & Bainbridge, 2002, p. 13).

The ways in which these four disciplines can come together and work together is explained in Figs. 7.1 and 7.2. As the authors point out, the relationship can be seen in the form of a tetrahedron, with each vertex representing a specific field, the lines showing the ways in which two disciplines connect together, each set of three fields represented by a surface while the union of all four fields is the volume of the tetrahedron. A large number of interdisciplinary projects have been chalked out and authors from different disciplines have detailed them in this volume. The areas of application would have far reaching implications, covering as they do, several fields, including, human cognition and communication, health and physical capabilities, group and societal outcomes, national security, science and education.

7.3.3 The Internet of Things (IoT) and Psychology of Technology

> ***The Internet of Things:*** *interconnectedness of objects and people through web-based technology*

In just a few years from now, we will not have the normal two or three remote controls in a household. Rather, we will have a basket of remote controls and according to the CEO of Cisco, the remote control market will soon be of the order of 19 trillion USD. What is the reason for this sudden onslaught of remote control devices in our lives? It is IoT or the Internet of Things, apparently at the zenith of the growth of convergent technologies and ranges from the use of wireless communication to the internet and from embedded systems to microelectromechanical systems. The time frame that is foreseen for IoT is given in the Fig. 7.3.

What would it enable? It will enable the establishment of the much talked about Big Data, which in turn will enable augmentation of a myriad other functions ranging from home automation to transportation, and from health management to environmental management.

By the year 2020, we expect nearly 26 billion devices on IoT, according to *Gartner Inc* (2013). We have been mentioning the *Pew Research Internet Project* time and again. According to one of their surveys, 83 % of the technology experts and internet users they targeted agreed that IoT will have widespread and beneficial effects by 2025 (*Pew Research and Internet Science and Technology,* 2014). The British Chancellor of the Exchequer, George Osborne has posited, that the Internet of Things is the next stage of the information revolution and is of the view that it will lead to an interconnectivity of everything from urban transport to medical devices to household appliances (GOV.IN, 2015).

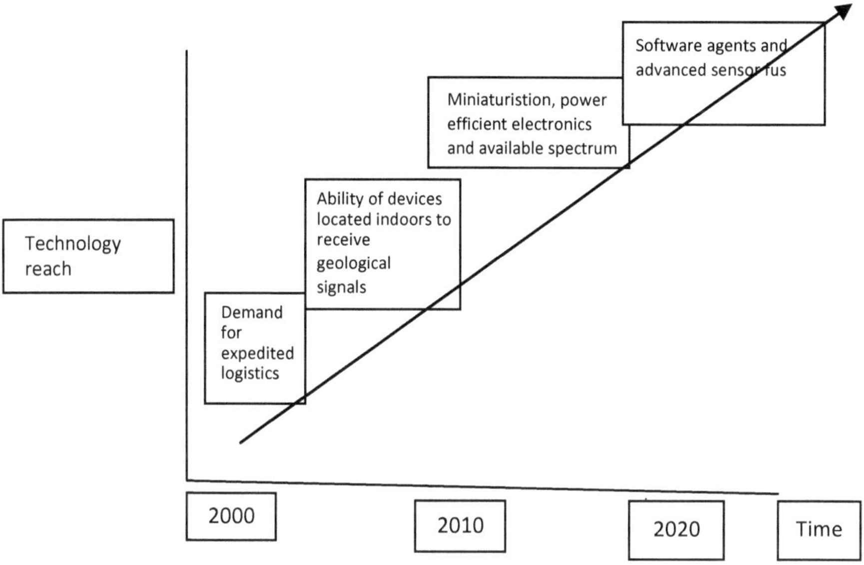

Fig. 7.3 The roadmap for Internet of Things. Source: adapted from SRI consulting business intelligence

7.4 Psychology, Technology, and Ethics

So, let us spend some time, thinking as a cognitive scientist would do. When technologists begin to use the power of the human brain in terms of a machine associating ideas at lightning speed and organizing it into a web form, it could well lead to the amplification of these ideas because of the interconnectedness between individuals. The description of convergent technologies and especially of the IoT shows that we are extremely near such interconnectedness. Such "bootstrapping," writes Engelhart (2000), could create an unending cycle of new information for the user. It was the thought of such a human–computer symbiosis in the form of a web of trails, that the term "knowledge workers" had been framed. Though, now outdated, this was a turning point in our conception of psychology of technology. A more mature view of this notion evolved in the form of a dynamic interface in which the user utilizes "the computer as a tool frames its appropriate form" (Glassman, 2012, p. 311). Thus, the core issue now is to link these pieces of information. Citing the example of a hand with five fingers, Bateson (1991) illustrated this point by emphasizing the crucial role that the four spaces between the fingers play, for it is with versatile use of these spaces that we are able to grasp and manipulate objects. Thus, in a scenario in which an interface with:

- The human being having the information
- The computer processing and managing such information to be given back to the individual
- Simple devices available to keep the level of efficiency

everything would become user-friendly. Unfortunately, this notion was lost with the arrival of personal computers. Developments in this new direction struggled with the frustration of users performing in wrong modes. One has only to browse on the internet to note that though considerable development has taken place in information technology during the past three decades, technologists have been obsessed with technology but have ignored the users. And, at the same time, when it comes down to the application and use of information technology, technologists often have no choice but to yield to business and market demands. Core, ethical questions, leading to psychological consequences of immense significance is rooted only in the later developments of any technology. One has only to imagine what could happen when information is received from individuals and managed by computers to communicate openly and freely, for example, say in the form of "Cognitive Commons" as stated by Dror and Harnad (2008). Will it not raise a plethora of ethical questions: is it safe to share information? What part of information do we need to share? Or, can we keep information neutral? Already, the question of net neutrality is being hotly debated in many parts of the world.

While many advantages have been pointed out for IoT and scientists and governments alike are ready to pump in both money and effort, it will also raise many a question regarding the ethicality of such interconnectedness, which will include among others, the invasion of privacy, challenges for security maintenance, harmful effects on children, environmental impacts in terms of the manufacture, use and disposal of the countless chip-based devices. With people concerned with issues such as "we will spy on you through your dishwasher" (Ackerman, 2012), "hackers remotely kill a jeep on the highway—with me in it" (Andy Greenberg, 2015) and "say goodbye to privacy" (Webb, 2015), the dangers that are lurking just around the corner are very real. Are we as human beings ready to handle the situation? IoT offers an immense opportunity to psychology of technology. It could come to the rescue, by helping both engineers and governments realize the implications of this technology and at the same time prepare the general public to use the technology in a meaningful manner.

In Chap. 6, we have pointed out how hedonomics is good economics. But is everything that a human being likes good for him? Do we not need to emphasize a need for making and using technology cautiously? As also mentioned in Chap. 6, we need to apply good chemistry, and one way in which it would be possible is if we start mimicking nature by using **Cradle to Cradle technology** (see Chap. 6 for more details).

Cradle-to-Cradle technology*: technology in which products are taken back by the manufacturer after their use is over*

Another ethical issue related to technology which seems to have even greater psychological significance is about our very existence and concerns questions such as "What does it mean to be human or what it is to be a human being?", as Nikolas Kompridis (2009) puts it. Technology, as we mentioned earlier, is likely to disturb our normative ways of life, the ways through which we inherited our culture, body, and mind. With the emergence of neuro-mediation technology, genetic engineering, and neurosurgery, the distinction between person and things, as philosopher Kant argued, might disappear or at the minimum, may raise ethical issues on, "having a body and being a body, between what is born and what is made, between organic and manufactured life" (Kompridis, p. 25). In other words, being at the disposal of technology would disturb our notion of "natality" and offer a challenge to our understanding of what it means to be human: with birth, is it one more life or is it a new life? The issue is best presented in the argument developed by Kompridis:

> "Just how essential to being human are our bodies? Enthusiasts of information technologies and biotechnologies like Kurzweil believe that these technologies will soon allow us to transcend just about all limitations imposed on us by the natural conditions of human embodiment. Whether this is achieved through genetic enhancement or by computerizing the body (e.g. by turning it into a wireless network), the success of this endeavor will confirm the belief that we can get along just fine without our bodies. Not just fine, in fact; better than fine" (p. 29).

The developments in the fields of biology, robotics, and nanotechnology are leading to genetic manipulation at a pace, so fast, that philosophers such as Jacques Derrida have begun to engage in apocalyptic thinking by asking if we would become unrecognizable as human beings. If our senses and cognition are parts of our awareness of our embodiment, it would constitute a sense of **intercorporeality** that is, being in control of our body. The argument is drawn on phenomenological lines on the basis of the notion of *having* a body or *being* a body (Table 7.1):

Given the ***having a body*** notion, we are heading in the direction of mixed creations—partly biological and partly mechanical. In this scenario, what would our self be like? What would a creation of this type look like, both mentally and physically? A cyborg? An anime? While we can design such a cyborg or anime to fit the existing nuances of a culture, the critical point is that we would be developing a notion of new technologies without clearly understanding its effects. In this regard, Rocco and Bainbridge's argument that given our lack of competence to foresee the impact of technology, it would be a smart move to make projections for a time frame of 10–20 years. Yes, it seems to be a pragmatic move to think of technology as an intervention, and it will help us to periodically evaluate our stake in "what it means to be a human being." In this way, the fiction or ghost of a counter-science will appear less imminent, at least for now. At the same time, it would not eliminate the psychological consequences resulting in the fear of unknown effects of technology and loss of freedom as cited in the surveillance example at the beginning of this chapter, and various other issues of human dignity. As we move away from our biological reality to a new era of artifacts dominated by technology, new questions regarding our survival will emerge to bring to the fore the distinction between good technology vs. bad technology, between good science vs. bad science, and issues of

Table 7.1 Differences between "having a body" and "being a body"

Having a body	Being a body
Lax self-instrumentation	Stiff self-instrumentation
Fusion permissible	Free will
Technologically determined	Naturally determined

normative existence, ethically perceived and practiced, individually or collectively. One example of the above is that of Artificial Intelligence.

Artificial Intelligence: *ability of machines to understand the environment and take actions to maximize its success*

One of the most amazing technologies of the last century is that of AI, that is **Artificial Intelligence**. Research on AI started in the 1950s but witnessed a drop in enthusiasm around the 1980s. Recent breakthroughs in computer sciences have once again brought it to its zenith with all major players in the industry from Google to Facebook to the Chinese giant, *Baidu* vying with each other to grab the major chunk of the AI pie. But what is AI? It has been defined as machine intelligence, or, the ability possessed by machines to take stock of its environment and take actions which will enable it to maximize its chances of success. From the early chess playing computer, we now have AI moving into a wide variety of spheres, ranging from education to industry. The future of AI looks bright, especially if one thinks of the countless ways in which it could resolve many of our problems related to say, nanotechnology and space technology (Kurzweil, 1999). The end result appears to be, as Kurzweil (2005) puts it, a final melding of human technology and human intelligence. According to a recent survey undertaken by Muller and Bostrom (2016), scientists have set a time line, with most respondents being of the view that the probability of AI reaching human ability around the decade 2040 is about 50 %, with about 90 % being of the view that we will probably have to wait till 2075.

While entrepreneurs are gung-ho about its significance, there are others who have voiced their concerns, including academicians such as Stephen Hawking and Nick Bostrom to people in the industry from Andrew Ng of *Baidu* to Luke Nosek of *Paypal* regarding the ways in which it could overpower the very minds that have created it. As Bostrom (2014) puts it, we will be able to conquer nature, but what about conflict and war? Will technology be able to automatically solve that?

In his book, *Superintelligence: Paths, Dangers, Strategies* (2014) Nick Bostrom has voiced concerns, many of which are existential in nature. According to him, we could be even heading towards the creation of a superintelligent machine. In other words, once we create a machine with general intelligence of the type we possess, it could generate, through a sequence of self-propelled iterations, higher and higher levels of intelligence with superintelligence as the end result. Are we ready for such an onslaught? With the field of machine ethics still in a very nascent stage, considerable thought is necessary, especially regarding machine morality and values and the future of AI could well depend on who controls it to whether we are able to control it at all (Hawking, Russell, Tegmark, & Wilczek, 2014).

Sensing the above issues in psychology, Gordon Allport, in his seminal writings at Harvard University during the 1960s (Allport, 1961), invited psychologists to understand the significance of the difference between the static status of a *human being* vs. studying the person as a dynamic *human becoming*. In doing this, he was trying to bolster arguments in favor of revamping the concept of personality in psychology. Remember, that this was the era in which psychoanalysis had already faded away from the scene of psychology. Behaviorism, Gestalt psychology, and cognitive psychology as initiated by the likes of Bruner and Miller, had revolutionized the scope of psychology by positing a highly mechanical and cognitive framework. However, though it was beginning to make its mark, it wasn't convincing enough. The Third force in psychology, as influenced by phenomenological approaches, appeared to be handy for those psychologists who wanted psychology to take its own course while studying the emerging experiences of individuals. The connectedness of human beings with machines opens a new vista of experiences in which technology is viewed as an extension of the self. In other words, the core issue in psychology remains the same, that is, the person. In the context of technology, the person becomes the user and more, given the direction of advancement in emerging technologies and its acceptance.

Among the four goals of psychology of technology, as stated in Chap. 1, the applicability goal invites a dialogue regarding the nature of practical applications of technology including its acceptance and desirability at the individual and community levels. As mentioned earlier, with the focus on labor cost cutting and usage of automated systems, we are placing demands on customers to learn technology even in situations in which they would not like to use it. Consider the range of problems and its desirability at the TSA security at airports where our bodies are searched and exposed through the X-ray machine. While no one would dispute the good intentions behind such security, in as much as it makes our travel safe, but it could also give rise to undesirable consequences, which at the minimum could be seen as an invasion of privacy. Therefore, the instrumental part of technology becomes suspect even though technology per se is claimed to be neutral.

Broadly speaking, the ethics of engineers involves the role of engineers as professionals in terms of their competence or skill and its demonstration, their ability to maintain confidentiality and responsibility and the implementation of quality control and safety. Whistle-blowing could also be added to this list. Those engineers who receive good training, undoubtedly, understand the ethics of their profession, but the psychological consequences of their professions place them in many unforeseen role conflicts. Such role conflicts get exacerbated, more often than not, when new technologies emerge. Let us examine the rationale underlying such an ethical dilemma. When scientists and engineers announce that they know how a product works, it is said to be the **descriptive base** of knowledge. On the other hand, when they prescribe a set of actions to a user to achieve his/her goal, they set up a **prescriptive form** of knowledge. A lot of water flows between the transitions from the descriptive knowledge base to the prescriptive set of actions (van Gorp, 2007; Vincenti, 1990). What would Albert Einstein be thinking after the mass destruction of life and material caused by the nuclear bomb? His knowledge base made him a towering intellectual of the previous century, but bad judgment caused the prescriptive set of actions to change its course.

> ***Descriptive knowledge base****: knowledge about how a product works*

> ***Prescriptive knowledge base****: telling the user how to operate the product*

Zwart, van de Poel, van Mil, and Brumsen (2006) contended that technological growth in our time is mostly contained in networks and not in the individuals. As such it raises specific issues in the context of the normative, prevalent ethical base. Further, if these specific issues are not addressed to in a timely manner, it might lead to the intensification of the existing ethical base. For example, with converging technologies, particularly in the context of nanotechnology known as nanoethics, we will require a fresh ethical debate to answer questions in several fields, for example, biotechnology, genetic engineering, and web-based cognitive enhancement. In the history of science, developments were never as fast as what we are seeing today. In the days of yore, we had enough time to think over and debate about the impact of the application of knowledge. Therefore, when a description was offered by a scientist and accepted by others without any challenge, it remained **epistemic** in nature. However, it becomes **nonepistemic** when existing cultural and social beliefs regulate it. According to Prilleltensky (1989), whether we use the epistemic or the non-epistemic approach, it is impossible to escape the values that influence our scientific approach and in the present context, it refers to behavior as affected by technology. As had been suggested by Kipnis (1994), this approach has one important value: if ethical issues are ignored by psychologists, the discipline would lose its significance. It is particularly important in the case of technology, for if non-epistemic values are adopted, it will be the influential, the rich and the powerful, who are generally in control of technology, who will start controlling the lives of ordinary people (Kool & Agrawal, 2006).

> ***Epistemic**: aimed at understanding how and what the product is in reality*

> ***Nonepistemic**: how society and culture perceives the product*

Prilleltensky has further argued that it is impossible for psychologists to merely describe a phenomenon in terms of what it appears to be, without at the same time subtly prescribing what we regard as desirable (p. 797). This issue becomes even more salient in the case of psychology of technology: should we limit ourselves to basic research and/or should we think about its applications, and/or should we examine its applicability even if it is applicable, and/or should we simply apply it directly (for details, see Chap. 1). Given the nature and scope of psychology of technology—technology being applied science for all practical purposes—we move towards the end of this chapter and this volume with an invitation to psychologists to become activists and not be sitting on the fence, simply as scientists, who do not want to take sides on social or moral issues. In short, psychology of technology, by its very nature, offers a platform where science and technology of behavior can appear more unified, making it a relatively stronger science of human behavior. The opportunity is there during this millennium, especially in the light of the fact that there is no better science than when we demonstrate the elements of a theory and its practice moving along in unison. Sensing this dilemma, Sherry Turkle (2004) remarked in her article, *How computers change the way we think:*

> "In the decades ahead, we need to rebuild the culture around informational technology. In that new sociotechnical culture, assumptions about the nature of mastery would be less absolute. The new culture would make it easier, not difficult to consider life in shades of gray, to see moral dilemmas in terms other than a battle of Good and Evil. For never has our world been more complex, hybridized, and global. Never have we so needed to have many contradictory thoughts and feelings at the same time. Our tools must help us accomplish that, not fight against us" (p. B26).

> ***Conscientization:*** *eliminating myths that cause confusion about a phenomenon*

We now focus on two specific issues related to the ethics of psychology of technology. The first is **conscientization** which involves eliminating myths that cause confusion about a phenomenon. In terms of technology, there were, are and always will be such myths. For example, when automobiles arrived, there were very few women drivers and if an accident occurred, a female driver was subjected to a strong attributional bias of personal disposition, in that, the accident was said to be caused by her lack of skills. Let us take a hypothetical example. So, if there were 100 male and 10 female drivers with accidents rates being 10 and 1 respectively, there was, in reality, no difference in terms of proportion of accidents between them, and yet owing to a false illusory correlation, women were being perceived as poor drivers. Compare such a line of biased thinking to what we see today. You will find that insurance companies charge lower rates of premium for female drivers. Not surprisingly, many families mention females as primary drivers to take advantage of these lower rates of insurance. As the number and variety of products flood the market during this century, there will a high probability that the intensity and frequency of such illusory correlations will increase, especially when these products are used by people not included in the normal sense of its use, for example, by those who are physically or mentally challenged or by older people. The social psychology of technology will, in its focus on conscientization, tend to bust such myths and mitigate confusions.

> ***Annunciation:*** *replacing the current system and developing an alternate arrangement*

> ***Battered women syndrome:*** *learned helplessness associated with tolerance of prolonged abuse by spouse*

The second ethical issue specific to psychology of technology deals with **annunciation**. It means replacing the current system and developing an alternate arrangement. Whereas conscientization is aimed at eliminating myths, annunciation goes a step beyond in the sense that an effort is generated to replace an existing argument. Should the **battered women syndrome** exist (learned helplessness amounting to tolerance of prolonged abuse by spouse), and, is it enough to merely explore and describe it? Is it not an obligation on the part of scientists to offer opportunities to raise the level of self-esteem of the abused person? New technologies will continuously batter unwilling learners, older people, and almost all those who are mentally and physically challenged. As technology is advancing, it has, without doubt, equipped us with many more choices, but at the same time it has increased our indecisiveness, and as James Surowiecki (2005) puts it, makes us "less happy in its relentless generation of newness" (p. 4). Quoting Woody Allen's joke that "the food may not be so great, but we want the portions to be as big as possible," Surowiecki draws a parallel between our craving for food and our craving for technology. We want more technology without really having a clear notion about its use. It is a complicated proposition which will continue to feature in our behavior.

Let us examine annunciation further. While no one will doubt that technology has helped industrial productivity by substituting manual work, it has also set limits to the autonomy of workers through increased surveillance and control (both possible only because of technology). A positive effect of technology is that it tends to make the workplace less formal yet more efficient and productive. Thus, while being in the annunciation mode, we may replace an existing arrangement, but the consequences of a new order initiated by technology must be examined proactively, not retrospectively. Therefore, Rocco and Bainbridge's suggestion of taking stock of technology in a 10–20 year time frame seems to be a reasonable idea.

In the section on ethics in Chap. 1, we raised concerns regarding operational psychology, that is, about the role of psychologists who are asked to team up with fellow engineers and others in the development of products and the degree of freedom afforded to them to judge the impact of the product in an independent fashion. Will they be perceived as adversarial or supportive when they give their independent report? Take, for instance, the ways in which law enforcement personnel view the role of psychologists in cases of an insanity plea or in determining the state of intentions of the suspect. Depending on whether the psychologist is able to follow in the footsteps of the prosecutor, her role is judged as being supportive or adversarial. As discussed in Chap. 1, even after its inception by the Office of Strategic Services, the significance of operational psychology did not go very far in mainstream psychology.

The many facets of ethics in psychological research cannot remain confined merely to the participating subject in the psychology laboratory (or in the case of person-machine in the context of psychology of technology). Rather, they are of vital significance to the threats caused to the society at large (Pittinsky & Diamante, 2015). The relevant and specific question before us is: what would be the role of psychologists in the twenty-first century when technology is used as an intervention, say, for teaching and treatment of physical and mental conditions, for our security or for any other significant aspect of our life? The chances are that because of the technological divide and the varying use and impact of technology across the world, the psychologist may become relegated to that of a mere bystander (Pittinsky & Diamante, 2015). Will we, as psychologists, be satisfied with such a role? Or, are we ready to take on technology head-on, working proactively with engineers and technology designers, working hand-in-glove with them and thereby helping to create a better, more sustainable world?

7.5 Technology, Psychology, and Culture

Woody Allen's joke on small portions of food mentioned above would most likely, not appeal to Amish people who prefer to live a very frugal life and avoid automobiles, computers, and other things considered essential in the twenty-first century. The Amish simply avoid technology, especially in its modern form. One advantage of this is that they are less perturbed by the choices that technology offers. Could this be the reason for why their happiness levels have been registered to be at a level similar to those found among the richest Forbes 400 people. From the history of technology we learn that while appreciation of technology has been considered the cornerstone for the development of a society, for example, as reflected in Samuel Butler's (1872) book, *Erewhon*, religions around the globe have been preaching the value of simplicity and stoicism.

The above goes to show that people in different cultures have differing views regarding the significance of technology. Some years back, Weil and Rosen (1994) conducted a study covering 38 universities in 23 countries. They were interested in measuring technological sophistication, measured in several ways, for example, the use of consumer technology, university computing systems, and computer ownership. The results of this study, albeit old, are of great significance, because, that was the time when computers were just beginning to show their impact. The sample from Japan strongly supported technology, although, at the same time, they preferred to work in their conventional ways. Both Israel and Singapore expressed a very positive view of computers. The general conclusion based on their study is likely to hold good even today, especially for cultures that have responded slowly to the computer revolution:

> "These findings suggest that the strongest criteria for students with no technophobia or low levels of technophobia include a culture that: (a) values technology, (b) integrates technology early in the formal educational system, and (c) has supportive political climate" (Weil & Rosen, 1994, p. 128).

For sure, technology has helped the flow of communication between people of different cultures. Internet and Skype have afforded new opportunities for growth of the formation of social groups and even communities, as we see on Facebook and Twitter and Whatsapp, or even in the world of virtual gaming. As mentioned in Chap. 6, users in individualistic and collectivist cultures differ in their preferences for technology, ranging from preferences for color to the variety of options that they want. While in Japan, robots have been employed to offer the human touch in the hospital setting, signaling that its collectivist culture is ready to move away from the mainstream of shared selves, such social robots would probably not be easily acceptable to many other collectivist cultures where substituting normal human interaction with technology-mediated interaction would mean the collapse of core human values. According to Rauterberg (2006), traditional, personal, and cooperative computing would have different implications for users in individual and collectivist cultures. Saariluoma and Oulasvirta (2010) write

> "Scientists working with human mind and society have naturally known for a long time that simple intuitions and lay science do not provide technologists with an accurate understanding of people. This is why psychology, sociology, anthropology, and other human sciences have developed sophisticated observations, concepts, and theories that make it easier for specialists to better understand the human being" (p. 325).

Even a decade back, while planning a vacation, would you have ever thought of seeing what your friends and acquaintances have liked? Today, this has become common place. We search for reviews not only from people we know but from even strangers. This seems to be the twenty-first century way of keeping up with the Joneses. Some data of how technology, and especially social media, is affecting vacation buying is presented in Box 7.6.

Box 7.6: Contagious Effects of Technology: How Technology Destroys Your Savings

According to a survey of more than 1000 Americans released this summer (2016) by the American Institute of Certified Public Accountants, roughly four in ten adults with a social media account (39 %) say that seeing other people's purchases and vacations on social media makes them look into a similar purchase or vacation. What's more, 11 % have taken a vacation or made a purchase in the last year after seeing someone's post about their vacation or purchase.

Moreover, 30 % of Americans say that social media has some influence on their purchasing decisions, with 5 % saying that it has a significant impact, as a 2014 Gallup poll found. Among millennials the numbers are even higher with roughly half saying that social media influences what they buy.

Some social media-related spending, of course, is driven by the fact that many brands advertise their goods and services on social media or pay celebrities and other influential people to post about them. By 2017, social network advertising spending is expected to hit nearly $36 billion, or roughly 16 % of all digital ads spending globally, up from about $24 billion in 2015,

Source: Hill (2016)

Looking at the significance of psychology of technology in the historical context, our final thought is rooted in the relationship between culture change and technology. The role of technology in human life has not always remained the same. With the passage of time, as cultures around the world underwent a variety of changes so did the importance of technology. For example, after the fall of the Roman Empire, growth in science, by and large, decelerated, a trend which has been observed in many other societies, too, in the course of history. Thereafter, with the Renaissance, intellectual horizons expanded once again, widening holistically with sciences, technologies, and humanities springing forward together. An intellectual of this era was a handy man of scholarship who could engage himself or herself in multiple tasks ranging from discovery to invention, being a carpenter to a writer, a scientist to a cleric, or even a combination of all. Over the years, human knowledge expanded and the scenario changed giving rise to what we now know as specialization of disciplines—each with their own principles and theories (Rocco & Bainbridge, 2002). No longer was the intellectual a handy man. Rather, each intellectual or scientist was isolated in his own ivory tower deeply involved in his narrow area of interest and knowledge. Technologies were developed in many parts of the world (some examples of which have been cited in earlier chapters), but since the cultures of the world remained loosely connected or some with only name sake contact, the gains of technology bore uneven effects around the globe. Today, however, vast cultural changes have taken place. With the explosive growth of information and communication technology, it has now become possible for cultures around the world to come together on a common platform, to interact and collaborate, and, as a result, to be ready to evaluate, from their own perspective, the impact of technology—it's good or bad aspects—with or without its own "web."

With the ways in which events are unfolding around the globe, we can probably expand our ability to envision the movement of human civilization, at the best, in a 10–20 years time frame (as suggested in the NSF report mentioned earlier), before the phenomenal growth of technology becomes elusive to the very mind that gave birth to it—and long before psychology would be able to claim its status as a mother.

Chapter Summary

Continuing from Chap. 6, this chapter starts out by pointing out favorable and unfavorable aspects of behavior in the context of interventional technology and the impact of the greater availability of free-market type of products. The focus is on, "where is the user in user technology?" and shows how psychology of technology will help us to understand the user as an actor, not merely as a factor. In short, the growth of psychology during the twenty-first century will continue to focus on two viewpoints: the Engineer's Psychology of Technology (EPT) and the Humanists' Psychology of Technology (HPT). **Second**, the growth of this subfield will become increasingly interdisciplinary with the advent of new techniques of research, for example, convergent technologies comprising inputs from nanotechnology, biotechnology, information technology, and cognitive science. **Third**, several social psychological explanations of behavior will need to be adapted in the context of technology, for example, how our attributions will change—say, attributing failures to the machine or its operators, treating robots and humans alike, blurring the boundaries between what is technologically determined vs. what is naturally determined, and more. Obviously, such issues will raise questions regarding "What does it mean to be human?", as Nikolas Kompridis put it. Undoubtedly, such ethical issues will continue to dominate the relationship between technology and human behavior in this millennium. The applicability issue, as discussed in Chap. 1, will, thus, often set limits to the role of technology. Overall, predicting behavior in the context of technology will become a challenge for the psychological sciences, with technology making it easier on the one hand but more complex on the other.

Suggestions for Further Reading

Kurzweil, R. (1999).*The age of spiritual machines: When computers exceed human intelligence.* New York: Penguin Books.

Roco, M. C., & Bainbridge, W. S. (2002).*Converging technologies for improving human performance: Nanotechnology, biotechnology, information technology and cognitive science.* NSF/DOC-sponsored report, Arlington, Virginia.

Salvendi, G. (2012). *Handbook of human factors and ergonomics* (4th ed.). New York: Wiley.

References

Abbott, L. F., & Nelson, S. B. (2000). Synaptic plasticity: Taming the beast. *Nature Neuroscience, 3*, 1178–1183.
Acampora, G., & Vitiello, A. (2013). Interoperable neuro-fuzzy services for emotion-aware ambient intelligence. *Neurocomputing, 122*, 3–12.
Achterhuis, H. (Ed.). (2001). *American philosophy of technology*. Bloomington, IL: Indiana University Press.
Ackerman, S. (2012, June 26). CIA Chief: We'll spy on you through your dishwasher. *Wired.*
Adams, J. A. (1971). A closed-loop theory of motor learning. *Journal of Motor Behavior, 3*, 111–149.
Adams, D. K., Sewell, M. A., Angerer, R. C., & Angerer, L. M. (2011). Rapid adaptation to food availability by a dopamine-mediated morphogenetic response. *Nature Communication, 2*, 592. doi:10.1038/ncomms1603.
Advisory Group on Non-Ionising Radiation (AGNIR). (2010). *Health effects of exposure to ultrasound and infrasound.* Report of the independent Advisory Group on Non-Ionizing Radiation, Health Protection Agency, UK.
AFP, Isumi. (2015, February 25). *In Japan, robot dogs are for life—And death.* Retrieved from www.ucanews.com.
Agarwal, R., & Prasad, J. (1998). A conceptual and operational definition of personal innovativeness in the domain of information technology. *Information Systems Research, 9*, 204–215.
Agrawal, R. (2001). *Stress in life and at work.* New Delhi, India: Sage.
Ahissar, M., & Hochstein, S. (2004). The reverse hierarchy theory of visual perceptual learning. *Trends in Cognitive Sciences, 8*, 457–464.
Alemseged, Z., Spoor, F., Kimbel, W. H., Bobe, R., Geerards, D., Reed, D., et al. (2006). A juvenile early hominin skeleton from Dikika, Ethiopia. *Nature, 443*, 296–301.
Allen Mouse Brain Connectivity Atlas. (2011). Retrieved from http://connectivity.brain-map.org.
Allford, J. (2015, Fall-Winter). Brain and mental health. *UCalgary Alumni Magazine.*
Allport, G. W. (1961). *Pattern and growth in personality*. New York: Holt, Rinehart and Winston.
Alsoufi, A., & Ali, H. (2014). Customers' perception of m-banking adoption in kingdom of Bahrain: An empirical assessment of an extended TAM model. *International Journal of Managing Information Technology, 6*, 547–550.
American Institute of Physics. (2013). https://www.aip.org/history/exhibits/laser/interviews/franken_laserexcited_interview.html.
American Psychological Associtaion. (2009). Obedience—Then and now. *American Psychologist, 64*(1), 1–11.

V.K. Kool, R. Agrawal, *Psychology of Technology*,
DOI 10.1007/978-3-319-45333-0

Ananova. (2003). *Young people could not live without their mobiles*. Retrieved from http://www.ananova.com/news/story/sm_779110.html?menuZ.

Anderson, C. A., & Dill, K. E. (2000). Video games and aggressive thoughts, feelings, and behavior in the laboratory and in life. *Journal of Personality and Social Psychology, 78*, 772–790.

Anderson, C. A., Dill, K. E. (2012, February 6). Why cognitive enhancement is in your future (and your past). *The Atlantic*.

Annett, J. (1995). Motor imagery: Perception or action? *Neuropsychologia, 33*, 1395–1417.

Annetta, L. (2010). The "I's" have it: A framework for serious educational game design. *Review of General Psychology, 14*, 105–112.

Arhippainen, L., & Tähti, M. (2003). Empirical evaluation of user experience in two adaptive mobile application prototypes. In *Proceedings of the 2nd International Conference on Mobile and Ubiquitous Multimedia (MUM)* (pp. 27–34). New York: ACM Press.

Arias-Carrion, O., & Poppel, E. (2007). Dopamine, learning, and reward-seeking behavior. *Acta Neurobiologiae Experimentalis, 2007*(67), 481–488.

Arthur, W. B. (2009). *The nature of technology*. New York: Free Press.

Ashbrook, D. L., Clawson, J. R., Lyons, K., Starner, T. E., & Patel, N. (2008). Quickdraw: The impact of mobility and on-body placement on device access time. In *Proceedings of CHI 2008* (pp. 219–222). New York: ACM Press.

Aslin, R. (2012). Infant eyes: A window on cognitive development. *Infancy, 1*, 126–140.

Atkinson, R. C., & Shiffrin, R. M. (1968). Human memory: A proposed system and its control processes. In K. W. Spence & J. T. Spence (Eds.), *The psychology of learning and motivation* (Vol. 2, pp. 89–195). New York: Academic.

Ayyagari, R. (2006). Examination of hedonism in TAM research. In *Proceedings of SAIS*, Paper 35.

Baddeley, A. D. (1966). Influence of depth on the manual dexterity of free divers: A comparison between open sea and pressure chamber testing. *Journal of Applied Psychology, 50*, 81–85.

Baddeley, A. D. (2000). The episodic buffer: A new component of working memory? *Trends in Cognitive Science, 4*, 417–423.

Baddeley, A. D. (2003). Working memory: Looking back and looking forward. *Nature Reviews. Neuroscience, 4*, 829–839.

Baddeley, A. D., & Hitch, G. (1974). Working memory. In G. H. Bower (Ed.), *The psychology of learning and motivation* (Vol. 8, pp. 47–89). New York: Academic.

Bailly, G., Müller, J., Rohs, M., Wigdor, D., & Kratz, S. (2012). ShoeSense: A new perspective on gestural interaction and wearable applicationsx. In *Proceedings of CHI 2012* (pp. 219–222). New York: ACM Press.

Bain, R. (1937). Technology and state government. *American Sociological Review, 2*, 860–874.

Baker, R. S. J. D., Gowda, S. M., Wixon, M., Kalka, J., Wagner, A. Z., Salvi, A., et al. (2012). Towards sensor-free affect detection in cognitive tutor algebra. In K. Yacef, O. Zaïane, H. Hershkovitz, M. Yudelson, & J. Stamper (Eds.), *Proceedings of the 5th International Conference on Educational Data Mining* (pp. 126–133). Retrieved from http://educationaldatamining.org/EDM2012/uploads/procs/Full_Papers/edm2012_full_1.pdf.

Baldwin, C. L. (2012). *Auditory cognition and human performance: Research and applications*. Boca Raton, FL: CRC.

Bandura, A. (1977). *Social learning theory*. Englewood Cliffs, NJ: Prentice Hall.

Bandura, A. (1986). *Social foundations of thought and action: A social cognitive theory*. Englewood Cliffs, NJ: Prentice-Hall.

Bandura, A. (1997). *Self-efficacy: The exercise of control*. Severn, MD: Freeman.

Bandura, A., Ross, D., & Ross, S. A. (1961). Transmission of aggression through imitation of aggressive models. *Journal of Abnormal Social Psychology, 63*, 575–582.

Barron, H. C., Dolan, R. J., & Behrens, T. E. (2013). Online evaluation of novel choices by simultaneous representation of multiple memories. *Nature Neuroscience, 16*, 1492–1498.

Bateson, G. (1991). *A sacred unity: Further steps to an ecology of mind*. New York: Cornelia and Michael Bessie, Harper Collins.

Batson, C. D., Batson, J. G., Todd, R. M., Brummett, B. H., Shaw, L. L., & Aldeguer, C. M. R. (1995). Empathy and the collective good: Caring for one of the others in a social dilemma. *Journal of Personality and Social Psychology, 68*, 619–631.

Battele, J. (2016, June 29). *Is AI the worst mistake in human history?* Retrieved from https://shift.newco.co/are-we-smart-enough-to-control-ai-eb23565f8a4#.db7izk2in.

Baumeister, R. F. (1984). Choking under pressure: Self-consciousness and paradoxical effects of incentives on skillful performance. *Journal of Personality and Social Psychology, 46*, 610–620.

Baylor, A. L. (2007). Pedagogical agents as a social interface. *Educational Technology, 47*, 11–14.

Beamish, D., Bhatti, S., Mackenzie, I. S., & Wu, J. (2009). Estimation of psychomotor delay from the Fitts' law coefficients. *Biological Cybernetics, 101*, 279–296.

Bee, M., & Micheyl, C. (2008). The cocktail party problem: What is it? How can it be solved? And why should animal behaviorists study it? *Journal of Comparative Psychology, 122*, 235–251.

Belleville, S. (2008). Cognitive training for persons with mild cognitive impairment. *International Psychogeriatrics, 20*, 57–66.

Belman, O. (2015, April 30). *'Skip' Rizzo honored for advances in virtual reality therapy*. Retrieved from https://news.usc.edu.

Bemelmans, R., Gelderblom, G. J., Jonker, P., & de Witte, L. (2012). Socially assistive robots in elderly care: A systematic review into effects and effectiveness. *Journal of the American Medical Directors Association, 13*, 114–120.

Benninger, D. H., Lomarev, M., Lopez, G., Wassermann, E. M., Li, X., Considine, E., et al. (2010). Transcranial direct current stimulation for the treatment of Parkinson's disease. *Journal of Neurology and Neurosurgical Psychiatry, 81*, 1105–1111.

Berkowitz, L. (1993). *Aggression: Its causes, consequences, and control*. New York: McGraw-Hill.

Berkun, S. (2000). *Fitts's law applied to the web*. Retrieved from www.scottberkun.com.

Bertucco, M., & Cesari, P. (2010). Does movement planning follow Fitts' law? *Neuroscience, 171*, 205–213.

Besecker, B. (2015). *5 top Virtual Reality & augmented reality trends for 2015*. Reteieved January 23, from www.marxentlabs.com.

Bethge, P., & Grolle, J. (2015, October 2). Jane Goodall interview: Even chimps understand sustainability. Interview conducted for *Spiegel* Online International.

Bhatt, R. S., & Quinn, P. C. (2011). How does learning impact development in infancy? The case of perceptual organization. *Infancy, 16*, 2–38.

Bier, N., Provencher, V., Gagnon, L., Van der Linden, M., Adam, S., & Desrosiers, J. (2008). New learning in dementia: Transfer and spontaneous use of learning in everyday life functioning. Two case studies. *Neuropsychological Rehabilitation, 18*, 204–235.

Billies, M. (2015). Surveillance threat as embodied psychological dilemma. *Peace and Conflict, 21*, 168–186.

Bispo, R., & Branco, R. (2008). Designing out stigma: The role of objects in the construction of disabled people's identity. In *Proceedings of the International Design & Emotion Conference.*

Bjorklund, D. F., & Pellegrini, A. D. (2010). Evolutionary perspectives on social development. In P. K. Smith & C. H. Hart (Eds.), *The Wiley-Blackwell handbook of childhood social development* (pp. 64–81). Oxford, England: Wiley-Blackwell.

Blascovich, J., & McCall, C. (2010). Attitudes in virtual reality. In J. P. Forgas, J. Cooper, & W. D. Crano (Eds.), *The psychology of attitudes and attitude change* (pp. 283–297). New York: Psychology Press.

Bläsing, B., Coogan, J., Biondi, J., Simmel, L., & Schack. T. (2014). Motor learning in dance using different modalities: Visual vs. verbal models. *Cognitive Processing*. Retrieved from http://dx.doi.org/10.1007/s10339-014-0632-2.

Bläsing, B., Tenenbaum, G., & Schack, T. (2009). The cognitive structure of movements in classical dance. *Psychology of Sport and Exercise, 10*, 350–360. doi:10.1016/j.psychsport.2008.10.001.

Bock, O., Schott, N., & Papaxanthis, C. (2015). Motor imagery: Lessons learned in movement science might be applicable for spaceflight. *Frontiers in Systems Neuroscience, 9*, 75.

Bohannon, J. (2015a, January 12). Your computer knows you better than your friends do. *Science AAAS*. Retrieved from www.sciencemag.org.

Bohannon, J. (2015b). The synthetic therapist. *Science, 349*, 250–251.

Bohannon, J. (2015c). Fears of an AI pioneer. *Science, 349*, 252.

Bombari, D., Schmid-Mast, M., Cañadas, E., & Bachmann, M. (2015). Studying social interactions through immersive virtual environment technology: Virtues, pitfalls, and future challenges. *Frontiers in Psychology, 6*, 1–11.

Bor, D. (2012). *The ravenous brain: How the new science of consciousness explains our insatiable search for meaning*. New York: Basic Books.

Bor, D., Duncan, J., Wiseman, R. J., & Owen, A. M. (2003). Encoding strategies dissociate prefrontal activity from working memory demand. *Neuron, 37*, 361–367.

Bor, D., & Owen, A. M. (2007). A common prefrontal-parietal network for mnemonic and mathematical recoding strategies within working memory. *Cerebral Cortex, 17*, 778–786.

Bor, D., & Seth, A. (2012). Consciousness and the prefrontal parietal network: Insights from attention, memory and chunking. *Frontiers in Psychology, 3*, 63.

Borsci, S., Kuljis, J., Barnett, J., & Pecchia, L. (2016). Beyond the user preference: Aligning the prototype design to the users' expectations in manufacturing & service industry. *Human Factors and Ergonomics, 26*, 16–39.

Bostrom, N. (2014). *Superintelligence: Paths, dangers, strategies*. New York: Oxford University Press.

Bostrom, N., & Roache, R. (2008). Ethical issues in human enhancement. In J. Ryberg, T. Petersen, & C. Wolf (Eds.), *New waves in applied ethics* (pp. 120–152). Basingstoke, UK: Palgrave Macmillan.

Botella, C., Garcia-Palacios, A., Banos, R. M., & Quero, S. (2009). Cybertherapy: Advantages, limitations, and ethical issues. *PsychNology Journal, 7*, 77–100.

Botti, S., & Iyengar, S. S. (2004). The psychological pleasure and pain of choosing: When people prefer choosing at the cost of subsequent outcome satisfaction. *Journal of Personality and Social Psychology, 87*, 312–326.

Boyd, D. (2006, February). *Identity production in a networked culture: Why youth heart MySpace*. Paper presented at the Annual Conference of the American Association for the Advancement of Science, St Louis, MO.

Boyd, L. A., & Linsdell, M. A. (2009). Excitatory repetitive transcranial magnetic stimulation to left dorsal premotor cortex enhances motor consolidation of new skills. *BMC Neuroscience, 10*, 72.

Boyd, L. A., & Winstein, C. J. (2004). Providing explicit information disrupts implicit motor learning after basal ganglia stroke. *Learning & Memory, 11*, 388–396.

Boyle, J. B., Kennedy, D. M., Wang, C., & Shea, C. H. (2014). The sine wave protocol: Decrease movement time without increasing errors. *Journal of Motor Behavior, 46*, 277–285.

Boyle, J. B., Kennedy, D. M., & Shea, C. H. (2012). Optimizing the control of high ID single degree of freedom movements: Re-thinking the obvious. *Experimental Brain Research, 223*, 377–387.

Boyle, J. B., Kennedy, D. M., & Shea, C. H. (2015). A novel approach to enhancing limb control in older adults. *Experimental Brain Research, 233*, 2061–2071.

Brackens, L. (2011). *Optimizing participation slows physical decline in seniors with mobility disabilities*. Retrieved from http//:mobilitymanagement.com.

Brandon, J. (2015, June 4). *Six amazing uses for virtual reality in business*. Retrieved from www.computerworld.com.

Braun, S. M., Beurskens, A. J., Schack, T., Marcellis, R. G., Oti, K. C., Scholos, J. M., et al. (2007). Is it possible to use the Structural Dimension Analysis of Motor Memory (SDA-M) to investigate representations of motor actions in stroke patients? *Clinical Rehabilitation, 21*, 822–832.

Braungart, M., & McDonough, W. (2002). *Cradle to cradle: Remaking the way we make things*. New York: North Point Press.

Braungart, M., & McDonough, W. (2013). *The upcycle: Beyond sustainability-designing for abundance*. New York: North Point Press.

Brembs, B., Lorenzetti, F. D., Reyes, F. D., Baxter, D. A., & Byrne, J. H. (2002). Operant reward learning in aplysia: Neuronal correlates and mechanisms. *Science, 296*, 1706–1709.

Brewster, S., Lumsden, J., Bell, M., Hall, M., & Tasker, S. (2003). Multimodal 'eyes-free' interaction techniques for wearable devices. In *Proceedings of CHI, 2003* (pp. 473–480). New York: ACM Press.

Broadbent, D. E. (1958). *Perception and communication*. Elmsford, NY: Pergamon Press.

Brooks, R. (2001). The relationship between matter and life. *Nature, 409*, 409–411.

Brown, M., Pope, N., & Voges, K. (2003). Buying or browsing? An exploration of shopping orientations and online purchase intention. *European Journal of Marketing, 37*, 1666–1684.

Brown, N. J., Read, D., & Summers, B. (2003). The lure of choice. *Journal of Behavioral Decision Making, 16*, 297–308.

Bruner, J. (1956). *A study of thinking*. Hoboken, NJ: Wiley.

Bruner, J. (1974). *Beyond the information given*. London: George Allen & Unwin.

Bruya, B. (2010). *Effortless attention*. Cambridge, MA: MIT Press.

Bryant, J. A., Sanders-Jackson, A., & Smallwood, A. M. K. (2006). IMing, text messaging and adolescent social networks. *Journal of Computer-Mediated Communications, 11*, 577–592.

Buchanan, A. (2011). Cognitive enhancement and education. *Theory and Research in Education, 9*, 145–162.

Buchanan, J. J., Park, J., & Shea, C. H. (2006). Target width scaling in a repetitive aiming task: Switching between cyclical and discrete units of action. *Experimental Brain Research, 175*, 710–725.

Budman, S. H. (2000). Behavioral health care dot-com and beyond: Computer mediated communications in mental health and substance abuse treatment. *American Psychologist, 55*, 1290–1300.

Burg, S., & Gorp, A. (2005). Understanding moral responsibility in the design of trailers. *Science and Engineering Ethics, 11*, 235–256.

Burkolter, D., & Kluge, A. (2011). Online consumer behavior and its relationship with socio-demographics, shopping orientations, need for emotion, and fashion leadership. *Journal of Business and Media Psychology, 2*, 20–28.

Bushman, B. J., & Anderson, C. A. (2002). Violent video games and hostile expectations: A test of the general aggression model. *Personality and Social Psychology Bulletin, 28*, 1679–1686.

Buss, D. M., Haselton, M. G., Shackelford, T. K., Bleske, A. L., & Wakefield, J. C. (1998). Adaptations, exaptations, and spandrels. *American Psychologist, 53*, 533–548.

Butler, S. (1872). *Erewhon*. Mineola, NY: Dover Thrift Edition.

Buzzi, M. C., Buzzi, M., Leporini, B., & Trujillo, A. (2015). Design of web-based tools to study blind people's touch based interaction with smartphones. In C. Stephanidis (Ed.), *HCII 2015 Posters, Part I*, CCIS 528 (pp. 7–12).

Cadwalla, C. (2014, February 22). Are the robots about to rise? Google's new director of engineering thinks so. *The Guardian*. Retrieved from www.theguardian.com.

Caeyenberghs, K., Wilson, P. H., van Roon, D., Swinnen, S. P., & Smits-Engelsman, B. C. (2009). Increasing convergence between imagined and executed movement across development: Evidence for the emergence of movement representations. *Developmental Science, 12*, 474–483.

Caflin, E. S., Krishnan, C., & Khot, S. P. (2015). Emerging treatments for motor rehabilitation after stroke. *Neurohospitalist, 5*, 77–88.

Cain, M. S., Landau, A. N., & Shimamura, A. P. (2012). Action video game experience reduces the cost of switching tasks. *Perception & Psychophysics, 74*, 641–647.

Campbell, J. (2009). Bayesian methods and universal Darwinism. *American Institute of Physics Conference Proceedings, 1193*, 40–47. doi:10.1063/1.3275642.

Caplovitz, G. P., & Kastner, S. (2009). Carrot sticks or joysticks: Video games improve vision. *Nature Neuroscience, 12*, 527–528.

Carlbring, P., & Andersson, G. (2006). Internet and psychological treatment. How well can they be combined? *Computers in Human Behavior, 22*, 545–553.

Carr, N. (2008, July/August). Is Google making us stupid? What the internet is doing to our brains. Retrieved from www.theatlantic.com.

Carrier, M. L., Cheever, N. A., Rosen, L. D., Benitez, S., & Chang, J. (2009). Multitasking across generations: Multitasking choices and difficulty ratings in three generations of Americans. *Computers in Human Behavior, 25*, 483–489.

Carrington, P., Hurst, A., & Kane, S. K. (2014). Wearables and chairables: Inclusive design of mobile input and output techniques for power wheelchair users. In *CHI '14 Proceedings of the SIGCHI Conference on Human Factors in Computing Systems* (pp. 3103–3112). New York: ACM Press.

Carson, E. (2015, March 10). *Nine industries using Virtual Reality*. Retrieved from www.techrepublic.com.

Cattani, G. (2008). Reply to Dew (2007)'s commentary: Pre-adoption, exaptation and technology speciation: A comment on Cattani (2006). *Industrial and Corporate Change, 17*, 585–596.

Cattell, R. B. (1963). Theory of fluid and crystallized intelligence: A critical experiment. *Journal of Educational Psychology, 54*, 1–22.

Cerritelli, B., Maruff, P., Wilson, P., & Currie, J. (2000). The effect of an external load on the force and timing components of mentally represented actions. *Behavioural Brain Research, 108*, 91–96.

Chabris, C., & Simons, D. (2011). *The invisible gorilla: And other ways our intuitions deceive us*. New York: Broadway Paperbacks.

Chang, R. S. (2009, July 1). *The spandrels of Dunkin Donuts, or how the Munchkin came to be*. Paper presented at the Convention of the Association for Psychological Science: Evolutionary Psychology.

Chase, C., Chin, D. B., Oppezzo, M., & Schwartz, D. L. (2009). Teachable agents and the protégé effect: Increasing the effort towards learning. *Journal of Science Education and Technology, 18*, 334–352.

Chatterjee, A., Chaubey, P., Martin, J., & Thakor, N. V. (2008). Quantifying prosthesis control improvements using a vibrotactile representation of grip force. In *Proceedings of the IEEE Region 5 Conference,* Kansan City, MO.

Chen, N. (1998). *Is that how you look: Body-image, beauty standards, and the creation of online bodies*. PhD prospectus, UCLA Department of Sociology.

Chhabra, M., & Jacobs, R. A. (2006). Properties of synergies arising from a theory of optimal motor behavior. *Neural Computation, 18*, 2320–2342.

Chiao, J. Y., & Blizinsky, K. D. (2010). Culture-gene coevolution of individualism collectivism and the serotonin transporter gene. *Proceedings of the Royal Society B: Biological Sciences, 277*, 529–537.

Chiappe, D., Conger, M., Liao, J., Caldwell, J. L., & Vu, K.-P. L. (2013). Improving multi-tasking ability through action videogames. *Applied Ergonomics, 44*, 278–284.

Chion, M. (1994). *Audio-vision: Sound on screen*. New York: Columbia University Press.

Chirchir, H., Kivell, T. L., Christopher, R. C. B., Jean-Jacques, J., & Richmond, B. G. (2014). The evolution of trabecular bone density in humans. In *The 83rd Annual Meeting of the American Association of Physical Anthropologists*, Calgary, AB.

Chollet, F., Cramer, S. C., Stinear, C., Kapelle, L. J., Baron, J. C., Weiller, C., et al. (2014). Pharmacological therapies in post stroke recovery: Recommendations for future clinical trials. *Journal of Neurology, 261*, 1461–1468.

Chow, A. Y., & Dickerson, C. (2009). Shoulder strength of females while sitting and standing as a function of hand location and force direction. In A. Y. Chow (Ed.), *Applied ergonomics*. Retrieved from www.elsevier.com/locate/apergo.

Christensen, M. G. (2008). On perceptual distortion measures and parametric modeling. In *Proceedings of Acoustics'08*. Société Francaise d'Acoustique.

Clark, A. (1993). *Sensory qualities*. Oxford, UK: Clarendon Press.

Clark, A. (2003). *Natural-born cyborgs: Minds, technologies, and the future of human intelligence*. New York: Oxford University Press.

Clark, B. C., Mahato, N. K., Nakazawa, M., Law, T. D., & Thomas, J. S. (2014). The power of the mind: The cortex as a critical determinant of muscle strength/weakness. *Journal of Neurophysiology, 112*, 3219–3226.

Clément, G., & Ngo-Anh, J. T. (2013). Space physiology II: Adaptation of the central nervous system to space flight-past, current, and future studies. *European Journal of Applied Physiology, 113*, 1655–1672.

Clubb, R., & Mason, G. (2003). Animal welfare: Captivity effects on wide-ranging carnivores. *Nature, 425*, 473–474.

CNN. (2013, July 7). *2 die, 305 survive after airliner crashes, burns at San Francisco airport*. Retrieved www.cnn.com.

Cocks, M., Moulton, C., Luu, S., & Cil, T. (2014). What surgeons can learn from athletes: Mental practice in sports and surgery. *Journal of Surgical Education, 71*, 262–269.

Cohen, R. G., & Sternad, D. (2009). Variability in motor learning: Relocating, channeling and reducing noise. *Experimental Brain Research, 193*, 69–83.

Coleman, G. W., Macaulay, C., & Newell, A. F. (2008). Sonic mapping: Towards engaging the user in the design of sound for computerized artifacts. In *5th Nordic Conference on Human-Computer Interaction: Building Bridges* (pp. 83–92). Lund, Sweden: ACM.

Collins, F. (2014, September 30). *BRAIN: Launching America's next moonshot*. Director's blog, NIH, Maryland.

Collins, K., Kapralos, B., & Kanev, K. (2014). *Smart table computer interaction interfaces with integrated sound*. Retrieved from http://www.researchgate.net/profile/Karen_Collins6/publication/260750965_Smart_Table_Computer_Interaction_Interfaces_with_Integrated_Sound/links/0c960537b45602df94000000.pdf.

Collins, S. H., Wisse, M., & Ruina, A. (2001). A 3-D passive dynamic walking robot with two legs and knees. *International Journal of Robotic Research, 20*, 607–615.

Congos, D. (2005). *9 Types of mnemonics for better memory*. The Learning Center Exchange.

Conti-Ramsden, G., Durkin, K., & Simtin, Z. (2010). Language and social factors in the use of cell phone technology by adolescents with and without specific language impairment (SLI). *Journal of Speech, Language, & Hearing Research, 53*, 196–208.

Cooper, M. (1981). *The best (and worst) technology predictions of all time*. Retreived April 29, 2016, from www.msn.com.

Corasaniti, N. (2010, September 14). How do people use their smartphones? *The New York Times BITS*.

Corballis, M. C. (2009). Mental time travel and the shaping of language. *Experimental Brain Research, 192*, 553–560.

Corbett, S. (2010, September 15). Learning by playing: Video games in the classroom. *New York Times*. Retrieved from www.nytimes.com.

Cornwall, W. (2015, October 15). *Can we build an 'Iron Man' suit that gives soldiers a robotic boost?* Retrieved from http://wwwsciencemag.org.

Cotelli, M., Calabria, M., Manenti, R., Rosini, S., Zanetti, O., Cappa, S. F., et al. (2011). Improved language performance in Alzheimer disease following brain stimulation. *Journal of Neurology, Neurosurgery & Psychiatry, 82*, 794–797.

Cottingham, S., Stoothoof, R., & Murdoch, D. (1984). *The philosophical writings of Descartes*. Cambridge, UK: Cambridge University Press.

Coulson, M., Barnett, J., Ferguson, C. J., & Gould, R. L. (2012). Real feelings for virtual people: Emotional attachments and interpersonal attraction in video games. *Psychology of Popular Media Culture, 1*, 176–184.

Cowie, D., Atkinson, J., & Braddick, O. (2010). Development of visual control in stepping down. *Experimental Brain Research, 202*, 181–188.

Craik, K. J. W. (1947/1948). Theory of the human operator in control systems. I. The operator as an engineering system. *British Journal of Psychology, 38*, 56–61.

Cropley, A. (2006). In praise of convergent thinking. *Creativity Research Journal, 18*, 391–404.
Csikszentmihalyi, M. (1975). *Beyond boredom and anxiety: Experiencing flow in work and play*. San Francisco: Jossey-Bass.
Csikszentmihalyi, M. (1990). *Flow: The psychology of optimal experience*. New York: Harper and Row.
Csikszentmihalyi, M. (1998). *Finding flow: The psychology of engagement with everyday life*. New York: Basic Books.
Csikszentmihalyi, M. (2014). *The systems model of creativity: The collected works of Mihaly Csikszentmihalyi*. Dordrecht, The Netherlands: Springer.
Curtis, D. (2012, October 3). 4 inches. Retrieved from https://dcurt.is/4-inches.
Cusack, W. F., Thach, S., Patterson, R., Acker, D., Kistenberg, R. S., & Wheaton, L. A. (2016). Enhanced neurobehavioral outcomes of action observation prosthesis training. *Neurorehabilitation and Neural Repair, 30*, 573–582. doi:10.1177/1545968315606992.
Daniels, A. C. (2000). *Bringing out the best in people: How to apply the astonishing power of positive reinforcement*. New York: McGraw-Hill.
Danzl, M. M., Chelette, K. C., Lee, K., Lykins, D., & Sawaki, L. (2013). Brain stimulation paired with novel locomotor training with robotic gait orthosis in chronic stroke: A feasibility study. *Neurorehabilitation, 33*, 67–76.
Daschmann, E. C., Goetz, T., & Stupnisky, R. H. (2011). Testing the predictors of boredom at school: Development and validation of the precursors to boredom scales. *British Journal of Educational Psychology, 81*, 421–440.
Davie, R., Panting, C., & Charlton, T. (2004). Mobile phone ownership and usage among pre-adolescents. *Telematics & Informatics, 4*, 359–373.
Davis, F. D. (1989). Perceived usefulness, perceived ease of use and user acceptance of information technology. *MIS Quarterly, 13*, 319–340.
Davis, R. A. (2001). A cognitive–behavioral model of pathological Internet use. *Computers in Human Behavior, 17*, 187–195.
Davis, F. D., Bagozzi, R. P., & Warshaw, P. R. (1992). Extrinsic and intrinsic motivation to use computers in the workplace. *Journal of Applied Social Psychology, 22*, 1109–1130.
Davis, M., & Whalen, P. J. (2001). The amygdala: Vigilance and emotion. *Molecular Psychology, 6*, 13–34.
Dawkins, R. (1983). Universal Darwinism. In D. S. Bendall (Ed.), *Evolution from molecules to man*. Cambridge, England: Cambridge University Press.
de Byl, P. (2015). A conceptual affective design framework for the use of emotions in computer game design. *Cyberpsychology: Journal of Psychosocial Research on Cyberspace, 9*, article 1. doi: 10.5817/CP2015-3-4.
de Peuter, A. (2014). *Development of an adaptive game by runtime adjustment of game parameters to EEG measurements*. Retrieved from http://www.few.vu.nl/~apr350/bachelor_project/Scriptie_Auke_de_Peuter_2512078.pdf.
Demirbilek, O., & Sener, B. (2003). Product design, semantics and emotional response. *Ergonomics, 46*, 1346–1360.
Dempster, W. T., & Gabel, F. (1959). The anthropometry of the manual work space for the seated subject. *American Journal of Physical Anthropometry, 17*, 289–317.
Demura, S., & Uchiyamal, M. (2009). Influence of cell phone email use on characteristics of gait. *European Journal of Sport Science, 9*, 303–309.
Denmead, K. (2015). *Sounds from the future: Aftershokz bone conducting headphones*. Retrieved January 28, from https://geekdad.com/2015/01/aftershokz/.
Dennett, D. D. (1991). *Consciousness explained*. New York: Little, Brown.
Dennis, C., Jayawardhena, C., Merrilees, W., & Wright, L. T. (2009). e-Consumer behaviour. *European Journal of Marketing, 43*, 1121–1139.
Desmet, P. M. A. (2002). *Designing emotion*. Unpublished PhD Thesis, Delft University of Technology, Delft, The Netherlands.
Desmet, P. M. A., & Hekkert, P. (2007). Framework of product experience. *International Journal of Design, 1*, 13–23.

Dew, N., Sarasvathy, S. D., & Venkataraman, S. (2004). The economic implications of exaptation. *Journal of Evolutionary Economics, 14*, 69–84.
Dewey, J. (1935). *Liberalism and social action*. New York: Putman.
Dickinson. E. (1921). *The complete poems of Emily Dickinson*. Boston: Little, Brown. (Original work published 1860)
Dickstein, R., & Deutsch, J. E. (2007). Motor imagery in physical therapist practice. *Physical Therapy, 87*, 942–953.
Diehl, S. (2002). *Erlebnisorientiertes internetmarketing, analyse, konzeption und umsetzung von internetshops aus verhaltenswissenschaftlicher perspective.* (Experience oriented internet marketing. Analysis, design and application of internet shops from the perspective of the behavioral sciences). Wiesbaden, Germany: Gabler.
Diener, E., & Biswas-Diener, R. (2002). Will money increase subjective well-being? A literature review and guide to needed research. *Social Indicators Research, 57*, 119–169.
D'Mello, S. K. (2013). A selective meta-analysis on the relative incidence of discrete affective states during learning with technology. *Journal of Educational Psychology, 105*, 1082–1099.
D'Mello, S. K., & Graesser, A. C. (2011). The half-life of cognitive-affective states during complex learning. *Cognition and Emotion, 25*, 1299–1308.
Dolan, R. (2003). Emotion, cognition, and behavior. *Science, 298*, 1191–1194.
Dooley, J. J., Pyzalski, J., & Cross, D. (2009). Cyberbullying versus face-to-face bullying: A theoretical and conceptual review. *Journal of Psychology, 217*, 182–188.
Drews, F. A., Yazdani, H., Godfrey, C. N., Cooper, J. M., & Strayer, D. L. (2009). Text messaging during simulated driving. *Human Factors, 51*, 762–770.
Dror, I. E., & Harnad, S. (Eds.). (2008). *Cognition distributed: How cognitive technology extends our minds*. Amsterdam: John Benjamins.
Drouin, M., & Landgraff, C. (2012). Texting, sexting, and attachment in college students' romantic relationships. *Computers in Human Behavior, 28*, 444–449.
Droumeva, M., & Mcgregor, L. (2012). Everyday listening to auditory displays: Lessons from acoustic ecology. In *Proceedings of the 18th International Conference on Auditory Display*, Atlanta, GA, June 18–21, 2012.
Droumeva, M., & Wakkary, R. (2010). Socio-ec(h)o: Focus, listening and collaboration in the experience of ambient intelligent environments. In *Proceedings of the 16th International Conference on Auditory Display* (pp. 327–334).
Duncan, A. (2010, March 3). Using technology to transform schools—Remarks by Secretary Arne Duncan. *Association of American Publishers Annual Meeting*, US Dept of Education.
Dupire, J., Gal, V., & Topol, A. (2009). Physiological player sensing: New interaction devices for video games. In S. Stevens & S. Saldamarco (Eds.), *Entertainment computing—ICEC 2008* (pp. 203–208). Berlin, Germany: Springer. Retrieved from http://dx.doi.org/10.1007/978-3-540-89222-9_25.
Dvorak, K. (2015). *Tufts Medical Center to use virtual reality to ease patient anxiety before surgery*. Retrieved December 10, from www.fiercehealthcare.com.
Dweck, C. S., & Molden, D. C. (2005). Self-theories: Their impact on competence motivation and acquisition. In A. J. Elliot & C. S. Dweck (Eds.), *Handbook of competence and motivation* (pp. 122–140). New York: Guilford.
Ehrlenspiel, F., Wei, K., & Sternad, D. (2010). Open-loop, closed-loop and compensatory control: Performance improvement under pressure in a rhythmic task. *Experimental Brain Research, 201*, 729–741.
Eichenbaum, H., & Cohen, N. J. (2001). *From conditioning to conscious recollection: Memory systems of the brain*. New York: Oxford University Press.
Eindhoven University of Technology, UG Program, Psychology and Technology. (2016). Eindhoven University Website. Retrieved from https://www.tue.nl/en/university/departments/industrial-engineering-innovation-sciences/education/undergraduate-programs/psychology-technology/.
Eisenbach, M., & Lengeler, J. W. (2004). *Chemotaxis*. London: Imperial College Press.

Elliot, D., Grierson, L. E. M., Hayes, S. J., & Lyons, J. (2011). Action representations in perception, motor control and learning: Implications for medical education. *Medical Education, 45*, 119–131.

Emara, T. H., Moustafa, R. R., Elnahas, N. M., Elganzoury, A. M., Abdo, T. A., Mohamed, S. A., et al. (2010). Repetitive transcranial magnetic stimulation at 1Hz and 5Hz produces sustained improvement in motor function and disability after ischaemic stroke. *European Journal of Neurology, 17*, 1203–1209.

Emmerson, S. (2007). *Living electronic music*. Aldershot, Hampshire: Ashgate.

Engelbart, D. C. (1962). *Augmenting human intellect: A conceptual framework*. Summary report. Washington, DC: Air Force Office for Scientific Research.

Engelhart, D. (2000). *Bootstrapping: Coevolution and the origins of personal computing*. Palo Alto, CA: Stanford University Press.

Entertainment Software Association. (2012). *Essential facts about the computer and video game industry*. Retrieved from www.theesa.com/facts/pdfs/ESA_EF_2012.pdf.

Ericsson, K. A., & Charness, N. (1994). Expert performance: Its structure and acquisition. *American Psychologist, 49*, 725–747.

Ericsson, K. A., Krampe, R. T., & Tesch-Romer, C. (1993). The role of deliberate practice in the acquisition of expert performance. *Psychological Review, 100*, 363–406.

Erikson, E. H. (1959). *Identity and the life cycle; selected papers, with a historical introduction by David Rapaport*. New York: International University Press.

Evans, G. W. (2006). Child development and the physical environment. *Annual Review of Psychology, 57*, 423–451.

Ewoldsen, D. R., Eno, C. A., Okdie, B. M., Velez, J. A., Guadagno, R. E., & DeCoster, J. (2012). Effect of playing violent video games cooperatively or competitively on subsequent cooperative behavior. *Cyberpsychology, Behavior, and Social Networking, 15*, 277–280.

Fallman, D. (2011). The new good: Exploring the potential of philosophy of technology to contribute to human-computer interaction. In *Proceedings of the International Conference on Human Factors in Computing Systems*, CHI 2011, Vancouver, BC, May 7–21, 2011.

Faulkner, X., & Culwin, F. (2005). When fingers do the talking: A study of text messaging. *Interacting with Computers, 17*, 167–185.

FDA (2013, December 13). *FDA letter to eNeura re de novo classification review*.

Felton, E. A., Radwin, R. G., Wilson, J. A., & Williams, J. C. (2009). Evaluation of a modified Fitts' law brain-computer interface target acquisition task in able and motor disabled individuals. *Journal of Neural Engineering, 6*, 056002.

Ferguson, C. J., & Garza, A. (2011). Call of (civic) duty: Action games and civic behavior in a large sample of youth. *Computers in Human Behavior, 27*, 770–775.

Ferguson, C. J., & Olson, C. K. (2013). Friends, fun, frustration and fantasy: Child motivations for video game play. *Motivation and Emotion, 37*, 154–164.

Finke, R. A. (1979). The functional equivalence of mental images and errors of movement. *Cognitive Psychology, 11*, 235–264.

Fishbein, M., & Ajzen, I. (1975). *Belief, attitude, intention and behavior: An introduction to theory and research*. Reading, MA: Addison-Wesley.

Fitch, W. T. (2011). The evolution of syntax: An exaptationist perspective. *Frontiers in Evolutionary Neuroscience, 3*, 9. doi:10.3389/fnevo.2011.00009.

Fitts, P. M. (1954). The information capacity of the human motor system in controlling the amplitude of movement. *Journal of Experimental Psychology, 47*, 381–391.

Fitts, P. M. (1964). Perceptual-motor skill learning. In A. W. Melton (Ed.), *Categories of human learning* (pp. 243–285). New York: Academic.

Fitts, P. M., & Peterson, J. R. (1964). Information capacity of discrete motor responses. *Journal of Experimental Psychology, 67*, 103–112.

Fitts, P. M., & Posner, M. I. (1967). *Human performance*. Belmont, CA: Brooks/Cole.

Fleming, S. M., Whiteley, L., Hulme, O. J., Sahani, M., & Dolan, R. J. (2010). Effects of category-specific costs on neural systems for perceptual decision-making. *Journal of Neurophysiology, 103*, 3238–3247.

Forlines, C., Vogel, D., & Balakrishnan, R. (2006). HybridPointing: Fluid switching between absolute and relative pointing with a direct input device. In *Proceedings of the 19th Annual ACM Symposium on User interface Software and Technology*, Montreux, Switzerland, October 15–18 (pp 211–220). UIST '06. New York: ACM Press.

Foster, P. (2013, July 5). George Orwell and climates of fear. *The Financial Post*. Retrieved from http://business.financialpost.com.

Fox, S. (2011). *Americans living with disability and their technology profile*. Pew Research Center's Internet & American Life Project.

Frank, C., Land, W. M., Popp, C., & Schack, T. (2014). Mental representation and mental practice: Experimental investigation on the functional links between motor memory and motor imagery. *PLoS One, 9*, e95175. doi:10.1371/journal.pone.0095175.

Frank, C., Land, W. M., & Schack, T. (2013). Mental representation and learning: The influence of practice on the development of mental representation structure in complex action. *Psychology of Sport and Exercise, 14*, 353–361.

Fredrickson, B. L. (2001). The role of positive emotions in positive psychology: The broaden-and-build theory of positive emotions. *American Psychologist, 56*, 218–226.

Fredrickson, B. L. (2008). Promoting positive affect. In M. Eid & R. J. Larsen (Eds.), *The science of subjective well-being* (pp. 449–468). New York: Guilford.

Fredrickson, B. L., & Cohn, M. A. (2008). Positive emotions. In M. Lewis, J. Haviland-Jones, & L. F. Barrett (Eds.), *Handbook of emotions* (Vol. 3, pp. 777–796). New York: Guilford.

Freeman, D. (2004). Creating emotion in games: The craft and art of emotioneering™. *Computers in Entertainment (CIE), 2*(3), article 15. Retrieved from http://dx.doi.org/10.1145/1027154.1027179.

Freitas, C., Mondragon-Llorca, H., & Pascual-Leone, A. (2011). Noninvasive brain stimulation in Alzheimer's disease: Systematic review and perspectives for the future. *Experimental Gerontology, 46*, 611–627.

Friedman, B. (2004). *Value sensitive design encyclopedia of human-computer interaction*. Great Barrington, MA: Berkshire.

Friedman, T. (2005). *The world is flat*. New York: Farrar, Straus and Giroux.

Fuller, B. (1895–1983). *A quote by Buckminster Fuller, 1895-1983*. Retrieved from www.qotd.org.

Gagnon, G., Schneider, C., Grondin, S., & Blanchet, S. (2011). Enhancement of episodic memory in young and healthy adults: A paired-pulse TMS study on encoding and retrieval performance. *Neuroscience Letters, 488*, 138–142.

Galvin, R. (2003). The making of the disabled identity: A linguistic analysis of marginalisation. *Disability Studies Quarterly, 23*, 149–178.

Ganesh, S., van Schie, H. T., de Lange, F. P., Thompson, E., & Wigboldus, D. H. J. (2011). How the human brain goes virtual: Distinct cortical regions of the person processing network are involved in self-identification with virtual agents. *Cerebral Cortex, 22*, 1577–1585.

Garcia, L., Nussbaum, M., & Preiss, D. D. (2011). Is the use of information and communication technology related to performance in working memory tasks? Evidence from seventh-grade students. *Computers & Education, 57*, 2068–2076.

Garcia-Rill, E. (2002). Focusing the possibilities of nanotechnology for cognitive evolution and human performance. In M. C. Roco & W. S. Bainbridge (Eds.), *Converging technologies for improving human performance: nanotechnology, biotechnology, information technology and cognitive science, NSF/doc-sponsored report*. Arlington, VA: National Science Foundation.

Gardner, J. E. (1991). Can the Mario Bros. help? Nintendo games as an adjunct in psychotherapy with children. *Psychotherapy: Theory, Research, Practice, Training, 28*, 667–670.

Gartner Inc. (2013, December 12). *Gartner says the Internet of Things installed base will grow to 26 billion units by 2020*. Stamford, CT: Gartner Inc. Retrieved from www.gartner.com.

Gartner Inc. (2014, January 22). *Gartner says by 2017, mobile users will provide personalized data streams to more than 100 apps and services every day*. Stamford, CT: Gartner Inc. Retrieved from www.gartner.com.

Gay, M. (2015). *The brain electric: The dramatic high tech race to merge minds with machines*. New York: Macmillan.

Gee, J. P. (2005). What would a state of the art instructional video game look like? *Innovate, 1*(6). Retrieved April 24, 2008, from http://www.innovateonline.info/index.php?view=article&id=80.
Gee, J. P. (2008). *Getting over the slump: Innovation strategies to promote children's learning*. New York: Joan Ganz Cooney Center.
Gentile, A. M. (1972). A working model of skill acquisition with application to teaching. *Quest, 17*, 3–23.
Gentile, A. M. (2000). Skill acquisition: Action, movement, and neuromotor processes. In J. A. Carr & R. B. Shephard (Eds.), *Movement science: Foundations for physical therapy in rehabilitation* (pp. 111–187). Rockville, MD: Aspen.
Gentile, D. A. (2009). Pathological video-game use among youth ages 8-18: A national study. *Psychological Science, 20*, 594–602.
Gentile, D. A., Anderson, C. A., Yukawa, S., Ihori, N., Saleem, M., Ming, L. K., et al. (2009). The effects of prosocial video games on prosocial behaviors: International evidence from correlational, longitudinal, and experimental studies. *Personality and Social Psychology Bulletin, 35*, 752–763.
Gerhardstein, P., Tse, J., Dickerson, K., Hipp, D., & Moser, A. (2012). The human visual system uses a global closure mechanism. *Vision Research, 71*, 18–27.
Gershman, S. G., Horvitz, E. J., & Tenenbaum, J. B. (2015). Computational rationality: A converging paradigm for intelligence in brains, minds, and machines. *Science, 349*, 273–278.
Ghose, T. (2013, October 28). Pediatricians: No more than 2 hours screen time daily for kids: The American Academy of Pediatrics' new guidelines also advise against TVs or Internet access in children's bedrooms. *Scientific American*. Retrieved from www.scientificamerican.com.
Gibbon, A. (2015). How human ancestors got a grip. *Science*. doi:10.1126/science.aaa6381.
Gilbert, D. T. (1991). How mental systems believe. *American Psychologist, 46*, 107–119.
Glassman, M. (2012). An era of webs: Technique, technology and the new cognitive (r)evolution. *New Ideas in Psychology, 30*, 308–318.
Gobet, F., Lane, P. C., Croker, S., Cheng, P. C., Jones, G., Oliver, I., et al. (2001). Chunking mechanisms in human learning. *Trends in Cognitive Science, 5*, 236–243.
Goldberg, M. E., Eggers, H. M., & Gluras, P. (1991). The ocular motor system. In E. R. Kandal et al. (Eds.), *Principles of neural science* (3rd ed., pp. 353–366). New York: Elsevier.
Goodall, J. (1986). *The chimpanzees of Gombe: Patterns of behavior*. Cambridge, MA: The Belknap Press of Harvard University Press.
Goodall, J. (1999). *Reason for hope: A spiritual journey*. New York: Warner Books.
Gordon, C. C., & Bradtmiller, B. (2012). Anthropometric change: Implications for office ergonomics. *Work, 41*, 4606–4611.
Gordon, C. C., Bradtmiller, B., &Ratnaparkhi, J. (1986). *Proposed sampling strategy for a U.S. Army anthropometric study (NATICK/TR-86)*. Natick, MA: U.S. Army Natick, Research Development and Engineering Center.
Gordon, J., & Ghez, C. (1987). Trajectory control in targeted force impulses. II. Pulse height control. *Experimental Brain Research, 67*, 241–252.
Gould, S. J., & Lewontin, R. C. (1979). The spandrels of San Marco and the Panglossian Paradigm: A critique of the adaptationist programme. *Proceedings of Royal Society: Biological Sciences, 205*, 581–598. doi:10.1098/rsbp.1979.0086.
Gould, S. J., & Vrba, E. S. (1982). Exaptation—A missing term in the science of form. *Paleobiology, 8*, 4–15.
GOV.IN. (2015, March 31). *Budget 2015: Some of the things we've announced*. GOV.UK.
Graesser, A. C., D'Mello, S. K., & Strain, A. C. (2014). Emotions in advanced learning technologies. In R. Pekrun & L. Linnen-Garcia (Eds.), *International handbook of emotions in education*. New York: Routledge.
Granic, I., Lobel, A., & Engels, R. C. M. E. (2013). The benefits of playing video games. *American Psychologist, 69*, 66–78.
Grant, K. W., VanWassenhove, V., & Poeppel, D. (2004). Detection of auditory (cross-spectral) and auditory-visual (cross-modal) synchrony. *Speech Communication, 44*, 43–53.

Gray, R. (2004). Attending to the execution of a complex sensorimotor skill: Expertise differences, choking, and slumps. *Journal of Experimental Psychology: Applied, 10*, 42–54.

Gray, W. D. (2012, September 24). *The cognitive science of natural interaction or why suboptimal performance results when human systems violate the design assumptions of nature.* Talk Presented at HFES Webinar.

Gray, J. R., Braver, T. S., & Raichle, M. E. (2002). Integration of emotion and cognition in the lateral prefrontal cortex. *Proceedings of the National Academy of Science, 99*, 4115–4120.

Gray, J. R., Chabris, C. F., & Braver, T. S. (2003). Neural mechanisms of general fluid intelligence. *Nature Neuroscience, 6*, 316–322.

Green, L. (2001). *Technoculture*. Crows Nest, Australia: Allen & Unwin.

Green, C. S., & Bavelier, D. (2012). Learning, attentional control, and action video games. *Current Biology, 22*, 197–206.

Green, D. M., & Swets, J. A. (1974). *Signal detection theory and psychophysics* (A reprint, with corrections of the original 1966 ed.). Huntington, NY: Robert E. Krieger Publishing. (Original work published 1966)

Greenberg, A. (2015, July 21). Hackers remotely kill a jeep on the highway—With me in it. *Wired.*

Grill-Spector, K., & Malach, R. (2004). The human visual cortex. *Annual Review of Neuroscience, 27*, 649–677.

Gucciardi, D. F., & Dimmock, J. A. (2008). Choking under pressure in sensorimotor skills: Conscious processing or depleted attentional resources? *Psychology of Sport and Exercise, 9*, 45–59.

Guilford, J. P. (1967). *The nature of human intelligence*. New York: McGraw-Hill.

Guimaraes, C., & Antunes, R. D. (2012). Hurtability: The lack of usability in products is hurting people and why manufacturers must be held accountable. *African Journal of Business Management, 26*, 7905–7914.

Hall, C. R., & Martin, K. A. (1997). Measuring movement imagery abilities: A revision of the Movement Imagery Questionnaire. *Journal of Mental Imagery, 21*, 143–154.

Halleck, T. (2015, January 14). *Google Inc. says self-driving car will be ready by 2020.* Retrieved from http://business.financialpost.com.

Hamel, M. F., & Lajoie, Y. (2005). Mental imagery: Effects on static balance and attentional demands of the elderly. *Aging Clinical and Experimental Research, 17*, 223–228.

Hancock, P. A., Pepe, A., & Murphy, L. L. (2005). Hedonomics: The power of positive and pleasurable ergonomics. *Ergonomics in Design, 13*, 8–14.

Hancock, P. A., & Szalma, J. L. (2003). The future of ergonomics. *Theoretical Issues in Ergonomics Sciences, 44*, 238–249.

Hardie, E., & Tee, M. Y. (2007). Excessive Internet use: The role of personality, loneliness and social support networks in Internet addiction. *Australian Journal of Emerging Technologies & Society, 5*, 34–47.

Hardy, L., Mullen, R., & Jones, G. (1996). Knowledge and conscious control of motor actions under stress. *British Journal of Psychology, 87*, 621–636.

Hardy, L., & Parfitt, G. (1991). A catastrophe model of a catastrophe model of anxiety and performance. *British Journal of Psychology, 82*, 163–178.

Harrelson, D. (2011, January 11). Defining virtual ethnography. *Cyber Anthropology.*

Harrison, M. A., Bealing, C. E., Salley, J. M. (2015). 2 TXT or not 2 TXT: College students' reports of when text messaging is social breach. *Social Science Journal, 52*(2). doi: 10.1016/j.soscij.2015.02.005.

Hart, T., Chaparro, B., & Halcomb, C. (2008). Evaluating websites for older adults: Adherence to senior-friendly guidelines and end-user performance. *Behaviour & Information Technology, 27*, 191–199.

Haslegrave, C. M., Tracy, M. F., & Corlett, E. N. (1997). Force exertion in awkward working postures: Strength capability while twisting or working overhead. *Ergonomics, 40*, 1335–1362.

Hassenzahl, M. (2004). The interplay of beauty, goodness, and usability in interactive products. *Human-Computer Interaction, 19*, 319–349.

Hatala, R. (2011). Practice makes perfect… sometimes. *Medical Education, 45*, 114–116.

Hatsopoulos, N., Joshi, J., & O'Leary, J. G. (2004). Decoding continuous and discrete motor behaviors using motor and premotor cortical ensembles. *Journal of Neurophysiology, 92*, 1165–1174.

Hawking, S., Russell, S., Tegmark, M., Wilczek, F. (2014, May 2). *Stephen Hawking: 'Transcendence looks at the implications of artificial intelligence—But are we taking AI seriously enough?'* Retrieved from www.independent.co.uk.

Hawks, J. (2013, July 1). How has the human brain evolved? *Scientific American*. Retrieved from http://www.scientificamerican.com/article/how-has-human-brain-evolved/.

Hedge, A. (2011a). *Anthropometry and workspace design*. Cornell University: DEA: 3250/6510.

Hedge, A. (2011b). Alternative workstations may be new but are they better? In M. M. Robertson (Ed.), *Ergonomics and health aspects of work with computers*. Heidelberg, Germany: Springer.

Heidegger, M. (1962). *Being and time* (J. Macquarrie & E. Robinson, Trans.). London: SCM Press.

Helander, M. G. (1997). The human actor's profession. In G. Salvendy (Ed.), *The handbook of human factors & ergonomics* (2nd ed., pp. 3–16). New York: Wiley.

Helander, M. G., & Tham, M. P. (2003). Hedonomics—Affective human factors design. *Ergonomics, 46*, 1269–1272.

Herrnstein, R. J., & Murray, C. (1994). *The bell curve*. New York: Free Press.

Heuer, H., & Sulzenbruck, S. (2009). Trajectories in operating a handheld tool. *Journal of Experimental Psychology: Human Perception and Performance, 35*, 375–389.

Hickman, L. (1990). *John Dewey's pragmatic technology*. Bloomington, IL: Indiana University Press.

Higgs, E., Light, A., & Strong, D. (Eds.). (2001). *Technology and the good life*. Chicago: University of Chicago Press.

Higuchi, T. (2002). Disruption of kinematic coordination under stress. *The Japanese Psychological Research, 42*, 168–177.

Hill, C. (2016, September, 13). *How your Facebook account can slowly destroy your finances*. Retrieved from http://www.msn.com.

Hills, T. T. (2004). *ARS-genetics: A genetic algorithm that evolves individual foragers*. Retrieved from http://ccl.northwestern.edu/netlogo/models/community/ARS-Genetics.

Hills, T. T. (2006). Animal foraging and the evolution of goal-directed cognition. *Cognitive Science, 30*, 3–41.

Hills, T. T. (2011). The evolutionary origins of cognitive control. *Topics in Cognitive Science, 3*, 231–237.

Hills, T. T., Brockie, P. J., & Maricq, A. V. (2004). Dopamine and glutamate control area-restricted search behavior in Caenorhabditis elegans. *The Journal of Neuroscience, 24*, 1217–1225.

Hills, T. T., & Butterfill, S. (2015). From foraging to autonoetic consciousness: The primal self as a consequence of embodied prospective foraging. *Current Zoology, 61*, 368–381.

Hills, T. T., & Dukas, R. (2012). The evolution of cognitive search. In P. M. Todd, T. Hills, & T. Robbins (Eds.), *Cognitive search: Evolution, algorithms, and the brain* (Vol. 9). Strüngmann Forum Reports. Cambridge, MA: MIT Press.

Hills, T. T., Todd, P. M., & Goldstone, R. L. (2008). Search in external and internal spaces: Evidence for generalized cognitive search processes. *Psychological Science, 19*, 676–682.

Hinckley, C. A., Alaynick, W. A., Gallarda, B. W., Hayashi, M., Hilde, K. L., Driscoll, S. P., et al. (2015). Spinal locomotor circuits develop using hierarchical rules based on motorneuron position and identity. *Neuron, 87*, 1008–1021.

Hodgson, G. M. (2005). Generalizing Darwinism to social evolution: Some early attempts. *Journal of Economic Issues, 39*, 899–914.

Hofstede, G. (1980). *Culture's consequences: International differences in work related values*. Beverly Hills, CA: Sage.

Holmes, P. S., & Collins, D. J. (2001). The PETTLEP approach to motor imagery: A functional equivalence model for sport psychologists. *Journal of Applied Sport Psychology, 13*, 60–83.

Hornyak, T. (2014, June 5). *Meet Pepper, the 'love-powered' humanoid robot that knows how you're feeling*. Retrieved from www.pcworld.com.

Hosking, S. G., Young, K. L., & Regan, M. A. (2009). The effects of text messaging on young drivers. *Human Factors, 51*, 582–592.

Hsee, C. K., & Tsai, C. I. (2008). Hedonics in consumer behavior. In C. P. Haugtvedt, P. M. Herr, & F. R. Kardes (Eds.), *Handbook of consumer psychology* (pp. 639–658). Mahwah, NJ: Erlbaum.

Hsiao, H., Long, D., & Snyder, K. (2002). Anthropometric differences among occupational groups. *Ergonomics, 45*, 136–152.

(2014, February 6). *It's ERWIN the friendly robot*. Retrieved from http://www.lincoln.ac.uk.

Hughes, T. P. (1989). *American genesis: A century of invention and technological enthusiasm, 1870-1970*. New York: Viking.

Huntley, J., Bor, D., Hampshire, A., Owen, A., & Howard, R. (2011). Working memory task performance and chunking in early Alzheimer's disease. *The British Journal of Psychiatry, 198*, 398–403.

Hussain, Z., & Griffiths, M. D. (2009). The attitudes, feelings, and experiences of online gamers: A qualitative analysis. *Cyberpsychology and Behavior, 12*, 747–753.

Huys, R., Jirsa, V. K., Studenka, B., Rheaume, N., & Zelaznik, N. N. (2008). Human trajectory formation: Taxonomy of movement based on phase flow topology. In A. Fuchs & V. K. Jirsa (Eds.), *Coordination: neural, behavioral and social dynamics (understanding complex systems)* (pp. 77–92). Berlin, Germany: Springer.

Iacovelli, A., & Valenti, S. (2008). Internet addiction's effect on likeability and rapport. *Computers in Human Behavior, 25*, 439–443.

Ifft, P. J., Lebedev, M. A., & Nicolelis, M. A. L. (2011). Cortical correlates of Fitts' law. *Frontiers in Integrative Neuroscience, 5*, 85.

Ihde, D. (1979). *Technics and praxis*. Dordrecht, The Netherlands: Reidel Publishing.

Ihde, D. (1990). *Technology and the lifeworld: From garden to earth*. Bloomington, IL: Indiana University Press.

Ihde, D. (1999). Technology and prognostic predicaments. *AI & Society, 13*, 44–51.

Ihde, D., & Selinger, E. (2003). *Chasing technoscience: Matrix for materiality*. Bloomington, IL: Indiana University Press.

IndianPsychologists Portal. (2014, February 1). *SHUT Clinic: An internet de-addiction centre right in the heart of Bangalore*.

Indvik, L. (2012, June 11). App store stats: 400 million accounts, 650,000 apps. *Mashable*.

Injury Prevention. (2012). *Injury Prevention*. doi: 10.1136/injuryprev-2012-040601. Retrieved from injuryprevention.bmj.com.

International Ergonomics Association. (2016). *Definition and domains of ergonomics*. Retrieved from http://www.iea.cc/whats/index.html.

Internet World Stats. (2016, June). *Internet users in the world by region: June 2016*. Retrieved from www.internetworldstats.com.

Ishizu, T., & Zeki, S. (2011). Toward a brain based theory of beauty. *PLoS One, 6*(7), e21852. doi:10.1371/journal.pone.0021852.

ISO. (2006). *General requirements for establishing international databases*. Geneva, Switzerland: International Organization for Standardization.

ISO 9241-400:2007. (2007). *Ergonomics of human–system interaction—Part 400: Principles and requirements for physical input devices*. Geneva, Switzerland: International Organization for Standardization.

ITU. (1998). *Recommendation BS. 1387: Method for objective measurement of perceived audio quality*. Geneva: International Telecommunications Union.

Iyengar, S. S. (2011). *The art of choosing. (Paperback)*. New York: Twelve-Hachette Book Group.

Iyengar, S. S., & Lepper, M. (2000). When choice is demotivating: Can one desire too much of a good thing? *Journal of Personality and Social Psychology, 76*, 995–1006.

Izquierdo, A., & Jentsch, D. J. (2012). Reversal learning as a measure of impulsive and compulsive behavior in addictions. *Psychopharmacology, 219*, 607–620.

Jackson, L. A., von Eye, A., Fitzgerald, H. E., Zhao, Y., & Witt, E. A. (2010). Self-concept, self-esteem, gender, race and information technology use. *Computers in Human Behavior, 26*, 323–328.

Jackson, L. A., Witt, E. A., Games, A. I., Fitzgerald, H. E., von Eye, A., & Zhao, Y. (2012). Information technology use and creativity: Findings from the children and technology project. *Computers in Human Behavior, 28*, 370–376.

Jacob, R. J. K. (1991). The use of eye movements in human-computer interaction techniques: What you look at is what you get. *ACM Transactions on Information Systems, 9*, 152–169.

James, W. (1890). *The principles of psychology* (2 vols). New York: Dover Publications. (Original work published 1950)

Janczyk, M., & Kunde, W. (2012). Visual processing for action resists similarity of relevant and irrelevant object features. *Psychonomic Bulletin Review, 19*, 412–417.

Janssen, I., Heymsfield, S. B., Wang, Z. M., & Ross, R. (2000). Skeletal muscle mass and distribution in 468 men and women aged 18-88 yr. *Journal of Applied Physiology, 89*, 81–88.

Jax, S. A., Rosenbaum, D. A., & Vaughan, J. (2007). Extending Fitts' Law to manual obstacle avoidance. *Experimental Brain Research, 180*, 775–779.

Jayawardhena, C., & Wright, L. T. (2009). An empirical investigation into e-shopping excitement: Antecedents and effects. *European Journal of Marketing, 43*, 1171–1187.

Jeannerod, M. (1994). The representing brain: Neural correlates of motor intention and imagery. *Behavioral and Brain Sciences, 17*, 187–245.

Jeannerod, M. (1995). Mental imagery in the motor context. *Neuropsychologia, 33*, 1419–1432.

Jeannerod, M. (1997). *The cognitive neuroscience of action*. Hoboken, NJ: Wiley-Blackwell.

Jeong, S., & Fishbein, M. (2007). Predictors of multimedia use and multitasking with media: Media factors and audience. *Media Psychology, 10*, 364–384.

Jeter, P. E., Dosher, B. A., Petrov, A., & Lu, Z. L. (2009). Task precision at transfer determines specificity of perceptual learning. *Journal of Vision, 9*, 1–13.

Jimison, H., Gorman, P., Woods, S., Nygren, P., Walker, M., Norris, S., et al. (2008). Barriers and drivers of health information technology use for the elderly, chronically ill, and underserved. *Evidence Report/Technology Assessment (Full Report), 175*, 1–1422.

John, Y. J., Bullock, D., Zikopoulos, B., & Barbas, H. (2013). Anatomy and computational modeling of networks underlying cognitive-emotional interaction. *Frontiers in Human Neuroscience, 7*, 101.

Johnson, A., & Proctor, R. (Eds.). (2013). *Neuroergonomics: A cognitive neuroscience applications*. New York: Palgrave Macmillan.

Jonas, H. (1979). Toward a philosophy of technology. *Hastings Center Report*, pp. 34–43.

Jones, J. G., & Hardy, L. (1989). Stress and cognitive functioning in sport. *Journal of Sports Sciences, 7*, 41–63.

Jordan, P. W. (1998). Human factors for pleasure in product use. *Applied Ergonomics, 29*, 25–33.

Jordan, M. I., & Mitchell, T. M. (2015). Machine learning: Trends, perspectives, and prospects. *Science, 349*, 255–260.

Jorgovanovic, N., Dosen, S., Djozic, D. J., Krajoski, G., & Farina, D. (2014). Virtual grasping: Closed-loop force control using electrotactile feedback. *Computational and Mathematical Methods in Medicine*, 120357, 13p. Retrieved from http://dx.doi.org/10.1155/2014/120357.

Kahneman, D. (1973). *Attention and effort*. Englewood Cliffs, NJ: Prentice-Hall.

Kahneman, D. (2011). *Thinking, fast and slow*. New York: Farrar, Straus & Giroux.

Kahneman, D., & Tversky, A. (1979). Prospect theory: An analysis of decision under risk. *Econometrica, 47*, 263–291.

Kakuda, W., Abo, M., Watanabe, S., Momsaki, R., Hashimoto, G., Nakayama, Y., et al. (2013). High-frequency rTMS applied over bilateral leg motor areas combined with mobility training for gait disturbance after stroke: A preliminary study. *Brain Injury, 27*, 1080–1086.

Kane, S. K., Wobbrock, J. O., & Ladner, R. E. (2011). Usable gestures for blind people: Understanding preference and performance. In *Proceedings of CHI 2011* (pp. 413–422). New York: ACM Press.

Kaneko, M., & Stryker, M. P. (2014). Sensory experience during locomotion promotes recovery of function in adult visual cortex. *eLife, 3*, e02798. Retrieved from http://dx.doi.org/10.7554/eLife.02798.

Kapp, E. (1877). *Grundlinien einer Philosophie der Technik: Zur Entstehungsgeschichte der Cultur aus neuen Gesichtspunkten*. Braunschweig, Germany: Westermann.

Karwoski, W. (2012). The discipline of human factors and ergonomics. In G. Salvendy (Ed.), *Handbook of human factors and ergonomics*. Hoboken, NJ: Wiley.

Kaye, H. S. (2000). *Computer and Internet use among people with disabilities* (Vol. 13). Washington, DC: US Department of Education, National Institute on Disability and Rehabilitation Research.

Kelion, L. (2013, July 3). Talking train window adverts tested by Sky Deutschland. *BBC News*.

Keller, P. E. (2012). Mental imagery in music performance: Underlying mechanisms and potential benefits. *Annals of the New York Academy of Sciences, 1252*, 206–213.

Kemp, N. (2011). Mobile technology and literacy: Effects across cultures, abilities and the lifespan. *Journal of Computer Assisted Learning, 27*, 1–3.

Kempf, D. S. (1999). Attitude formation from product trial: Distinct roles of cognition and affect for hedonic and functional products. *Psychology & Marketing, 16*, 35–50.

Khalid, H. M. (2005, July 22). Computerized automotive technology reconfiguration (CATER) system for reverse engineering and mass customization. In *EuroMalaysia IST Call 5 FP6 Proposal, Damai Sciences*, Kuala Lumpur, Malaysia.

Khalid, H. M. (2006). Embracing diversity in user needs for affective design. *Ergonomics, 37*, 409–418.

Khalid, H. M., & Helander, M. G. (2006). Customer emotional needs in product design. *Concurrent Engineering, 14*, 197–206.

Khalighy, S., Green, G., Scheepers, C., & Whittet, C. (2015). Quantifying the qualities of aesthetics in product design using eye-tracking technology. *International Journal of Industrial Ergonomics, 49*, 31–43.

Kim, M. (2015). *The good and the bad of escaping to Virtual Reality*. Retrieved February 18, from www.theatlantic.com.

Kim, J., & Forsythe, S. (2007). Hedonic usage of product virtualization technologies in online apparel shopping. *International Journal of Retail & Distribution Management, 35*, 502–514.

Kipnis, D. (1994). Accounting for the use of behavior technologies in psychology. *American Psychologist, 49*, 165–172.

Kirk, W. (2015). *Tiny dancers: Can ballet bugs help us build better robots?* Retrieved from releases.jhu.edu.

Kivell, T. L., & Skinner, M. (2015). Early human ancestors used their hands like modern humans. *Science, 347*, 395–399.

Kleinsorge, T., Schmidtke, V., Gajewski, P., & Heuer, H. (2003). The futility of explicit knowledge of a sequence of tasks. *European Journal of Cognitive Psychology, 15*, 455–469.

Kobayashi, M., Hutchinson, S., Théoret, H., Schlaug, G., & Pascual-Leone, A. (2004). Repetitive TMS of the motor cortex improves ipsilateral sequential simple finger movements. *Neurology, 62*, 91–98.

Kochanska, G., & Aksan, N. (2004). Development of mutual responsiveness between parents and their young children. *Child Development, 75*, 1657–1676.

Koffka, K. (1935). *Principles of gestalt psychology*. London: Lund Humphries.

Kohlberg, L. (1958). *The development of modes of thinking and choices in years 10 to 16*. Ph.D. Dissertation, University of Chicago.

Kohlberg, L. (1976). Moral stages and moralization: The cognitive-developmental approach. In T. Lickona (Ed.), *Moral development and behavior: Theory, research and social issues*. New York: Holt, Rinehart and Winston.

Kohler, W. (1925). *The mentality of apes* (Ella Winter, Transl.). New York: Harcourt, Brace and World. (Original work published 1917)

Kolb, B., & Gibb, R. (2011). Brain plasticity and behavior in the developing brain. *Journal of Canadian Academy of Child & Adolescent Psychiatry, 20*, 265–276.

Kompridis, N. (2009). Technology's challenge to democracy: What of the human? *Parrhesia, 8*, 22–33.

Kool, V. K. (1980). Short term recall of linear and curvilinear movements by blind and sighted subjects (Chapter #27). In R. S. Nickerson (Ed.), *Attention and performance* (Vol. VIII). Mahwah, NJ: Erlbaum.
Kool, V. K. (2008). *Psychology of nonviolence and aggression*. New York: Macmillan Palgrave.
Kool, V. K., & Agrawal, R. (2006). *Applied social psychology: A global perspective*. New Delhi, India: Atlantic Books.
Kool, V. K., & Agrawal, R. (2009). The psychology of nonkilling. In J. E. Pims (Ed.), *Toward a nonkilling paradigm*. Honolulu: Center for Global Nonkilling.
Kool, V. K., & Agrawal, R. (2012). From empathy to altruism: Is there an evolutionary basis for nonkilling? In D. Christie & J. E. Pims (Eds.), *Psychology of nonkilling*. Honolulu: Center for Global Nonkilling.
Kool, V. K., & Agrawal, R. (2015). On using experimental designs. In D. S. Hegde (Ed.), *Essays on research methodology*. New Delhi, India: Springer.
Kool, V. K., & Singh, S. (1988). Tactual and motor memory of blind people. In M. M. Gruenberg, P. E. Morris, & R. N. Sykes (Eds.), *Practical aspects of memory*. London: Wiley.
Kopko, N., Huang, S., Belliveau, J. W., Raij, T., Tengshe, C., & Ahveninen, J. (2012). Neuronal representations of distance in human auditory cortex. *PNAS, 109*, 111019–111024.
Kothiyal, K., & Tettey, S. (2001). Anthropometry for design for the elderly. *International Journal of Occupational Safety and Ergonomics, 7*, 15–34.
Kotler, P., Kartajaya, H., & Setiawan, I. (2010). *Marketing 3.0: From products to customers to the human spirit*. Hoboken, NJ: Wiley.
Kowalczyk, L. (2011). *Suit over cardiac monitor settled: MGH patient died despite alarms*. Retrieved November 28, from https://www.bostonglobe.com.
Kromand, D. (2007). Avatar categorization. In *Proceedings from DiGRA 2007: Situated play* (pp. 400–406), Tokyo, Japan.
Kruisselbrink, L. D., & Van Gyn, G. H. (2011). Task characteristics and the contextual interference effect. *Perceptual and Motor Skills, 113*, 19–37.
Kuhn, T. S. (1962). *The structure of scientific revolutions*. Chicago: University of Chicago Press.
Kurzweil, R. (1999). *The age of spiritual machines*. New York: Viking Books.
Kurzweil, R. (2005). *The singularity is near*. New York: Viking Books.
Lacourse, M. G., Orr, E. L., Cramer, S. C., & Cohen, M. J. (2005). Brain activation during execution and motor imagery of novel and skilled sequential hand movements. *NeuroImage, 27*, 505–519.
Laible, M., Grieshammer, S., Seidel, G., Rijntjes, M., Weiller, C., & Hamzei, F. (2012). Association of activity changes in the primary sensory cortex with successful motor rehabilitation of the hand following stroke. *Neurorehabilitation and Neural Repair, 26*, 881–888.
Lam, W. K., Maxwell, J. P., & Masters, R. (2009). Analogy learning and the performance of motor skills under pressure. *Journal of Sport and Exercise Psychology, 31*, 337–357.
Lamberg, E. M., & Muratori, L. M. (2012). Cell phones change the way we walk. *Gait and Posture, 35*, 688–690.
Landers, R. N., & Lounsbury, J. W. (2006). An investigation of Big Five and narrow personality traits in relation to Internet usage. *Computers in Human Behavior, 22*, 283–293.
Lanier, J. (2010). *You are not a gadget*. New York: Vintage Books.
Lanier, J. (2013). *Who owns the future?* New York: Simon and Schuster.
Lanier, J. (2015). *Digital future*. Retrieved April 2, from www.c-span.org.
Latour, B. (1987). *Science in action: How to follow scientists and engineers through society*. Cambridge, MA: Harvard University Press.
Latour, B. (1993). *We have never been modern* (Catherine Porter, Transl.). Cambridge, MA: Harvard University Press.
Latour, B. (2013). *An inquiry into modes of existence: An anthropology of the moderns* (Catherine Porter, Transl.). Cambridge, MA: Harvard University Press.
Lavrinc, D. (2012, April 24). Exclusive: Google expands its autonomous fleet with hybrid Lexus RX450h. *Wired*.

Law, J., Masters, R. S. W., Bray, S. R., Eves, F. F., & Bardswell, I. (2003). Motor performance as a function of audience affability and metaknowledge. *Journal of Sport and Exercise Psychology, 25*, 484–500.

Law, D. M., Shapka, J. D., Domene, J. F., & Gagné, M. H. (2012). Are cyberbullies really bullies? An investigation of reactive and proactive online aggression. *Computers in Human Behavior, 28*, 664–672.

Lefaucheur, J. P., Andre-Obadia, N., Antal, A., Ayache, S. S., Beaken, C., Benninger, D. H., et al. (2014). Evidence based guidelines on the therapeutic use of repetitive transcranial surgery (rTMS). *Clinical Neurophysiology, 125*, 2150–2206.

Lenhart, A. (2015, April 9). *Teens, social media & technology overview, 2015*. Pew Research Center.

Lenhart, A., Kahne, J., Middaugh, E., Macgill, A. R., Evans, C., & Vitak, J. (2008). *Teens, video games, and civics: teens' gaming experiences are diverse and include significant social interaction and civic engagement*. Pew Internet & American Life Project. Retrieved from the http://www.pewinternet.org/ Reports/2008/Teens-Video-Games-and-Civics.aspx.

Lenhart, A., Madden, M., & Hitlin, P. (2005). *Teens and technology: Youth are leading the transition to a fully wired and mobile nation*. Pew Internet and American Life Project.

Leotti, L. S., Iyengar, S. S., & Ochsner, K. (2010). Born to choose: The origins and value of the need for control. *Trends in Cognitive Sciences, 14*, 457–463.

Leuthardt, E. C., Schalk, G., Wolpaw, J. R., Ojemann, J. G., & Moran, D. (2004). A brain-computer interface using electrocorticographic signals in humans. *Journal of Neural Engineering, 1*, 63–71.

Levy, P. (2013). Beyond kansei engineering: The emancipation of kansei design. *International Journal of Design, 7*, 83–94.

Lévy, P., Lee, H., & Yamanaka, T. (2007). On kansei and kansei design: A description of Japanese design approach. In S. Poggenpohl (Ed.), *Proceedings of the 2nd World Conference on Design Research*. London, UK: Design Research Society.

Lewis, T. (2015). *When will Virtual Reality headsets stop making people sick?* Retrieved March 13, from www.livescience.com.

Li, D. D., Liau, A. K., & Khoo, A. (2013). Player-avatar identification in video gaming: Concept and measurement. *Computers in Human Behavior, 29*, 257–263.

Li, S., & Swanson, P. (Eds.). (2014). *Engaging language learners through technology integration*. Hershey, PA: IGI Global.

Liang, G. K. (2010). An empirical test of mnemonic devices to improve learning in elementary accounting. *The Journal of Education for Business, 85*, 349–358.

Liao, C. M., & Masters, R. S. W. (2001). Analogy learning: A means to implicit motor learning. *Journal of Sport Sciences, 19*, 307–319.

Liao, C. M., & Masters, R. S. W. (2002). Self-focused attention and performance failure under psychological stress. *Journal of Sport & Exercise Psychology, 24*, 289–305.

Liddell, H. G., & Scott, R. (1980). *A Greek-English Lexicon* (Abridgedth ed.). Oxford, UK: Oxford University Press.

Liebana-Cabanillas, F. J., Sanchez-Fernandez, J., & Munoz-Leiva, F. (2014). Role of gender on acceptance of mobile payment. *Industrial Management & Data Systems, 114*, 220–240.

Lin, L. (2009). Breadth-biased versus focused cognitive control in media multitasking. *Proceedings of the National Academy of Science, 106*, 1.

Lin, P. (2015). *A user-centric process for designing mobile apps*. Retrieved March 17, from www.empiricalworks.com.

Lin, C. H., Wu, A. D., Udompholkul, P., & Knowlton, B. J. (2010). Contextual interference effects in sequence learning for young and older adults. *Psychology of Aging, 25*, 929–939.

Lindow, M. (2004). *Seeking riches from the poor*. Retrieved from www.wired.news.com.

Lindstrom, M. (2011). *Brandwashed: Tricks companies use to manipulate our minds and persuade us to buy (Hardcover)*. New York: Crown Business.

Ling, R. (2004). *The mobile connection—The cell phone's impact on society*. San Francisco: Morgan Kaufmann.

Liu, Y. (2003). The aesthetic and the ethic dimensions of human factors and design. *Ergonomics, 46*, 1293–1305.

Liu, Q., Liu, S., Kodama, L., Driscoll, M. R., & Wu, M. (2012). Two dopaminergic neurons signal to the dorsal fan-shaped body to promote wakefulness. *Current Biology, 22*, 2114–2123.

Loeb, G. (2012). Technology and older persons issue overview. *The Journal of Community Informatics, 8*(1). Retrieved from http://cijournal.net/index.php/ciej/article/view/904.

Loevenich, P., & Lingenfelder, M. (2004). Substitution assessment of online shopping by means of the forced switching approach. In K.-P. Wiedmann, H. Buxel, T. Frenzel, & G. Walsh (Eds.), *Konsumentverhalten im Internet*. Wiesbaden, Germany: Gabler.

Loram, I. D., Maganaris, C. N., & Lakie, M. (2006). Use of ultrasound to make noninvasive in vivo measurement of continuous changes in human muscle contractile length. *Journal of Applied Physiology, 100*, 1311–1323.

Luber, B., & Lisanby, S. H. (2014). Enhancement of human cognitive performance using transcranial magnetic stimulation (TMS). *NeuroImage, 85*, 961–970.

Luczak, H., Roetting, M., & Schmidt, L. (2003). Let's talk: Anthropomorphization as means to cope with stress of interacting with technical devices. *Ergonomics, 46*, 1361–1374.

Luppicini, R. (2013). *Handbook of research on technoself: Identity in a technological society*. Hershey, PA: IGI Global.

Lupton, D. (2013). Quantifying the body: Monitoring and measuring health in the age of health technologies. *Critical Public Health, 23*, 393–403.

Machizawa, M. G., Goh, C. W. C., & Driver, J. (2012). Human visual short-term memory precision can be varied at will when the number of retained items is low. *Psychological Science, 6*, 554–559.

MacKenzie, I. S., & Teather, R. J. (2012). Fitts tilt: The application of Fitts' law to tilt based interaction. In *Proceedings of the 8th Nordic Conference on Human-Computer Interaction, ACM NordiCHI 2012* (pp. 568–577). ISBN 978-1-1482-4.

Magill, R. A. (2011). *Motor learning and control: Concepts and applications* (9th ed.). New York: McGraw Hill.

Magsamen-Conrad, K., Billotte-Verhoff, C., & Greene, K. (2014). Technology addiction's contribution to mental wellbeing: The positive effect of online social capital. *Computers and Human Behavior, 40*, 23–30.

Mahan, S. (2013, February 7). *Self-driving car test: Steve Mahan*. YouTube.

Mai, N., & Marquardt, C. (1996). The neglected behavior: Kinematic analysis of the writing movements in the writer's cramp. In H. S. Reinecker & D. Schmelzer (Eds.), *Behavioral therapy, self regulation and management*. Gottingen, Germany: Hogrefe.

Malamed, C. (2013). *Chunking information for instructional design*. Retrieved from theelearningcoach.com.

Malle, B. F., Scheutz, M., Arnold, T., Voiklis, J., & Cusimano, C. (2015). Sacrifice one for the good of many? People apply different moral norms to human and robot agents. In *HRI '15: Proceedings of the Tenth Annual ACM/IEEE International Conference on Human-Robot Interaction* (pp. 117–124). New York: ACM.

Malouin, F., & Richards, C. L. (2010). Mental practice for relearning locomotor skills. *Physical Therapy, 90*, 240–251.

Malta, S. (2008). Intimacy and older adults: A comparison between online and offline romantic relationships. In T. Marjoribanks, J. Barraket, J.-S. Chang, A. Dawson, et al. (Eds.), *Re-imagining sociology. Refereed conference proceedings*. Melbourne, Australia: The Australian Sociological Association.

Mandryk, R. L., & Atkins, M. S. (2007). A fuzzy physiological approach for continuously modeling emotion during interaction with play technologies. *International Journal of Human-Computer Studies, 65*, 329–347.

Mantyla, T. (2013). Gender differences in multitasking reflect spatial ability. *Psychological Science, 24*, 514–520. doi:10.1177/0956797612459660.

Marcus, S. (1974, January 13). Obedience to authority: An experimental view by Stanley Milgram. *New York Times Book Review*.

Marek, T., Karwowski, W., & Rice, V. (Eds.). (2010). *Advances in understanding human performance: Neuroergonomics, human factors design, and special populations*. Boca Raton, FL: CRC.

Marsden, P., & Hollnagel, E. (1996). Human interaction with technology: The accidental user. *Acta Psychologia, 91*, 345–358.

Martin, M. W., & Schinzinger, R. (2005). *Ethics in engineering*. Boston, MA: McGraw-Hill.

Maslow, A. H. (1943). A theory of human motivation. *Psychological Review, 50*, 370–396.

Masters, R. S. W., & Maxwell, J. (2008). The theory of reinvestment. *International Review of Sport and Exercise Psychology, 1*, 160–183.

Maxmen, A. (2010). Video games and the second life of science class. *Cell, 141*, 201–203.

Maynard, A. E., Subrahmanyam, K., & Greenfield, P. M. (2005). Technology and the development of intelligence: From the loom to the computer. In R. J. Sternberg & D. D. Preiss (Eds.), *Intelligence and technology: The impact of tools on the nature and development of human abilities* (pp. 29–54). Mahwah, NJ: Lawrence Erlbaum.

Mazzoni, P., Hristova, A., & Krakauer, J. W. (2007). Why don't we move faster? Parkinson's disease, movement vigor, and implicit motivation. *Journal of Neuroscience, 27*, 105–116.

McCabe, J. A. (2011). *Integrating mnemonics into psychology instruction*. Baltimore: OTRP.

McCreery, M. P., Krach, K. S., Schrader, P. G., & Boone, R. (2012). Defining the virtual self: Personality, behavior, and the psychology of embodiment. *Computers in Human Behavior, 28*, 976–983.

McDaniel, B. T., & Coyne, S. M. (2014). Technoference: The interference of technology in couple relationships and implications for women's personal and relational well-being. *Brigham Young University Release*, 2014.

McGinn, R. E. (2010). What's different, ethically, about nanotechnology? Foundational questions and answers. *Nanoethics, 4*, 115–128.

McGonigal, J. (2011). *Reality is broken: Why games make us better and how they can change the world*. New York: Penguin Press.

McGregor, I., Lepltre, G., Turner, P., & Flint, T. (2010). Soundscape mapping: A tool for evaluating sounds and auditory environments. In *Proceedings of the 16th International Conference on Auditory Display* (pp. 237–244).

McGurk, H., & MacDonald, J. (1976). Hearing lips and seeing voices. *Nature, 264*, 746–748.

McKhann, G. M. (2002). Neurology: Then, now, and in the future. *Archives of Neurology, 59*, 1369–1373.

McNab, F., Varrone, A., Farde, L., Jucaite, A., Bystritsky, P., Forssberg, H., et al. (2009). Changes in cortical dopamine D1 receptor binding associated with cognitive training. *Science, 323*, 800–802.

McPherron, S. P., Alemseged, Z., Marean, C. W., Wynn, J. G., Reed, D., Geraads, D., et al. (2010). Evidence for stone tool assisted consumption of animal tissues before 3.39 million years ago at Dikika Ethiopia. *Nature, 466*, 857–860.

Medina, J., Jax, S. A., & Coslett, H. B. (2009). Two-component models of reaching: Evidence from deafferentation in a Fitts' law task. *Neuroscience Letters, 451*, 222–226.

Melkerson, M. N. (2008). *Special premarket 510(k) notification for NeuroStar® TMS Therapy System for major depressive disorder*. Washington, DC: Food and Drug Administration.

Menon, S., & Kahn, B. (2002). Cross-category effects of induced arousal and pleasure on the internet shopping experience. *Journal of Retailing, 78*, 31–40.

Mercier, M. R., Foxe, J. J., Fiebelkorn, I. C., Butler, J. S., Schwartz, T. H., & Molholm, S. (2013). Auditory-driven phase reset in visual cortex: Human electrocorticography reveals mechanisms of early multisensory integration. *NeuroImage, 79*, 19–29.

Mercola, J. (2015, January 15). *Neuroplasticity studies reveal your brain's amazing malleability*. Retrieved from http://articles.mercola.com/.

Merriam Webster Dictionary. (2007). *Definition of technology*. Retrieved February 16, from www.merriam-webster.com/dictionary/technology.

Milgram, S. (1965). Some conditions of obedience and disobedience to authority. *Human Relations, 18*, 57–76.

Milgram, S. (1974). *Obedience to authority: An experimental view*. New York: Harper Collins.

Millan, J. del R. (2015, March 30). Intelligent neuroprostheses mimic natural motor control. *CNS 2015 Press Release*. Retrieved from www.cognitiveneurosociety.org.

Miller, G. A. (1956). The magical number seven, plus or minus two: Some limits on our capacity for processing information. *Psychological Review, 101*, 343–352.

Miller, G. A., Galanter, E., & Pribram, K. H. (1960). *Plans and the structure of behavior*. New York: Holt, Rinehart, & Winston.

Minsky, M. (2007). *Emotion machine: Common sense thinking, artificial intelligence and the future of the human mind*. New York: Simon and Schuster.

Mirabella, G., Pani, P., & Ferraina, S. (2008). Context influences on the preparation and execution of reaching movements. *Cognitive Neuropsychology, 25*, 996–1010.

Mishkin, M., Ungerleider, L. G., & Macko, K. A. (1983). Object vision and spatial vision: Two cortical pathways. *Trends in Neurosciences, 6*, 414–417.

Mitcham, C. (1994). *Thinking through technology: The path between engineering and philosophy*. Chicago: University of Chicago Press.

Molenbroek, J., & Bruin, R. (2005). Enhancing the use of anthropometric data. In D. de Waard et al. (Eds.), *Human factors in design, safety and management* (pp. 289–297). Maastricht, The Netherlands: Shaker Publishing.

Monks, K. (2014, May 29). *Bone conduction: Get used to the voices in your head*. Retrieved from www.cnn.com.

Moore, C. D., Cohen, M. X., & Ranganath, C. (2006). Neural mechanisms of expert skills in visual working memory. *Journal of Neuroscience, 26*, 11187–11196.

Moran, A., Guillot, A., MacIntyre, T., & Collet, C. (2012). Re-imagining motor imagery: Building bridges between cognitive neuroscience and sport psychology. *British Journal of Psychology, 103*, 224–247.

Morgan, K. N., & Tromberg, C. T. (2007). Sources of stress in captivity. *Applied Animal Behaviour Science, 102*, 262–302.

Morris, T., Spittle, M., & Watt, A. (2005). *Imagery in sport*. Champaign, IL: Human Kinetics.

Muller, V. C., & Bostrom, N. (2016). Future progress in artificial intelligence: A survey of expert opinion. In V. C. Müller (Ed.), *Fundamental issues of artificial intelligence*. Berlin, Germany: Springer.

Muratori, L. M., Lamberg, E. M., Quinn, L., & Duff, S. V. (2013). Applying principles of motor learning and control to upper extremity rehabilitation. *Journal of Hand Therapy, 26*, 94–103.

Murdoch, E., Osterreicher, A., Guse, L., Roger, K., & Thompson, G. (2013). *Use of social commitment robots in the care of elderly people with dementia: A literature review*. Retrieved November 9, 2012, from http://dx.doi.org/10.1016/j.maturitas.2012.10.015.

Muri, J., Winter, S. L., & Challis, J. H. (2008). Changes in segmental inertial properties with age. *Journal of Biomechanics, 41*, 1809–1812.

Murphie, A., & Potts, J. (2003). *Culture and technology*. London: Palgrave.

Murphy, L. L. (2005). *An hedonomic evaluation of pleasurable human-technology experience: The effect of exposure and aesthetics on the experience of flow*. A dissertation submitted in partial fulfillment of the requirements for the degree of Doctor of Philosophy, University of Central Florida, Orlando, FL.

Murphy, C. (2011). *Why games work—The science of learning*. Retrieved November 28, from cmmurphy@alionscience.com.

Murray, E. A., Wise, S. P., & Rhodes, S. E. V. (2011). What can different brains do with reward? In J. A. Gottfried (Ed.), *Neurobiology of sensation and reward*. Boca Raton, FL: CRC. Retrieved from http://www.ncbi.nlm.nih.gov/pubmed/22593912.

Myers, D. (2013). *Social psychology*. New York: McGraw-Hill.

Nada, E., Khater, A., & Saeed, A. (2014). Value of early intervention for hearing impairment of language and speech acquisition. *The Egyptian Journal of Otolarygology, 30*, 237–242.

Naeini, H. S., & Mostowfi, S. (2015). The role of pleasure criteria in product design: An integrated approach in ergonomics and hedonomics (a review). *Research in Psychology and Behavioral Sciences, 3*, 39–50.
National Institute for Occupational Safety and Health (NIOSH). (2013).*The state of the national initiative on prevention through design*. Publication No. 2014-123. Cincinnati, OH: U.S. Department of Health and Human Services, Centers for Disease Control and Prevention/ National Institute for Occupational Safety and Health, DHHS (NIOSH).
National Institute for Occupational Safety and Health (NIOSH). (2015). *Anthropometric study of U.S. truck drivers: Methods, summary statistics, and multivariate accommodation models*. DHHS (NIOSH). Publication No. 2015-116. Cincinnati, OH: National Institute for Occupational Safety and Health, DHHS (NIOSH).
Nees, M. A., & Walker, B. N. (2011). Auditory displays for in-vehicle technologies. *Reviews of Human Factors and Ergonomics, 7*, 58–99.
Neisser, U. (1967). *Cognitive psychology*. Englewood Cliffs, NJ: Prentice Hall.
Nelson, G. (1994). *Chairs*. New York: Acanthus Press.
Nelson, R. R. (2007). Universal Darwinism and evolutionary social science. *Biology and Philosophy, 22*, 73–94.
Nettersheim, A., Hallschmid, M., Born, J., & Diekelmann, S. (2015). The role of sleep in motor sequence consolidation: Stabilization rather than enhancement. *The Journal of Neuroscience, 35*, 6696–6702.
Neves, B. B., & Amaro, F. (2012). Too old for technology? How the elderly of Lisbon use and perceive ICT. *Journal of Community Informatics, 8*(1).
Nie, N. H., Hillygus, D. S., & Erbring, L. (2002). Internet use, interpersonal relations, and sociability. In B. Wellman & C. Haythornthwaite (Eds.), *The Internet in everyday life* (pp. 215–243). Malden, MA: Blackwell.
Nieoullon, A. (2002). Dopamine and the regulation of cognition and attention. *Progress in Neurobiology, 67*, 53–83.
Nieoullon, A., & Coquerel, A. (2003). Dopamine: A key regulator to adapt action, emotion, motivation and cognition. *Current Opinion in Neurology, 2*, 3–9.
Nilsen, D. M., & Dirusso, T. (2014). Using mirror therapy in the home environment: A case report. *American Journal of Occupational Therapy, 68*, e84–e89.
Nimrod, G. (2013). Applying gerontographics in the study of older internet users. *Journal of Audience and Reception Studies, 10*, 46–64.
NIOSH. (2013). *The state of the national initiative on prevention through design*. Publication No. 2014-123. Cincinnati, OH: U.S. Department of Health and Human Services, Centers for Disease Control and Prevention/National Institute for Occupational Safety and Health, DHHS (NIOSH).
Noe, A. (2004). *Action in perception*. Cambridge, MA: MIT Press.
Noe, A. (2009). *Out of our heads: Why you are not your brain and other lessons from the biology of consciousness*. New York: Hill and Wang.
Norman, D. A. (1990). *The design of everyday things* (1st ed.). New York: Knopf Doubleday.
Norman, D. A. (1993). *Things that make you smart*. New York: Perseus Books.
Norman, D. A. (1999). *The invisible computer*. Cambridge, MA: MIT Press.
Norman, D. A. (2003). Prologue: three teapots. Draft manuscript for Norman, D. A. (2004). *Emotional design: Why we love (or hate) everyday things*. New York: Basic Books.
Norman, D. A. (2004). *Emotional design: Why we love (or hate) everyday things*. New York: Basic Books.
Norman, D. A. (2005). *Emotional design: Why we love (or hate) everyday things* (Paperback Edition). New York: Perseus Books.
Norman, D. A. (2008). Simplicity is not the answer. *ACM Digital Library, 15*, 45–46.
Nowak, E. (1996). The role of anthropometry in design of work and life environments of the disabled population. *International Journal of Industrial Ergonomics, 17*, 113–121.
Nuccitelli, M. (2014). *Cyber bullying: Cyber bullying tactics and examples*. Retreived from www.ipredator.com.

Nuclear Regulatory Commission. (2014). *Backgrounder on the Three Mile Island accident.* Rockville, MD: Nuclear Regulatory Commission

O'Callaghan, C. (2010). *Sounds: A philosophical theory.* New York: Oxford University Press.

Obama, B., & Biden, J. (2013). *Remarks by the president and the vice president on gun violence.* Retrieved January 16, from http://www.whitehouse.gov/photos-and-video/video/2013/01/16/president-obama-introduces-plan-reduce-gun-violence#transcript.

Ogburn, W. F. (1922). *Social change with respect to culture and original nature.* New York: B. W. Huebsch.

Oh, S. W., Harris, J. A., Ng, L., Winslow, B., Cain, N., Mihalas, S., et al. (2014). A mesoscale connectome of the mouse brain. *Nature, 508*, 207–214.

Olesen, P. J., Westerberg, H., & Klingberg, T. (2004). Increased prefrontal and parietal activity after training of working memory. *Nature Neuroscience, 7*, 75–79.

Omura, D. T., Clark, D. A., Samuel, A. D. T., & Horvitz, R. H. (2012). Dopamine signaling is essential for precise rates of locomotion by C. elegans. *PLoS, 7*(6), e38649. doi:10.1371/journal.pone.0038649.

Oomen, P. (2014, February 6). Psychoacoustics: An introduction. *Red Bull Music Magazine.*

Ophir, E., Nass, C., & Wagner, A. D. (2009). Cognitive control in media multitaskers. *Proceedings of the National Academy of Sciences, 106*, 15583–15587.

Oron-Gilad, T., & Hancock, P. A. (2009). From ergonomics to hedonomics: Trends in human factors and technology. In Y. Amichai-Hamburger (Ed.), *Technology and psychological well-being* (pp. 131–147). New York: Cambridge University Press.

Orr, E. S., Sisic, M., Ross, S., Arsenaeault, J. M., & Orr, R. R. (2009). The influence of shyness on the use of Facebook in an undergraduate sample. *Cyberpsychology and Behavior, 12*, 337–340.

Ortiz-Catalan, M., Hakansson, B., & Branemark, R. (2014). An osseointegrated human–machine gateway for long-term sensory feedback and motor control of artificial limbs. *Science Translational Medicine, 6*, 257.

Orwell, G. (1949). *1984.* New York: Penguin Books.

Osborne, C. (2011, February 12) 'Phone on the table' students: Driven by social status. *iGeneration.* Retrieved from http://www.znet.com/blog/igeneration/phone-on-the-table-students-driven-by-social-status/8376.

Oskamp, S., & Schultz, W. P. (1998). *Applied social psychology.* Upper Saddle River, NJ: Prentice Hall.

Oskamp, S., & Spacaman, S. (Eds.). (1990). *People's reactions to technology: In factories, offices, and aerospace (Claremont Symposium on Applied Social Psychology).* New York: Sage.

Otto Bock Sensor Hand Speed. (2013). Retrieved from http://www.ottobock.de/cps/rde/xchg/ob_com_en/hs.xsl/3652.html.

Palarea, R. E. (2007). Operational psychology: An emerging discipline. *American Psychology-Law Society News, 27*, 9–11.

Papacharissi, Z., & Rubin, A. M. (2000). Predictors of Internet use. *Journal of Broadcasting & Electronic Media, 44*, 175–196.

Parasuraman, R. (2011). Neuroergonomics: Brain, cognition, and performance at work. *Current Directions in Psychological Sciences, 20*, 181–186.

Parasuraman, R., & Rizzo, M. (Eds.). (2007). *Neuroergonomics: The brain at work.* New York: Oxford University Press.

Parducci, A. (1965). Category judgment: A range-frequency model. *Psychological Review, 72*, 407–418.

Parducci, A. (1995). *Happiness, pleasure, and judgment: The contextual theory and its applications.* Mahwah, NJ: Erlbaum.

Parisi, T. (2015). *Learning virtual reality.* Sebastopal, CA: O'Reilly Media.

Parker-Pope, T. (2010, June 6). An ugly toll of technology: Impatience and forgetfulness. *The New York Times.*

Patall, E. A., Cooper, H., & Robinson, J. C. (2008). The effects of choice on intrinsic motivation and related outcomes: A meta-analysis of research findings. *Psychological Bulletin, 134*, 270–300.

Patrick, R., Kring, J., McBride, M., & Letowski, T. (2012). Conduction equivalency ratios: A means for comparing the frequency response of bone and air conduction auditory displays. In *Proceedings of the Human Factors and Ergonomics Society, 56th Annual Meeting*.

Pavol, M. J., Owings, T. M., & Grabiner, M. D. (2002). Body segment inertial parameter estimation for the general population of older adults. *Journal of Biomechanics, 35*, 707–712.

Pekrun, R., Goetz, T., Daniels, L. M., Stupnisky, R. H., & Perry, R. H. (2010). Boredom in achievement settings: Exploring control–value antecedents and performance outcomes of a neglected emotion. *Journal of Educational Psychology, 102*, 531–549.

Pekrun, R., & Linnenbrink-Garcia, L. (2014). *International handbook of emotions in education (Educational psychology handbook)* (1st ed.). New York: Routledge.

Pellis, S. M., & Pellis, V. C. (2007). Rough-and-tumble play and the development of the social brain. *Current Directions in Psychological Science, 16*, 95–98.

Peña, J., Hancock, J., & Merola, N. (2009). The priming effects of avatars in virtual settings. *Communication Research, 36*, 838–856.

Penn State. (2015, April 1). Texting too tempting for college students even when inappropriate. *ScienceDaily*. Retrieved from <www.sciencedaily.com/releases/2015/04/150401115748.htm>.

Perrow, C. (1984). *Normal accidents: Living with high-risk technologies*. New York: Basic Books.

Perry, L. S. (2010). *Designing the workplace for the aging workforce: How to use ergonomics to improve the workplace design*. Retrieved from http://195.28.226.39/NR/rdonlyres/AEFC0FF5-EE0B-4765-B5D4-640D99E1412/0/Designingtheworkplacefortheagingworkforce.pdf.

Perry, L. S. (2012, June). Standing up: Redesigning the workplace to address obesity. *Professional Safety.* Retrieved from www.asse.org.

Pessoa, L. (2008). On the relationship between emotion and cognition. *Nature, 9*, 148–158.

Pew Internet and American Life Project. (2012). *Older adults and internet use for the first time, half of adults ages 65 and older are online*. Retrieved from http://pewinternet.org/reports/2012.

Pew Research Center. (2013). *Majority of Americans expect NASA astronauts to land on Mars by 2050*. Retrieved June 21, from www.pewresearch.org.

Pew Research Center. (2014). *The web at 25*. Retrieved February 27, from http://www.pewinternet.org/2014/02/25/the-web-at-25-in-the-u-s.

Pew Research Center, Internet Science and Technology. (2014). *Main report: An indepth look at expert responses*. Retrieved May 14, from www.pewresearch.org.

Phelps, E. A. (2006). Emotion and cognition: Insights from studies of the human amygdala. *Annual Review of Psychology, 57*, 27–53.

Piaget, J. (1954). *The construction of reality in the child.* New York: Basic Books.

Picard, R. W. (1997). *Affective computing*. Cambridge, MA: MIT Press.

Pidgeon, N. (2011). In retrospect: Normal accidents. *Nature, 477*, 404–405.

Pierce, P. (2009). Social anxiety and technology: Face-to-face communication versus technological communication among teens. *Computers in Human Behavior, 25*, 1367–1372.

Pijnappels, M., Bobbart, M. F., & van Dieen, J. H. (2005). Control of support limb muscles in recovery after tripping in young and older subjects. *Experimental Brain Research, 160*, 326–333.

Pijpers, J. R., Oudejans, R. R. D., & Bakker, F. C. (2005). Anxiety-induced changes in movement behaviour during the execution of a complex whole-body task. *The Quarterly Journal of Experimental Psychology, Section A, 58*, 421–445.

Pittinsky, T. L., & Diamante, N. (2015, October). Going beyond fun in STEM. *Phi Delta Kappan* (pp. 47–51).

Poels, K., Van Den Hoogen, W., Ijsselsteijn, W., & De Kort, Y. (2012). Pleasure to play, arousal to stay: The effect of player emotions on digital game preferences and playing time. *CyberPsychology, Behavior, and Social Networking, 15*, 1–6.

Poeter, D. (2013, July 25). *U.K. scientists cook up Mars 'Concept Mission*.' Retrieved from www.pcmag.com.

Pohang University of Science & Technology (POSTECH). (2015). Virtual K-pop dance teacher developed to make dance learning easier. Pohang University of Science & Technology (POSTECH); *ScienceDaily*. Retrieved October 22, from www.sciencedaily.com/releases/2015/10/151022103842.htm.

Poljac, E., de Wit, L., & Wagenas, J. (2012). Perceptual wholes can reduce the conscious accessibility of the parts. *Cognition, 123*, 308–312.

Popescue, A., Broekens, J., & van Somerenet, M. (2013). GAMYGDALA: An emotion engine for games. *Journal of IEEE Transactions on Affective Computing, 5*, 32–45.

Popova, M. (2012). *The science of chunking, working memory and how pattern recognition fuels creativity*. Retrieved from https://www.brainpickings.org/2012/09.

Porter, C. (2013). *Translation of An inquiry into modes of existence, an anthropology of the moderns by Bruno Latour*. Boston: Harvard University Press.

Posner, M. I. (1986). *Chronometric explorations of mind*. New York: Oxford University Press.

Pradeep, A. K. (2010). *The buying brain: Secrets for selling to the subconscious mind*. Hoboken, NJ: Wiley.

Prensky, M. (2012). *From digital natives to digital wisdom: Hopeful essays for 21st century learning*. Thousand Oaks, CA: Corwin Press.

Prescott, T. J. (2007). Forced moves or good tricks in design space? Landmarks in the evolution of neural mechanisms for action selection. *Adaptive Behavior, 15*, 9–13.

Previc, F. (2009). *The dopaminergic mind in human evolution and history*. Cambridge: Cambridge University Press.

Prilleltensky, I. (1989). Psychology and the status quo. *American Psychologist, 44*, 795–802.

Pylatiuk, C., Kargov, A., Schulz, S., & Döderlein, L. (2006). Distribution of grip force in three different functional prehension patterns. *Journal of Medical Engineering and Technology, 30*, 176–182.

Pyszczysnki, T., Sheldon, S., & Greenberg, J. (2003). *In the wake of 9/11: The psychology of terror*. Washington, DC: American Psychological Association.

Raghanti, M. A., Stimpson, C. D., Marcinkiewicz, J. L., Erwin, J. M., Hof, P. R., & Sherwood, C. C. (2008). Cortical dopaminergic innervation among humans, chimpanzees, and macaque monkeys: A comparative study. *Neuroscience, 155*, 203–220.

Rainie, H. L. (2004). Foreward. In P. N. Howard & S. Jones (Eds.), *Society online: The internet in context* (pp. xi–xiv). Thousand Oaks, CA: Sage.

Rauterberg, M. (2006). From personal to cultural computing: How to assess a cultural experience. In G. Kemper & P. von Hellberg (Eds.), *uDayIV—Information nutzbar machen*. Lengerich, Germany: Pabst Science.

Ravaja, N., Turpeinen, M., Saari, T., Puttonen, S., & Keltikangas-Järvinen, L. (2008). The psychophysiology of James Bond: Phasic emotional responses to violent video game events. *Emotion, 8*, 114–120.

Rebok, G. W., Ball, K., Guey, L. T., Jones, R. N., Kim, H. Y., King, J. W., et al. (2014). Ten-year effects of the ACTIVE cognitive training trial on cognition and everyday functioning in older adults. *Journal of the American Geriatric Society, 62*, 16–24.

Reeves, B., & Nass, C. (1998). *The media equation: How people treat computers, television and new media like real people and places*. Cambridge: Cambridge University Press.

Reeves, B., & Voelker, D. (1993). *Effects of audio-video asynchrony on viewer's memory, evaluation of content and detection ability*. (Research Report). Los Gatos, CA: Pixel Instruments.

Reid, D. J., & Reid, F. J. M. (2004). Text or talk? Social anxiety, loneliness, and divergent preferences for cell phone use. *CyberPsychology and Behavior, 10*, 424–435.

Report of the Schalk Lab. (2015). *Brain-computer interfacing*. Retrieved from www.schalklab.org.

Rettie, R. (2007). Texters not talkers: Phone call aversion among mobile phone users. *PsychNology Journal, 5*, 33–57.

Richtel, M. (2010, June 6). Attached to technology and paying a price. *The New York Times*.

Rizzo, S. (2011). Symposium on innovative technologies for psychological intervention, consultation and training. In *APA Annual Convention*, Washington, DC.

Rizzo, S., Buckwalter, G., Forbell, E., Reist, C., Difede, J., Rothbaum, B. O., et al. (2013). Virtual Reality applications to address the wounds of war. *Psychiatric Annals, 43*, 123–138.

Robinette, K., & Daanen, H. A. M. (2003) Lessons learned from CAESAR: A 3-D anthropometric survey. In *Proceedings of the XVth Triennial Congress of the International Ergonomics Association, Ergonomics in the Digital Age*, 24–29 August, paper 00730.

Robinson, L. (2007). The cyberself: The self-ing project goes online, symbolic interaction in the digital age. *New Media & Society, 9*, 93–110.

Robinson, T. E., & Kolb, B. (2004). Structural plasticity associated with exposure to drugs of abuse. *Neuropharmacology, 47*, 33–46.

Rocco, M. C., & Bainbridge, W. S. (Eds.) (2002). *Converging technologies for improving human performance: Nanotechnology, biotechnology, information technology and cognitive science*. NSF/DOC-sponsored report. Arlington, VA: National Science Foundation.

Roeser, S. (2012). Moral emotions as guide to acceptable risk. In S. Roeser, R. Hillerbrand, P. Sandin, & M. Peterson (Eds.), *Handbook of risk theory: Epistemology, decision theory, ethics, and social implications of risk*. New York: Springer.

Rohmert, W., & Jenik, P. (1971). Isometric muscular strength in women. In R. J. Shephard (Ed.), *Frontiers of fitness*. Springfield, IL: Thomas.

Rosegrant, J. (2011). Technologically altered reality inside the therapist's office. *Psychoanalytic Psychology, 29*, 226–240.

Rosen, L. D. (2010). *Rewired: Understanding the iGeneration and the way they learn*. New York: Palgrave Macmillan.

Rosen, L. D. (2011, August 6). Social networking's good and bad impacts on kids. Plenary talk at the *119th Annual Convention of the American Psychological Association*, Washington, DC.

Rosen, L. D. (2012). *idisorder: Understanding our obsession with technology and overcoming its hold on us*. New York: Palgrave MacMillan.

Rosen, L. D., Cheever, N., & Carrier, L. M. (Eds.). (2015). *Wiley handbook of psychology, technology and society*. Hoboken, NJ: Wiley.

Rosen, L. D., Lim, A. F., Carrier, L. M., Lara-Ruiz, J. M., Mendoza, J. S., & Rokkum, J. (2014). Media and technology use predicts ill-being among children, preteens and teenagers independent of the negative health impacts of exercise and eating habits. *Computers and Human Behavior, 35*, 364–375.

Rosenbaum, D. A. (2005). The Cinderella of psychology: The neglect of motor control in the science of mental life and behavior. *American Psychologist, 60*, 308–317.

Rosenbaum, D. A., Cohen, R., Dawson, A., Jax, S. A., Meulenbroek, R. G., van der Wel, R., et al. (2009). The posture-based motion planning framework: New findings related to object manipulation, moving around obstacles, moving in three spatial dimensions, and haptic tracking. In D. Sternad (Ed.), *Progress in motor control–A multidisciplinary perspective* (pp. 485–498). Berlin, Germany: Springer.

Rosenbaum, D. A., Meulenbroek, R. J., Vaughan, J., & Jansen, C. (2001). Posture-based motion planning: Applications to grasping. *Psychological Review, 108*, 709–734.

Roseth, C. J., Saltarelli, A. J., & Glass, C. R. (2011). Effects of face-to-face and computer-mediated constructive controversy on social interdependence, motivation, and achievement. *Journal of Educational Psychology, 103*, 804–820.

Rothbart, M. K., & Posner, M. I. (2015). The developing brain in a multitasking world. *Developmental Review, 1*, 42–63.

Rothbaum, B. O., Hodges, L., Smith, S., Lee, J. H., & Price, L. (2000). A controlled study of virtual reality exposure therapy for the fear of flying. *Journal of Consulting and Clinical Psychology, 68*, 1020–1026.

Rowan, C. (2009, Summer). Technology overuse on child sensory development. *Newsletter of the British Columbia Society of Occupational Therapists*.

Rowan, C. (2013, July 29). The impact of technology on the growing child. *The Huffington Post*. Retrieved from www.huffingtonpost.com.

Rowlands, I., Nicholas, D., Williams, P., Huntington, P., Fieldhouse, M., Gunter, B., et al. (2008). The Google generation: The information behaviour of the researcher of the future. *Aslib Proceedings, 60*, 290–310.

Russoniello, C. V., O'Brien, K., & Parks, J. M. (2009). EEG, HRV and psychological correlates while playing Bejeweled II: A randomized controlled study. In B. K. Wiederhold & G. Riva (Eds.), *Annual review of cybertherapy and telemedicine 2009: Advance technologies in the behavioral, social and neurosciences* (Vol. 7, pp. 189–192). Amsterdam, The Netherlands: Interactive Media Institute and IOS Press.

Ryan, V. (2013). *Ergonomics*. Retrieved from www.technologystudent.com.

Ryan, T. M., & Shaw, C. N. (2014). Gracility of the modern homo sapiens skeleton is the result of decreased biomechanical loading. *Proceedings of the National Academy of Sciences, 112*, 372–377.

Saad, G. (2012). Nothing in popular culture makes sense except in the light of evolution. *Review of General Psychology, 16*, 109–120.

Saariluoma, P., & Oulasvirta, A. (2010). User psychology: Re-assessing the boundaries of a discipline. *Psychology, 1*, 317–328.

Sackett, A. M., Meyvis, T., Nelson, L. D., Converse, B. A., & Sackett, A. L. (2010). You're having fun when time flies: The hedonic consequences of subjective time progression. *Psychological Science, 21*, 111–117.

Salk Institute for Biological Studies. (2015, September 2). Scientists see motor neurons 'walking' in real time. *ScienceDaily*. Retrieved from www.sciencedaily.com/releases/2015/09/150902123617.htm.

Salthouse, T. A. (1991). Expertise as the circumvention of human processing limitations. In K. A. Ericsson & J. Smith (Eds.), *Toward a general theory of expertise: Prospects and limits*. Cambridge, UK: Cambridge University Press.

Salvendy, G. (Ed.). (2005). *Handbook of human factors and ergonomics*. Hoboken, NJ: Wiley.

Salvendy, G. (2007). *Handbook of industrial engineering: Technology & operations management*. Hoboken, NJ: Wiley.

Salvendy, G. (Ed.). (2012). *Handbook of human factors and ergonomics* (4th ed.). Hoboken, NJ: Wiley.

Samejima, K., Ueda, K., & Kimura, M. (2005). Representation of action-specific reward values in the striatum. *Science, 310*, 1337–1340.

Sanchez-Franco, M. J., Ramos, A. F. V., & Velicia, F. A. M. (2009). The moderating effect of gender on relationship quality and loyalty toward Internet service providers. *Information and Management, 46*, 196–202.

Sanchez-Franco, M. J., & Rondan-Cataluña, F. J. (2010). Virtual travel communities and customer loyalty: Customer purchase involvement and web site design. *Electronic Commerce Research and Applications, 9*, 171–182.

Sarlegna, F., Blouin, J., Bresciani, J., Bourdin, C., Vercher, J., & Gauthier, G. M. (2003). Target and hand position information in the online control of goal-directed arm movements. *Experimental Brain Research, 151*, 524–535.

Saunders, I., & Vijayakumar, S. (2011). The role of feed-forward and feedback processes for closed-loop prosthesis control. *Journal of Neuroengineering and Rehabilitation, 8*, 60.

Savioja, P., Liinsuo, M., & Koskinen, H. (2013). User experience: Does it matter in complex systems? *Cognition, Technology and Work, 16*, 429.

Sawaguchi, T. (2001). The effects of dopamine and its antagonists on directional delay-period activity of prefrontal neurons in monkeys during an oculomotor delayed-response task. *Neuroscience Research, 41*, 115–128.

Schack, T., & Hackfort, D. (2007). Action-theory approach to applied sport psychology. In G. Tenenbaum & R. C. Eklund (Eds.), *Handbook of sport psychology* (pp. 332–351). Hoboken, NJ: Wiley.

Schafer, R. M. (1977). *The tuning of the world*. New York: Knopf.

Schell, J. (2008). *The art of game design: A book of lenses*. Burlington, MA: Morgan Kauffman.

Schenk, T. T., Walther, E. U. E., & Mai, N. N. (2000). Closed- and open-loop handwriting performance in patients with multiple sclerosis. *European Journal of Neurology, 7*, 269–279.

Schmid, G., Thielmann, A., & Ziegler, W. (2009). The influence of visual and auditory information on the perception of speech and non-speech oral movements in patients with left hemisphere lesions. *Clinical Linguistics & Phonetics, 23*, 208–221.

Schmidt, R. A. (1975). A schema theory of discrete motor skill learning. *Psychological Review, 82*, 225–260.

Schmidt, R. A., & Donald, T. D. (2005). *Motor control and learning: A behavioral emphasis*. Champaign, IL: Human Kinetics.

Schmidt, R. A., & Lee, T. D. (2005). *Motor control and learning: A behavioral emphasis*. Champaign, IL: Human Kinetics.

Schneiderman, B. (2000). Universal usability. *Communications of the ACM, 43*, 84–91.

Schott, N., Frenkel, M.-O., Korbus, H., & Francis, K. L. (2013). Mental practice in orthopedic rehabilitation: Where, what and how? A case report. *Science & Motricité, 82*, 93–103.

Schuch, C. P., Balbinot, G., Boos, M., Peyré-Tartaruga, L. A., & Susta, D. (2011). The role of anthropometric changes due to aging on human walking: Mechanical work, pendulum and efficiency. *Biology of Sport, 2011*(28), 165–170.

Schultz, D., & Schultz, S. E. (2009). *Psychology and work today: An introduction to industrial and organizational psychology*. Upper Saddle River, NJ: Prentice-Hall. (Original work published 1998)

Schummer, J. (2001). Aristotle on technology and nature. *Philosophia Naturalis, 38*, 105–120.

Schumpeter, J. A. (1942). *Capitalism, socialism and democracy*. New York: Harper.

Schütte, S., Eklund, J., Axelsson, J., & Nagamachi, M. (2004). Concepts, methods and tools in Kansei engineering. *Theoretical Issues in Ergonomics Science, 5*, 214–231.

Schwartz, B. (2004). *The paradox of choice: Why more is less*. New York: Harper Perennial.

Schwebel, D. C., Stavrinos, D., Byington, K. W., Davis, T., O'Neal, E. E., & de Jong, D. (2012). Distraction and pedestrian safety: How talking on the phone, texting, and listening to music impact crossing the street. *Accident Analysis and Prevention, 45*, 266–271.

Schweitzer, J., & Driever, W. (2009). Development of the dopamine systems in zebrafish. *Advances in Experimental Medicine and Biology, 651*, 1–14.

ScienceDictionary.com. (2007). *Meaning of the word science*.

Scott, K. (2012, July 29). Individual relationships with technology. *AMCIS 2012 Proceedings*. Paper 25. Retrieved from http://aisel.aisnet.org/amcis2012/proceedings/HCIStudies/.

Segev-Jacubovski, O., Herman, T., Yogev-Seligmann, G., Mirelman, A., Giladi, N., & Hausdorff, J. M. (2011). The interplay between gait, falls and cognition: Can cognitive therapy reduce fall risk? *Expert Review of Neurotherapeutics, 11*, 1057–1075.

Seligman, M. E. P. (1972). Learned helplessness. *Annual Review of Medicine, 23*, 407–412.

Seligman, M. E. P., & Csikszentmihalyi, M. (2000). Positive psychology: An introduction. *American Psychologist, 55*, 5–14.

Seymour, N. (2008). VR to OR: A review of the evidence that virtual reality simulation improves operating room performance. *World Journal of Surgery, 32*, 182–188.

Sharkey, A. J. C., & Sharkey, N. E. (2012). Granny and the robots: Ethical issues in robot care for the elderly. *Ethics and Information Technology, 14*, 27–40.

Shastry, V. M. (2013, April 21). Seeing the world after a gap of thirteen years. *The Hindu*.

Shenoy, P., Miller, K. J., Ojemann, J. G., & Rao, R. P. N. (2007). Generalized features for electrocorticographic BCIs. *IEEE Transactions on Biomedical Engineering, 55*, 273–280.

Sherry, J., Lucas, K., Greenberg, B., & Lachlan, K. (2006). Video game uses and gratifications as predicators of use and game preference. In P. Vorderer & J. Bryant (Eds.), *Playing video games: Motives, responses, and consequences* (pp. 213–224). Mahwah, NJ: Lawrence Erlbaum.

Shiffrin, R. M., & Schneider, W. (1977). Controlled and automatic human information processing: II. Perceptual learning, automatic attending, and a general theory. *Psychological Review, 84*, 127–190.

Shimizu, Y., Sadoyama, T., Kamijo, M., Hosoya, S., Hashimoto, M., Otani, T., et al. (2004). On-demand production system of apparel on the basis of engineering. *International Journal of Clothing Science and Technology, 16*, 32–42.

Shinohara, K., & Wobbrock, J. O. (2011). In the shadow of misperception: Assistive technology use and social interactions. *Proceedings of CHI 2011* (pp. 705–714). New York: ACM Press.

Sidaway, B., Ahn, S., Boldeau, P., Griffin, S., Noyes, B., & Pelletier, K. (2008). A comparison of manual guidance and knowledge of results in the learning of a weight-bearing skill. *Journal of Neurologic Physical Therapy, 32*, 32–38.

Silvia, P. J. (2009). *Confusion and interest: The role of knowledge emotions in aesthetic experience*. Unpublished manuscript, University of North Carolina at Greensboro.

Simon, H. (1969). *The sciences of the artificial*. Cambridge, MA: MIT.

Simon, H. (1991). Bounded rationality and organizational learning. *Organization Science, 2*, 125–134.

Skierskowski, D., & Wood, R. M. (2012). To text or not to text? The importance of text messaging among college-aged youth. *Computers in Human Behavior, 28*, 744–756.

Skinner, B. F. (1959). *Pigeons in a pelican*. Paper presented at the American Psychological Association, Cincinnati, OH. Retrieved from www.bfskinner.org/org/project-pigeon.

Skinner, B. F. (1983). *A matter of consequence*. New York: Knopf.

Skolimowski, H. (1966). The structure of thinking in technology. *Technology and Culture, 7*, 371–383.

Slonje, R., Smith, P. K., & Frisen, A. (2013). The nature of cyberbullying, and strategies for prevention. *Computers in Human Behavior, 29*, 26–32.

Smith, M. A. (2012a, September 14). *Creativity and IQ, part I: What is divergent thinking? How is it helped by sleep, humor and alcohol?* Retrieved from www.thecreativitypost.com.

Smith, P. K. (2012b). Cyberbullying and cyber aggression. In S. R. Jimerson, A. B. Nickerson, M. J. Mayer, & M. J. Furlong (Eds.), *Handbook of school violence and school safety: International research and practice* (pp. 93–103). New York: Routledge.

Smith, B., Bi, X., & Zhai, S. (2015). Optimizing touchscreen keyboards for gesture typing. In *Proceedings of CHI 2015—The SIGCHI Conference on Human Factors in Computing Systems* (pp. 3365–3374). New York: ACM Press.

Smith, P. K., Mahdavi, J., Carvalho, M., Fisher, S., Russell, S., & Tippett, N. (2008). Cyberbullying: Its nature and impact in secondary school pupils. *Journal of Child Psychology and Psychiatry, 49*, 376–385.

Smith, D., Vertegaal R., Sohn, C., & Cheng, D. (2005). ViewPointer: Lightweight calibration-free eye tracking for ubiquitous handsfree deixis. In *Proceedings of ACM Symposium on User Interface Software Technology (UIST)*. New York: ACM Press.

Snyder, J. L., Gregg, M. K., Weintraub, D. M., & Alain, C. (2012). Attention, awareness and the perception of auditory scenes. *Frontiers in Psychology, 3*, 15.

Sorkin, J. D., Muller, D. C., & Andres, R. (1999). Longitudinal change in height of men and women: implications for interpretation of the body mass index: The Baltimore longitudinal study of aging. *American Journal of Epidemiology, 150*, 969–977.

Spahiu, T., Shehi, E., & Piperi, E. (2015). Anthropometric studies: Advanced 3D method for taking anthropometric data in Albania. *International Journal of Innovative Research in Science, Engineering and Technology, 4*, 2136–2142.

Sperling, G. (1960). The information available in brief visual presentations. *Psychological Monographs: General and Applied, 74*, 1–29.

Staal, M. A., & Greene, C. (2015). An examination of "adversarial" operational psychology. *Peace and Conflict, 21*, 264–268.

Stanney, K. M., Mourant, R. R., & Kennedy, R. S. (1998). Human factors: Issues in virtual environments: A review of the literature. *Presence, 7*, 327–351.

Stepp, C. E., An, Q., & Matsuoka, Y. (2012). Repeated training with augmentative vibrotactile feedback increases object manipulation performance. *PLoS One, 7*, e32743.

Stiegler, B. (1998). *Technics and time, 1: The fault of Epimetheus*. Stanford, CA: Stanford University Press.

Stöckel, T., Hughes, C. M., & Schack, T. (2012). Representation of grasp postures and anticipatory motor planning in children. *Psychological Research, 76*, 768–776.

Strebel, J., O'Donnel, K., & Myres, J. G. (2004). Exploring the connection between frustration and consumer choice behavior in a dynamic decision environment. *Psychology & Marketing, 21*, 1059–1076.

Strong, D., & Higgs, E. (2000). Borgmann's philosophy of technology. In E. S. Higgs, A. Light, & D. Strong (Eds.), *Technology and the good life*. Chicago: University of Chicago Press.

Suler, J. (2004). The online disinhibition effect. *Cyberpsychology & Behavior, 7*, 321–326.

Sulzenbruck, S., & Heuer, H. (2012). Enhanced mechanical transparency during practice impedes open loop control of a complex tool. *Experimental Brain Research, 218*, 283–294.

Surowiecki, J. (2005). *The wisdom of crowds: Why the many are smarter than the few and how collective wisdom shapes business, economies, societies and nations*. New York: Anchor Books.

Sweetser, P., & Wyeth, P. (2005). GameFlow: A model for evaluating player enjoyment in games. *Computers in Entertainment, 3*, Article 3A.

Takacs, B. (2006). Cognitive, mental and physical rehabilitation using a configurable virtual reality system. *The International Journal of Virtual Reality, 5*, 1–12.

Takeuchi, A. H., & Hulse, S. H. (1993). Absolute pitch. *Psychological Bulletin, 113*, 345–361.

Takeuchi, N., Tada, T., Toshima, M., Chuma, T., Matsuo, Y., & Ikoma, K. (2008). Inhibition of the unaffected motor cortex by 1 Hz repetitive transcranical magnetic stimulation enhances motor performance and training effect of the paretic hand in patients with chronic stroke. *Journal of Rehabilitation Medicine, 40*, 298–303.

Tan, D. P., Liu, Q. Y., Koshiya, N., Gu, H., & Alkon, D. (2006). Enhancement of long-term memory retention and short-term synaptic plasticity in cbl-b null mice. *PNAS, 103*, 5125–5130.

Tang, Y., Tang, R., & Posner, M. I. (2013). Brief meditation training induces spontaneous smoking reduction. *Proceedings of the National Academy of Sciences of the United States of America, 110*, 13971–13975.

Taylor, S. A. (2009). Reconciling satisfaction, emotions, attitudes, and ambivalence within consumer models of judgment and decision making: A cautionary tale. *Journal of Consumer Satisfaction, Dissatisfaction & Complaining Behavior, 21*, 41–65.

Temme, S., Brunet, P., & Keele, D. B. Jr. (2009). *Practical measurement of loudspeaker distortion using a simplified auditory perceptual model*. Audio Engineering Society Convention Paper Presented at the 127th Convention, October 9–12, New York.

Thaxton, S., & Rajulu, S. (2007). Anthropometric accommodation in space suit design. In *Tenth Annual Applied Ergonomics Conference*, Dallas, Texas.

Thayer, S. E., & Ray, S. (2006). Online communication preferences across age, gender, and duration of Internet use. *Cyberpsychology and Behavior, 9*, 432–440.

The Economic Times. (2015, May 27). Being so heavily dependent on data, can we imagine a day without it? *The Economic Times*.

The Guardian. (2013, October 11). Mobile apps revenues tipped to reach $26 bn in 2013. *The Guardian*.

The New York Times. (2012, July 2). Oxygen problems on F-22 elude the air force's fixes. *The New York Times*.

The New York Times. (2013, May 17). Lawmakers show concerns about Google's new glasses. *The New York Times*.

The Time. (2015, August 16). Inside the box. *The Time*.

The Times of India. (2013, July 4). I change my watch three times a day: Saif Ali Khan. *The Times of India*.

The UCLA Internet Report. (2000). *Surveying the digital future*. UCLA Center for Communication Policy.

The UCLA Internet Report. (2002). *Surveying the digital future, year two*. UCLA Center for Communication Policy.

Theilig, S., Podubecka, J., Bosl, K., Wiederer, R., & Nowak, D. A. (2011). Functional neuromuscular stimulation to improve severe hand dysfunction after stroke: Does inhibitory rTMS enhance therapeutic efficiency? *Experimental Neurology, 230*, 149–155.

Thompson, L., & Cupple, J. (2008). Seen and not heard? Text messaging and digital sociality. *Social and Cultural Geography, 9*, 95–108.

Thompson, R. H., Ménard, A., Pombal, M., & Grillner, S. (2008). Forebrain dopamine depletion impairs motor behavior in lamprey. *European Journal of Neuroscience, 27*, 1452–1460.

Thompson, L. L., Rivera, F. P., Ayyagari, R. C., & Ebel, B. E. (2012). Impact of social and technological distraction on pedestrian crossing behaviour: An observational study. *Injury Prevention, 19*, 232–237.

Thorndike, E. L. (1898). Animal intelligence: An experimental study of the associative processes in animals. *Psychological Monographs: General and Applied, 2*, i-109.

Thurlow, C. (2003). Generation Txt? The sociolinguistics of young people's text-messaging. *Discourse Analysis*. Retrieved from http://extra.shu.ac.uk/daol/articles/v1/n1/a3/thurlow2002003-01.html.

Toffler, A. (1980). *The third wave*. New York: Bantam Books.

Tokunaga, R. S. (2010). Following you home from school: A critical review and synthesis of research on cyberbullying victimization. *Computers in Human Behavior, 26*, 277–287.

Tolman, E. C. (1948). Cognitive maps in rats and men. *The Psychological Review, 55*, 189–208.

Toney, A. P., & Thomas, B. H. (2007). *Modeling reach for use in user interface design*. Paper presented at the Eighth Australian User Interface Conference (AUIC2007), Ballarat, Australia.

Tongeren van, M. C. (2013). *Grenzen van het hoorbare: Over de meerstemmigheid van het lichaam*. Doctoral Thesis, Leiden University, Leiden, Germany. Retrieved from https://openaccess.leidenuniv.nl/handle/1887/20611.

Touch Bionic i-Limb. (2013). http://www.touchbionics.com/.

Trainor, L. J. (2015). The origins of music in auditory scene analysis and the roles of evolution and culture in musical creation. *Philosophical Transactions of the Royal Society of London. Series B, Biological Sciences*. doi: 10.1098/rstb.2014.0089.

Treger, I., Aidinof, L., Lehrer, H., & Kalichman, L. (2012). Constraint-induced movement therapy alters cerebral blood flow in subacute post-stroke patients. *American Journal of Physical Medicine & Rehabilitation, 91*, 804–809.

Trent, G. T., Bai, M., Glick, D., Annin, P., & Keene-Osborn, S. (1999). Searching for answers: Shooting at Columbine High School, Colorado. *Newsweek, 133*, 30.

Tretriluxana, J., Runnarong, N., Tretluxana, S., Prayoonwiwat, N., Vachalathiti, R., & Winstein, C. (2013). Feasibility investigation of the Accelerated Skill Acquisition Program (ASAP): Insights into reach-to-grasp coordination of individuals with postacute stroke. *Top Stroke Rehabilitation, 20*, 151–160.

Truax, B. (2001). *Acoustic communication* (2nd ed.). Norwood, NJ: Ablex.

Tun, J. K., & Kneebone, R. (2011). Bridging worlds: Applying the science of motor learning to clinical education. *Medical Education, 45*, 111–114.

Tuomainen, J., Andersen, T. S., Tiippana, K., & Sams, M. (2005). Audio-visual speech perception is special. *Cognition, 96*, 13–22.

Turing, A. M. (1950). Computing machinery and intelligence. *Mind, 49*, 433–460.

Turkle, S. (1984). *The second self: Computers and the human spirit*. New York: Simon & Schuster.

Turkle, S. (1995). *Life on the screen: Identity in the age of the internet*. New York: Simon & Schuster.

Turkle, S. (1997). Seeing through computers: Education in a culture of simulation. *The American Prospect*, no. 31.

Turkle, S. (2004). How computers change the way we think. *The Chronicle of Higher Education, 50*(B), 26.

Turkle, S. (2011a). *Alone together: Why we expect more from technology and less from each other*. New York: Basic Books.

Turkle, S. (2011b). *Introduction to alone together*. New York: Basic Books.

Turkle, S., Taggart, W., Kidd, C. D., & Daste, O. (2006). Relational artifacts with children and elders: The complexities of cybercompanionship. *Connection Science, 18*(4).

Turner, P., & Turner, S. (2013). Emotional and aesthetic attachment to digital artifacts. *Cognition and Work, 15*, 403–414.

UN. (2004). *World population prospects: The 2004 revision*. New York: United Nations Organization.

UN. (2010). *Human development report 2010, UNDP*. New York: Palgrave MacMillan.

Underwood, E. (2015). *Typing using just your thoughts just got a lot faster*. Retrieved November 11, from www.sciencemag.org.

Uttal, D. H., Meadow, N. G., Tipton, E., Hand, L. L., Alden, A. R., Warren, C., et al. (2013). The malleability of spatial skills: A meta-analysis of training studies. *Psychological Bulletin, 139*, 352–402.

Valkenburg, P. M., & Peter, J. (2011). Online communication among adolescents: An integrated model of its attraction, opportunities, and risks. *Journal of Adolescent Health, 48*, 121–127.
Van Cleemput, K. (2010). I'll see you on IM, text, or call you: A social network approach of adolescents' use of communication media. *Bulletin of Science, Technology & Society, 30*, 75–85.
van den Hoven, M. J., & Weckert, J. (Eds.). (2008). *Information technology and moral philosophy*. Cambridge, UK: Cambridge University Press.
van Gorp, T. J. (2006). *Emotion, arousal, attention and flow: Chaining emotional states to improve human-computer interaction*. Master's Degree Project, Faculty of Environmental Design, University of Calgary.
van Gorp, A. (2007). Ethical issues in engineering design processes: Regulative frameworks for safety and sustainability. *Design Studies, 28*, 117–131.
van Osselaer, S. M. J., Alba, J. W., & Manchanda, P. (2004). Irrelevant information and mediated intertemporal choice. *Journal of Consumer Psychology, 14*, 257–270.
Vandebosch, H., & Van Cleemput, K. (2008). Defining cyberbullying: A qualitative research into the perceptions of youngsters. *CyberPsychology & Behavior, 11*, 499–503.
Vapenstad, C., & Buzink, S. N. (2013). Procedural virtual reality simulation in minimally invasive surgery. *Surgical Endoscopy, 27*, 364–377.
Vaziri, P. M., Bahrpeyma, F., Firoozabadi, M., Forough, B., Hatef, B., Sheikhhoseini, R., et al. (2014). Low frequency repetitive transcranial magnetic stimulation to improve motor function and grip force of upper limbs of patients with hemiplegia. *Iran Red Crescent Medical Journal, 16*, e13579.
Velentzas, K., Heinen, T., Tenenbaum, G., & Schack, T. (2010). Functional mental representation of volleyball routines in German youth female national players. *Journal of Applied Sport Psychology, 22*, 474–485.
Ventura, M., Shute, V., & Zhao, W. (2013). The relationship between video game use and a performance-based measure of persistence. *Computers & Education, 60*, 52–58.
Vezina, K. (2011, July 14). Evidence suggests that the Internet changes how we remember. *MIT Technology Review*. Retrieved from www.technologyreview.com.
Vicario, D. S., & Ghez, C. (1984). The control of rapid limb movement in the cat. IV. Updating the ongoing isometric responses. *Experimental Brain Research, 55*, 134–144.
Vickers, J. N., & Williams, A. M. (2007). Performing under pressure: The effects of physiological arousal, cognitive anxiety, and gaze control in biathlon. *Journal of Motor Behavior, 39*, 381–394.
Vincenti, W. G. (1990). *What engineers know and how they know it: Analytical studies from aeronautical history*. Baltimore: Johns Hopkins University Press.
Vogel, G. (2015). Chimps destroy documentary drone with twig tools. *Science*, doi: 10.1126/science.aad1712.
Volkow, N. D., Wang, G. J., Telang, F., Fowler, J. S., Logan, J., Wong, C., et al. (2008). Sleep deprivation decreases binding of (11C) raclopride to dopamine D2/D3 receptors in the human brain. *Journal of Neuroscience, 28*, 8454–8461.
Wade, L. (2015). Birth of the moralizing gods. *Science, 349*, 918–922.
Wagner, N., Hassanein, K., & Head, M. (2010). Computer use by older adults: A multi-disciplinary review. *Computers in Human Behavior, 26*, 870–882.
Wagster, J., Tan, J., Wu, Y., Biswas, G., & Schwartz, D. L. (2007). Do learning by teaching environments with metacognitive support help students develop better learning behaviors? In *Proceedings of the 29th Annual Meeting of the Cognitive Science Society*.
Wai, J., Lubinski, D., Benbow, C. P., & Steiger, J. H. (2010). Accomplishment in science, technology, engineering, and mathematics (STEM) and its relation to STEM educational dose: A 25-year longitudinal study. *Journal of Educational Psychology, 102*, 860–871.
Walsh, S. P., White, K. M., & Young, R. M. D. (2009). The phone connection: A qualitative exploration of how belongingness and social identification relate to mobile phone use amongst Australian youth. *Journal of Community & Applied Social Psychology, 19*, 225–240.

Wan, C. Y., & Huon, G. F. (2005). Performance degradation under pressure in music: An examination of attentional processes. *Psychology of Music, 33*, 155–172.
Wang, R. Y., Tseng, H. Y., Liao, K. K., & Wang, C. J. (2012). rTMS combined with task-oriented training to improve symmetry of interhemispheric corticomotor excitability and gait performance after stroke: A randomized trial. *Neurorehabilitation and Neural Repair, 26*, 222–230.
Warren, T. (2012). *How researchers hacked into Stephen Hawking's brain*. Retrieved June 25, from http://cosmiclog.nbcnews.com.
Warwick, K., Gasson, M., Hutt, B., Goodhew, I., Kyberd, P., Andrews, B., et al. (2003). The application of implant technology for cybernetic systems. *Archives of Neurology, 60*, 1369–1373.
Waugh, N. C., & Norman, D. A. (1965). Primary memory. *Psychological Review, 72*, 89–104.
Webb, G. (2015, February 15). Say goodbye to privacy. *WIRED*.
Webster, B. R., Celnik, P. A., & Cohen, L. G. (2006). Noninvasive brain stimulation in stroke rehabilitation. *NeuroRx, 3*, 474–481.
Wedell, D. H., & Parducci, A. (1988). The category effect in social judgment: Experimental ratings of happiness. *Journal of Personality and Social Psychology, 55*, 341–356.
Wei, R., Wang, H., Han, J., Zhao, S., Dumbleton, J., Agarwal, P., et al. (2015). Chitosan-decorated doxorubicin-encapsulated nanoparticle targets and eliminates tumor reinitiating cancer stem-like cells. *ACS Nano, 9*, 5725–5740.
Weil, M. M., & Rosen, L. D. (1994). The psychological impact of technology. *Computers and Society, 24*, 3–9.
Weil, M. M., & Rosen, L. D. (1998). *TechnoStress: Coping with technology @WORK @HOME @ PLAY*. Hoboken, NJ: Wiley.
Weisskirch, R. S., & Delevi, R. (2011). Sexting and adult romantic attachment. *Computers in Human Behavior, 27*, 1697–1701.
Werner, N. E., Bumpus, M. F., & Rock, D. (2010). Involvement in Internet aggression during early adolescence. *Journal of Youth and Adolescence, 39*, 607–619.
Wertheimer, M. (1923). Untersuchungen zur Lehre von der Gestalt, II. (Investigations in Gestalt Theory: II. Laws of organization in perceptual forms). *Psychologische Forschung, 4*, 301–350.
Wheaton, L. (2015). *Opposites don't attract when learning how to use a prosthesis*. Retrieved from www.sciencedaily.com/releases/2015/10/151008173519.html.
Willis, S. L. (2004). Technology and learning in current and future older cohorts. In R. W. Pew & S. B. Van Hemel (Eds.), *National Research Council (US) Steering Committee for the Workshop on Technology for Adaptive Aging*. Washington, DC: National Academies Press.
Wills, S. (2013, July 17). PayPal accidentally credits man $92 quadrillion. *CNN*. Retrieved from www.cnn.com.
Wilson, M. (2008). From processing efficiency to attentional control: A mechanistic account of the anxiety-performance relationship. *International Review of Sport and Exercise Psychology, 1*, 184–201.
Winston, M., & Edelbach, R. (2014). *Society, ethics and technology*. Boston: Wadsworth/Cenagage Learning.
Witt, J. K., Linkenauger, S. A., Bakdash, J. Z., & Proffitt, D. R. (2008). Putting to a bigger hole: Golf performance relates to perceived size. *Psychonomic Bulletin & Review, 15*, 581–585.
Witt, E. A., Massman, A. J., & Jackson, L. A. (2011). Trends in youth's videogame playing, overall computer use, and communication technology use: The impact of self-esteem and the Big Five personality factors. *Computers in Human Behavior, 27*, 763–769.
Witt, J. K., & Proffitt, D. R. (2005). See the ball, hit the ball: Apparent ball size is correlated with batting average. *Psychological Science, 16*, 937–938.
Wolpaw, J. R., & Mcfarland, D. J. (2004). Control of a two-dimensional movement signal by a noninvasive brain-computer interface in humans. *PNAS, 101*, 17849–17854.
Woodworth, R. S. (1899). The accuracy of voluntary movements. *Psychological Review Monograph Supplement, 3*, 114.
Woodworth, R. S. (1918). *Dynamics of behavior*. Unpublished manuscript.
Woodworth, R. S. (1958). *Dynamics of behavior*. New York: Henry, Holt.

Woolaston, V. (2015, September 30). *Controlling a computer with your mind: Paralysed patients move on-screen cursor using just their brain waves*. Retrieved from www.dailymail.co.uk.

Wronska, N., Garcia-Zapirain, B., & Mendez-Zorrilla, A. (2015). An iPad-based tool for improving the skills of children with attention deficit disorder. *International Journal of Environmental Research and Public Health, 12*, 6261–6280.

Wu, J., Yang, J., & Honda, T. (2010). Fitts' law holds for pointing movements under conditions of restricted visual feedback. *Human Movement Science, 29*, 882–892.

Wulf, G., Chiviacowsky, S., Schiller, E., & Ávila, L. T. G. (2010). Frequent external focus feedback enhances motor learning. *Frontiers in Psychology, 11*, 2010.

Wulf, G., & Mornell, A. (2008). Insights about practice from the perspective of motor learning: A review. *Music Performance Research, 2*, 1–25.

Yantis, S. (2010, August 15). In M. Richtel, Outdoors and out of reach, studying the brain. *The New York Times*.

Yantis, S., & Serences, J. T. (2003). Cortical mechanisms of space-based and object-based attentional control. *Current Opinion in Neurobiology, 13*, 187–193.

Yee, N., Bailenson, J. N., & Ducheneaut, N. (2009). The Proteus effect: Implications of transformed digital self-representation on online and offline behavior. *Communication Research, 36*, 285–312.

Yun, M. H., Han, S., Hong, S., & Kim, J. (2003). Incorporating user satisfaction into the look-and-feel of mobile phone design. *Ergonomics, 46*, 1423–1440.

Zegas, A. (2013, July 5). We're closer to George Orwell than we may think. *Star-Ledger*.

Zeki, S. (2008). *Splendors and miseries of the brain: Love, creativity, and the quest for human happiness*. Oxford: Blackwell.

Zeki, S. (2009, November 24). Statement on neuroesthetics. *Neuroesthetics*. Retrieved from http://www.neuroesthetics.org/statement-on-neuroesthetics.php.

Zelaznik, H. N., Croxall, R., & Vanhooser, J. (2011). Performance is not related to perception of target width in Fitts' Law. *BIO Web of Conferences, 1*, 00100.

Zendel, B. R., Tremblay, C. D., Belleville, S., & Peretz, I. (2014). The impact of musicianship on the cortical mechanisms related to separating speech from background noise. *Journal of Cognitive Neuroscience, 12*, 1–16.

Zhang, Y., Lu, H., & Bargmann, C. I. (2005). Pathogenic bacteria induce aversive olfactory learning in Caenorhabditis elegans. *Nature, 438*, 179–184.

Zhang, H., Xu, L., Wang, S., Xie, B., Guo, J., Long, Z., et al. (2011). Behavioral improvements and brain functional alterations by motor imagery training. *Brain Research, 1407*, 38–46.

Zhang, H., Xu, L., Zhang, R., Hui, M., Long, Z., Zhao, X., et al. (2012). Parallel alterations of functional connectivity during execution and imagination after motor imagery learning. *PLoS ONE, 7*, e36052.

Zhou, L., Dai, L., & Zhang, D. (2007). Online shopping acceptance model: A critical survey of consumer factors in online shopping. *Journal of Electronic Commerce Research, 8*, 41–62.

Zhuang, Z., Landsittel, D., Benson, S., Roberge, R., & Shaffer, R. (2010). Facial anthropometric differences among gender, ethnicity, and age groups. *Annals of Occupational Hygiene, 54*, 391–402.

Zwart, S. D., Poel, van de, I. R., Mil, van, H. G. J., & Brumsen, M. (2006). A network approach for distinguishing ethical issues in research and development. *Science and Engineering Ethics, 12*, 663–684.

Index

V.K. Kool, R. Agrawal, *Psychology of Technology*,
DOI 10.1007/978-3-319-45333-0

MIX
Papier aus verantwortungsvollen Quellen
Paper from responsible sources
FSC® C105338

If you have any concerns about our products,
you can contact us on
ProductSafety@springernature.com

In case Publisher is established outside the EU,
the EU authorized representative is:
Springer Nature Customer Service Center GmbH
Europaplatz 3, 69115 Heidelberg, Germany

Printed by Libri Plureos GmbH
in Hamburg, Germany